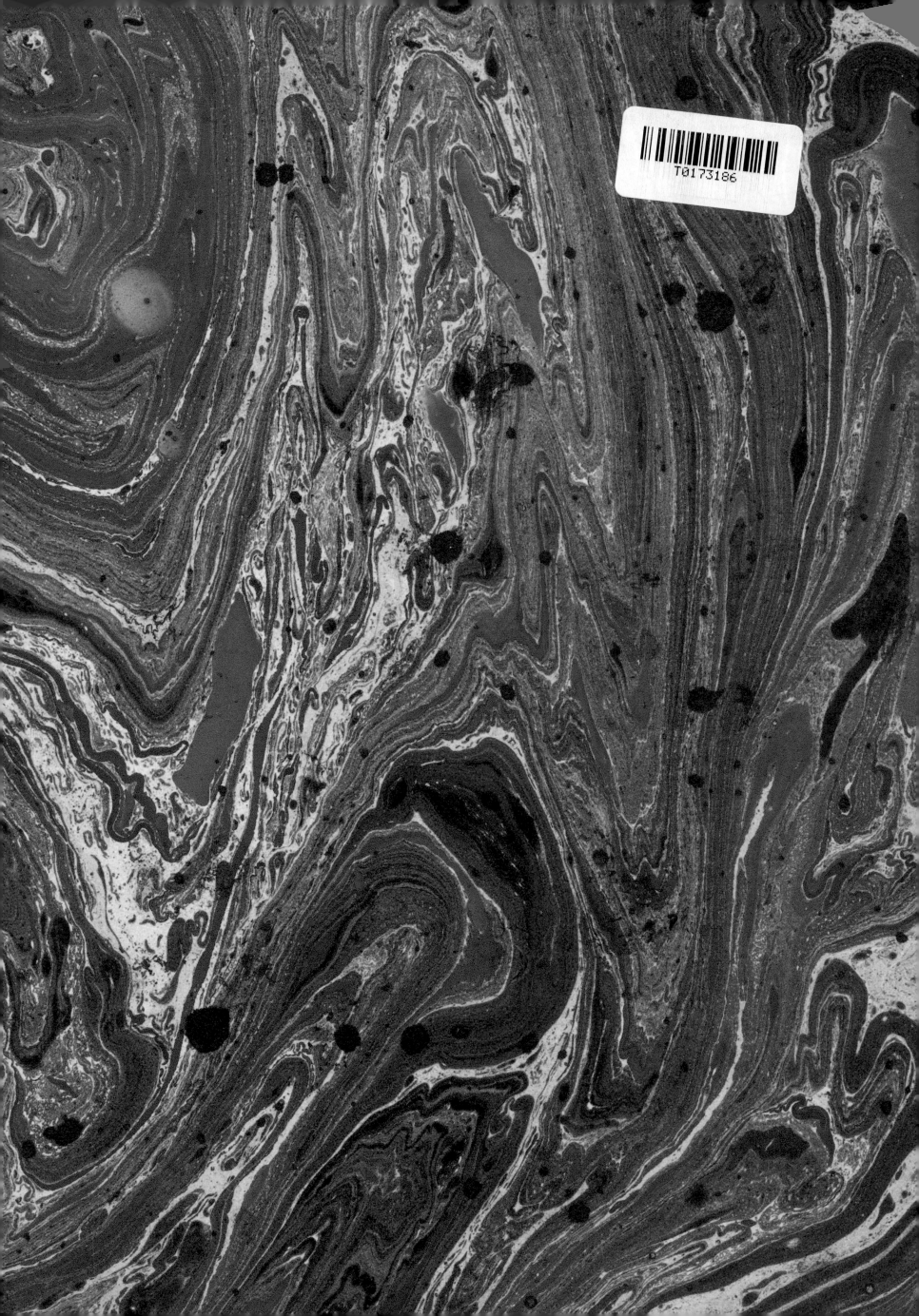

PHÆNOMENA

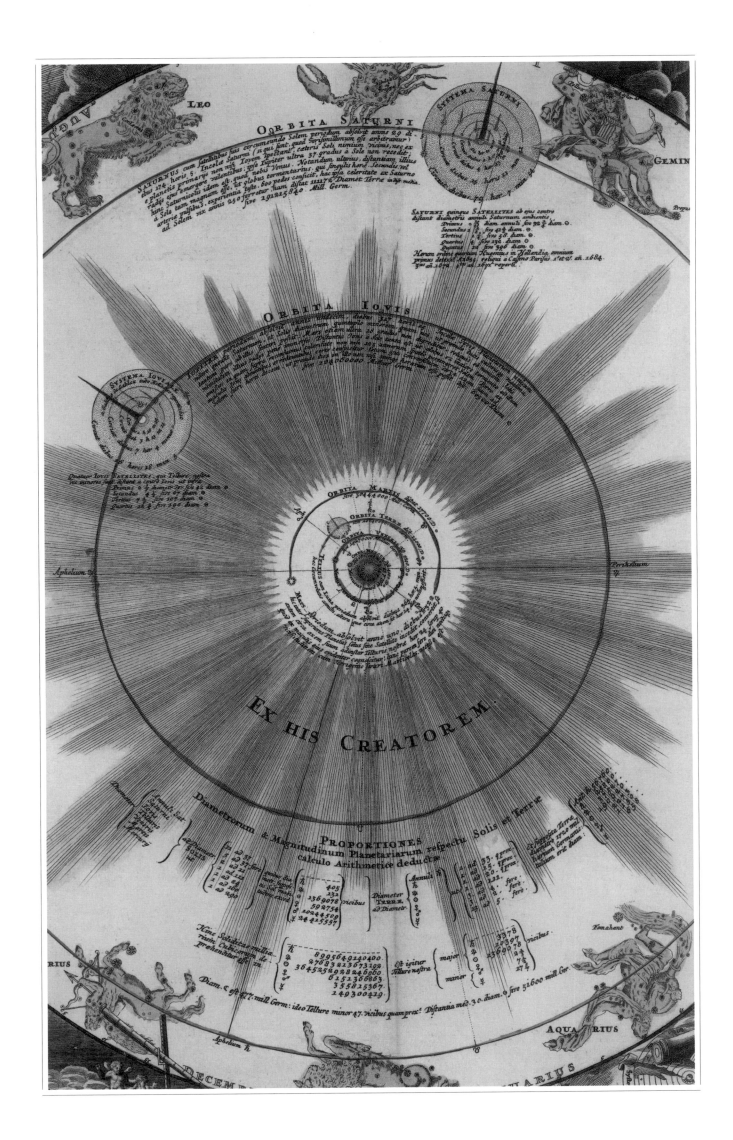

PHÆNOMENA

DOPPELMAYR'S
CELESTIAL ATLAS

TEXT BY
GILES SPARROW

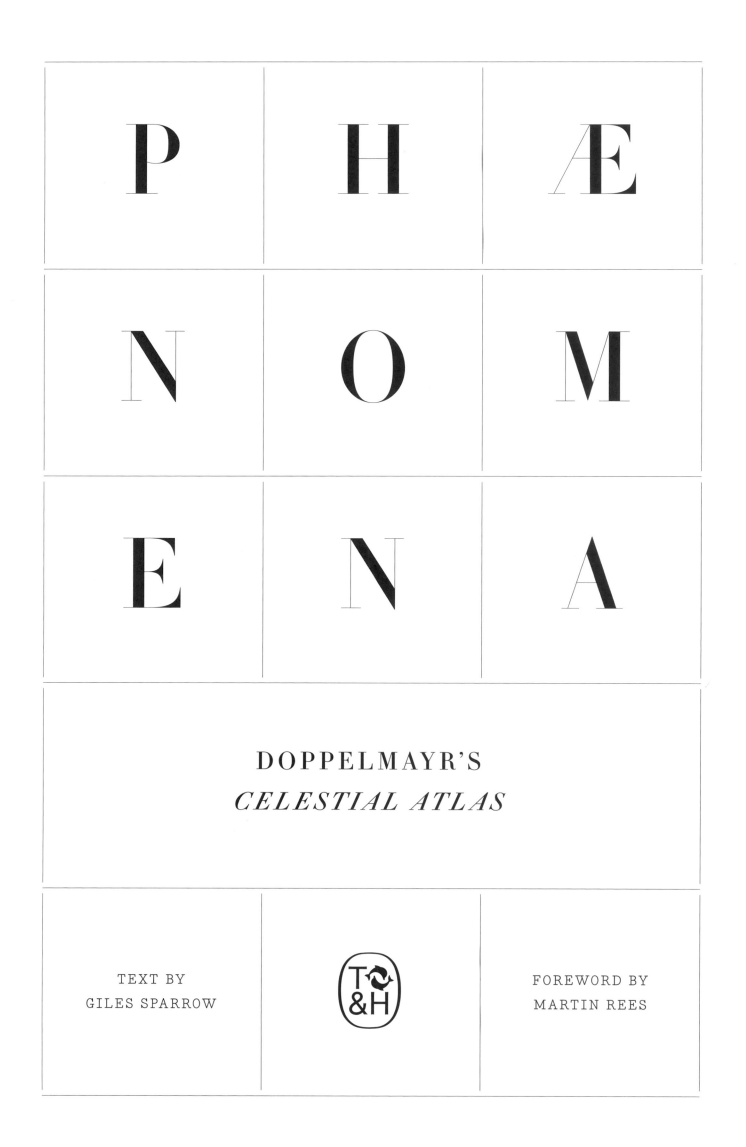

FOREWORD BY
MARTIN REES

TABLE OF CONTENTS.

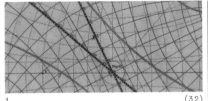

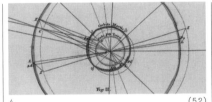

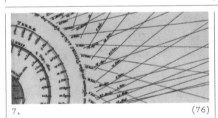

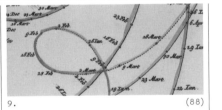

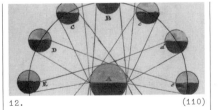

TABLE OF CONTENTS.

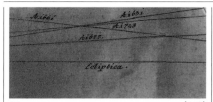

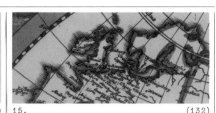

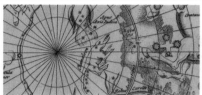

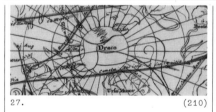

FOREWORD.

Martin Rees, Astronomer Royal.

The starry sky is the most universal feature of our environment. The 'vault of heaven' has been viewed with awe throughout human history – the still-familiar names of some constellations are vestiges of that early wonderment. That the Moon orbits the Earth – and is closer to us than the Sun – has been known since antiquity. The ancients also wondered at the perplexing motions of the planets against the backdrop of the 'fixed stars'. And Ptolemy's (*c.* 100–170 CE) geocentric cosmology dates from the second century CE.

Over the centuries that followed, geocentric cosmology was 'improved', to take account of successively more accurate data, by adding epicycles and equants, and a more complex variant was proposed by the great Tycho Brahe (1546–1601). Nicolaus Copernicus's (1473–1543) heliocentric system eventually triumphed, losing some of its simplicity after Johannes Kepler (1571–1630) discovered planetary orbits to be ellipses rather than circles. Rival 'models' survived because at the time there was no deeper understanding of why the planets should trace out any particular paths. That understanding came only with Isaac Newton (1643–1727), whose laws of gravity, published in 1687, showed that the elliptical shapes of the orbits, and their speeds, were direct consequences of the Sun's gravitational pull.

In the *Atlas Coelestis* Johann Gabriel Doppelmayr (1677–1750) masterfully presents a panorama of what was believed about the cosmos in the early 18th century. The splendid ornamentation of his illustrations reflects the cultural breadth and aesthetic sense of his intended readers. The fact that a work of this style was widely appreciated testifies to the range of cultural interests among the intelligentia. The early members of London's Royal Society, and their counterparts elsewhere in the world, were polymaths. They were steeped in the literary culture of the time but fascinated by the cosmos and by travellers' tales of remote continents. The maps and globes depicted in this book reveal the still-vague delineation of the Earth's southern hemisphere at that time. Doppelmayr's sky maps portray the most prominent stars, the tracks of comets, and the orbits of Saturn's and Jupiter's moons, themselves 'cosmic clocks', observable in principle by navigators.

It proved a real challenge to measure the actual scale of the solar system (rather than relative sizes of different orbits). The best method required observation of a rare phenomenon, the transit of Venus across the face of the Sun. Government-funded expeditions were made to Pacific islands to observe the transits of 1761 and 1769.

FIG. 1.
Dutch artist Olivier van Deuren's *A Young Astronomer* of about 1685 depicts its subject equipped with some of the key tools of Enlightenment astronomy – star charts, a celestial globe and a small quadrant for measuring angles and positions in the sky. The painting is one of several contemporary Dutch depictions of similar figures, reflecting the fascinations of an age when understanding of the heavens was being transformed by precise measurement.

FIG. 1.

There were speculations – even in the 16th century by Giordano Bruno (1548–1600) – that stars were 'other suns'. Their faint light implied that they must be hugely more remote than any of the planets; it was only much later that such distances were corroborated. In the 18th century there was a realization – due to Immanuel Kant (1724–1804), inspired by Thomas Wright (1711–86) – that the Milky Way was an assemblage of stars configured in a disc or sheet. It became apparent only in the 1920s that we live in a galaxy harbouring over one hundred billion stars, orbiting a central hub, and that some faint and fuzzy 'nebulae', such as Andromeda, are systems fully the equal of our own galaxy. There are billions of such galaxies within range of our modern telescopes.

Until the mid 19th century it was widely believed that the stars consisted of a 'fifth essence' – quite different from the earth, air, fire and water that constituted the chemistry of the material world. Discovering what made them shine with characteristic colours had to await the deployment of the spectrograph by Joseph von Fraunhofer (1787–1826) and his successors – first to sunlight and then to the stars. It was not until the 20th century that was it recognized that stars were mainly powered by nuclear fusion (turning hydrogen into helium, helium into carbon, and so on), and that they had a life cycle.

A similar atlas created by a modern-day Doppelmayr would depict glowing clouds – like the Orion Nebula – where new stars are forming, and where stellar deaths are signalled by supernovae. These colossal explosions fling 'processed' material back into space. There it mixes with the nebula gas, which later condenses into new stars. Our Sun was one such star. We are the ashes of long-dead stars – or, less romantically, we are 'nuclear waste' from the fuel that makes stars shine. Our bodies contain atoms forged from pristine hydrogen in many different stars, throughout the Milky Way, which lived and died more than five billion years ago – before our Sun formed. We are more intimately connected to the stars than astrologers could ever have conceived.

Some of Doppelmayr's most remarkable illustrations depict how the solar system would appear if viewed not from Earth but from another planet. One wonders whether it ever crossed his mind that 300 years later we would have sent probes to these other worlds, and would have mapped their surfaces more precisely than the Earth's southern hemisphere had been in his lifetime.

What are the transformative advances that would have fascinated Doppelmayr and his readers if they were alive today? They would

FIG. 2.

certainly have been amazed to learn of the huge age of the cosmos. The 'chronologies' of 18th-century thinkers were restricted – in some cases by theological arguments. In the 19th century, Charles Darwin (1809–82) and the geologists argued that the Earth must be at least tens of millions of years old, but in the 20th century it was recognized that nuclear energy could fuel the Sun and stars for billions of years, and that the light from the most distant observed galaxies had taken more than ten billion years to journey towards us. Today the origin of our cosmos can be traced back to an era 13.8 billion years ago, when everything was compressed into a dense fireball hotter than the centres of stars.

I think they would also have been fascinated by the new perspective on the possibility of alien life. They would be disappointed that space probes reveal the Moon and planets as arid and uninhabited – contrary to speculations by scientists such as Christiaan Huygens (1629–95) and William Herschel (1738–1822). But they would be elated by the discovery, within the last twenty-five years, that most stars are orbited by retinues of planets, just as our Sun is orbited by the Earth and the other planets of our solar system. There are many millions of 'exo-planets' in our galaxy that resemble the Earth. But is there life on them? This is, to me at least, the most exciting and challenging question in the whole of science, deeply relevant to humanity's significance and destiny. A successful quest for alien life will require a synthesis of many sciences – mathematics, physics, chemistry and biology – and an engagement with cultural studies. It will also mean a revival of the spirit of wonder and mystery that motivated the polymaths who Doppelmayr celebrates in the *Atlas Coelestis*.

FIG. 2.
The techniques of astronomy have advanced beyond recognition since the era of the *Atlas Coelestis*, allowing modern stargazers to extract astonishing levels of information from the light of distant planets, stars and galaxies, frequently without ever looking through a telescope. Yet images of the cosmos – such as this distant galaxy viewed by the Hubble Space Telescope – still have the power to capture the imagination in ways that data alone cannot.

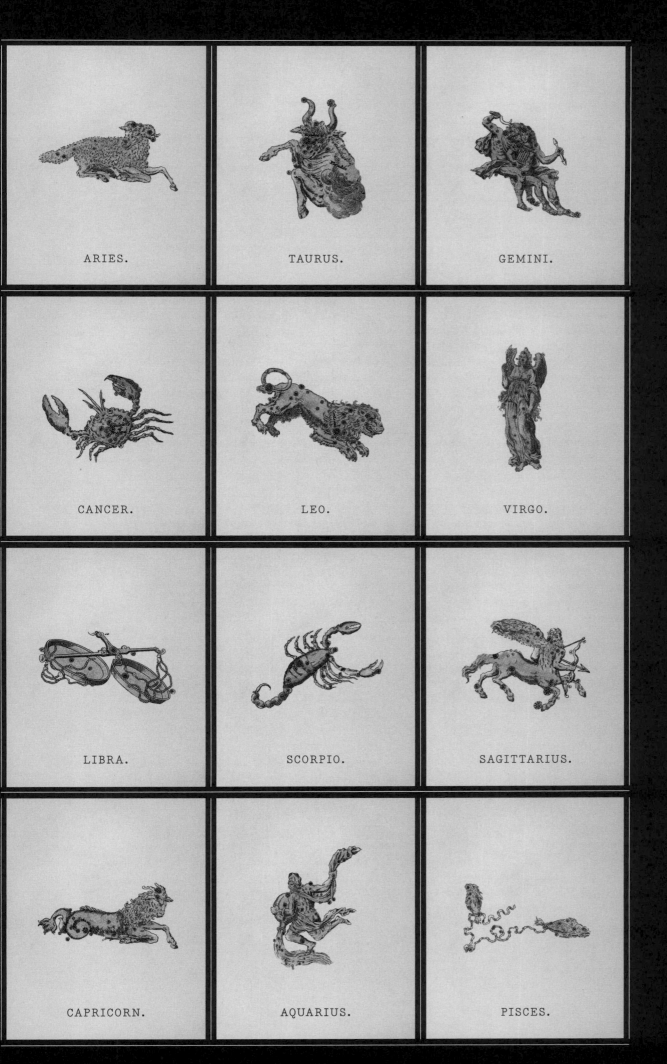

ARIES.

TAURUS.

GEMINI.

CANCER.

LEO.

VIRGO.

LIBRA.

SCORPIO.

SAGITTARIUS.

CAPRICORN.

AQUARIUS.

PISCES.

INTRODUCTION

The *Atlas Coelestis* of Johann Gabriel Doppelmayr (1677-1750) is one of the most spectacular artistic and scientific achievements of astronomy from the European Enlightenment. Across thirty remarkable plates, it gathers together and explains countless aspects of astronomical science as it was known at that time, ranging from the motions of the planets to the timing of eclipses, the passage of comets and the properties of distant stars. Published by the great cartographic house known as Homann Heirs in Nuremberg, Germany, in 1742, it collates illustrations created over the preceding decades for previous world atlases with many produced especially for the project. Together, they provide an unrivalled insight into the Enlightenment view of the cosmos: a world that had shaken off many of the wrong-headed theories that had persisted since classical times, but for whom many questions remained unanswered.

From the distance of the 21st century, Doppelmayr's time feels comfortably removed from the great turmoil that had swept through astronomy in the late 16th and early 17th centuries. When we consider the Copernican Revolution, which uprooted Earth from its privileged place at the centre of the cosmos and transformed it into one of several planets orbiting the Sun, it is all too easy to think of it as beginning with Nicolaus Copernicus's (1473-1543) treatise *On the Revolutions of the Heavenly Spheres*, published in the year of his death, and culminating with the trial of Galileo Galilei (1564-1642) before the Inquisition in 1633. History is written by the winners, and it is often assumed that, despite his condemnation by the Church, Galileo's discoveries and arguments settled the matter in the mind of all rational thinkers from that point onwards.

The truth, of course, is more complex, and the *Atlas Coelestis*, published almost exactly two centuries after Copernicus, hints at some of that complexity with its lengthy explorations of alternative systems of the universe.

When telescopic discoveries such as the Moon-like phases of Venus and the satellites orbiting Jupiter fatally undermined the case for a cosmos in which absolutely everything orbited Earth, the collapse took with it an entire system of physics that had stood mostly unchallenged for 2,000 years – and which had just as many applications on Earth as it did in the heavens. Medieval Islamic and European astronomers might have fretted at the increasingly obvious failings of the cosmological system erected by the ancients, but most assumed the solution lay in further refinement rather than wholesale replacement.

It is little wonder, then, that many unresolved questions lingered. What controlled the shape of planets' orbits around the Sun, and the periods in which they orbited? Could the details of orbits be modelled with enough accuracy to predict planetary motions? What was the true scale of the universe, and the true nature of the planets? Above all, if the classical laws of physics – in which materials naturally fell towards the centre of Earth and the universe, and thereby found their orderly place – were swept away, what should replace them?

Today, we understand the answer to that last question to be gravitation, an attractive force exerted by all heavy objects in proportion to their mass. Described in monumental detail by English mathematician and astronomer Isaac Newton (1643-1727) in his 1687 *Mathematical Principles of Natural Philosophy* (often known, from its abbreviated Latin title, as the *Principia*), the theory of gravitation has stood the test of time. Despite a paradigm shift in our understanding of its causes courtesy of Albert Einstein (1879-1955), it remains an elegant

FIG. 1.
The frontispiece of *Harmonia Macrocosmica*, a star atlas produced in 1660 by Dutch-German cartographer Andreas Cellarius, depicts key figures in the debate about the nature of the universe attending on Urania, the Greek muse of astronomy. They include Tycho Brahe (front left), Nicolaus Copernicus (front right) and Ptolemy of Alexandria (back row, left).

FIG. 1.

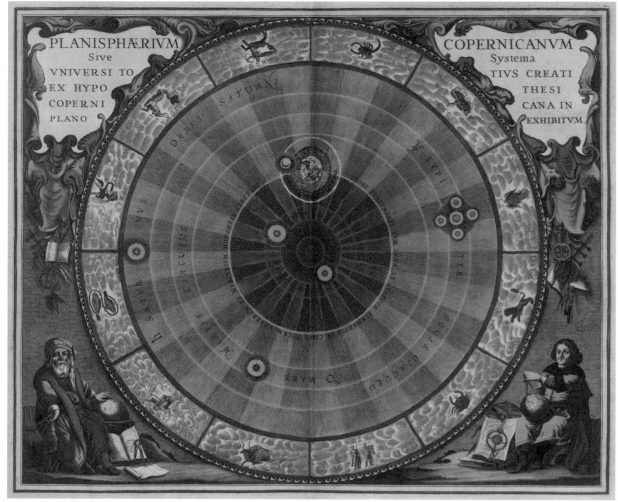

FIG. 2.

FIG. 2.
A plate from Cellarius's atlas displays the Copernican model of the universe in plan view. The Sun is at the centre, circled by the planets and ringed by the sphere of the stars (represented by the traditional constellations of the zodiac). Both Earth and Jupiter are depicted with their accompanying satellites.

description of the way that bodies behave in all but the most extreme circumstances.

It is easy to imagine, then, Newton's *Principia* being greeted with relief by the educated minds of the time as the longed-for solution to countless astronomical problems. But the truth is that Newtonian physics was slow to catch on – particularly in mainland Europe – and these questions remained open for far longer than we might now imagine.

In part, this was due to suspicions about Newton's methods; at the time, many natural philosophers were devoted to French mathematician and philosopher René Descartes's (1596–1650) mechanical model of the cosmos, in which all of space was filled with moving matter, and motion was passed from one body to another through direct physical contact. In contrast, Newton's argument for a mysterious attraction capable of influencing the behaviour of distant objects separated by empty voids of space came a little too close to the occult forces (in the word's original sense of hidden and undetectable) from which the rationalist natural philosophers of the Enlightenment were trying to move away.

Furthermore, while the predictive success of the Copernican system as modified by Johannes Kepler (1571–1630) convinced most astronomers that it must be right, there were still valid reasons for doubt, based not only on dogma but also on what seemed like solid observational grounds: namely, the absence of evidence from astronomical tests that were supposed to support it. Even Doppelmayr's younger German contemporary Tobias Mayer (1723–62), although a convinced Copernican himself, justifiably lamented: 'Where has anyone ever…given a thorough proof of the Copernican worldview? Everywhere it is either already presupposed as truth, or accepted as an arbitrary proposition…Should a treatise of astronomy be called clear if its fundamental thesis, the basis of all astronomical truths, remains unproven?'

What is more, there was a significant rival theory, courtesy of the renowned pre-telescopic astronomer Tycho Brahe (1546–1601). The so-called Tychonic system (introduced in detail on Plate 3) placed the planets in orbit around the Sun, but still kept the Moon and Sun in orbit around a static Earth. For many, this appeared to offer the best of both worlds – or at least a rational model in keeping with the observed facts of the time. The spectre of Tycho's system haunts the pages of the *Atlas*, not only as an alternative cosmology fresh in recent memory, but also as a tool for understanding the loops and circuits of celestial objects observed by Earthbound stargazers.

* * *

Doppelmayr was born in Nuremberg, a free city in the Holy Roman Empire, which at the time encompassed modern Germany, Austria, Czechia and neighbouring lands. During the late medieval period, Nuremberg's position at the crossroads of trade routes between northern and southern, eastern and western Europe had made it prosperous, and thriving business combined with religious tolerance put it at the heart of the late German Renaissance. The great painter and engraver Albrecht Dürer (1471–1528) remains the city's most famous son; along with his contemporaries, he created a tremendous flourishing in Germanic arts and culture that still lingered and shaped Doppelmayr's own work almost two centuries later.

Several more or less coincidental aspects of Nuremberg's role in the so-called Northern Renaissance put it close to the heart of the ongoing debate between rival cosmologies. A loose chain of cause and effect began in the 15th century, when the city's geographical location and growing status allowed it to gain renown for the manufacture of astronomical instruments. Monopolies on certain metals, access via trade routes to other materials, and a canny eye for industrial infrastructure supported by the city's patrician governing council (including, for example, the world's first wire-pulling mill) had made the city a centre of metalworking since the 14th century. When demand for the complex hardware needed for contemporary stargazing began to grow (for reasons explored in the Prelude to the *Atlas*; see pages 26–9), the city's skilled artisans were happy to meet it, turning out industrial quantities of instruments from sundials to quadrants, astrolabes and globes.

In turn, this reputation for craftsmanship drew leading astronomers of the day to Nuremberg, establishing a scientific lineage that Doppelmayr would surely have felt looking over his shoulder. Nicholas of Cusa (1401–64), an influential cleric and philosopher of the 15th century, travelled there to purchase his instruments when he turned his attention to the heavens in around 1444. More significantly, however, in 1471, Johannes Müller of Königsberg (1436–76) settled in Nuremberg at the age of thirty-five after a peripatetic early career.

Better known by his Latin monicker of Regiomontanus, Müller was already internationally renowned for both his own instrument-making prowess and his writings on subjects ranging from astrology and trigonometry to calendar reform and algebra. In Nuremberg, he set out to fulfil a promise made to his late friend and mentor Georg von Peuerbach (1423–61) to publish the latter's commentary on the venerable astronomy of Ptolemy, whose ideas had dominated Western astronomy for some 1,300 years. Partnering with a successful local merchant, Bernhard Walther (c. 1430–1504), he established what was effectively the world's first scientific publishing house. Its first product (and the first astronomical textbook to be published using the printing

FIGS. 3–5.
These astrolabes are typical of the high-quality astronomical instruments produced in Nuremberg in the 15th and 16th centuries. *From left to right*: a 16th-century brass astrolabe by Georg Hartmann, who pioneered the standard-ized production of the instruments in bulk; a 1538 instrument by Johann Wagner, designed primarily for astrological use; a board-and-paper astrolabe by Hartmann, sold as printed paper sheets for home construction.

FIG. 3.

FIG. 4.

FIG. 5.

FIG. 6.

FIG. 7.

FIG. 6.
This view of the
Egidienplatz,
Nuremberg, was
drawn by Johann
Andreas Graff in
1682. It shows the
Romanesque church
of Saint Egidien on the
right, the *Aegidianum*
next to it, and the
grand Pellerhaus
mansion at the far end.
The original church
was destroyed by fire
in 1696 and replaced
by a new baroque
building.

FIG. 7.
Graff created this
view of Nuremberg's
baroque Neuer Bau
square (now the
Maxplatz) in 1693.
It features the recently
installed fountain,
the Tritonbrunnen.

press, then barely a generation old) was Peuerbach's *New Theory of the Planets* (1472).

Although Regiomontanus's career was cut short by his untimely death in 1476, his legacy was continued by Walther, thus ensuring Nuremberg's growing stature as a hub for new ideas in astronomy, astrology and mathematics. In 1515, Dürer himself even made a significant contribution, through the production of the first star charts to be printed in Europe (see pages 172–3).

Nuremberg's reputation was further cemented in the 16th century by the establishment of the city's Protestant school (known as the *Egidiengymnasium* or *Aegidianum*, after its original location in a former monastery next to the church of Saint Egidien) in 1526. While the project largely focused on education in key attributes of Renaissance humanism, such as the classical languages, oratory and rhetoric, it attracted the renowned 'mathematicus' Johannes Schöner (1477–1547) as its first professor of mathematics. Schöner was already a publisher in his own right, and also a skilled cartographer, globe maker and astronomer. In 1538, he played host to Wittenberg professor Georg Joachim Rheticus (1514–74), with whom he discussed the ideas of Rheticus's own former tutor, Copernicus. Five years later, this led to the publication of Copernicus's famous *On the Revolutions of the Heavenly Spheres* at Nuremberg through the publishing house, or *Offizin*, of Johannes Petreius (c. 1497–1550).

Nuremberg's publishers would continue to act as a clearing house for new astronomical ideas throughout the following decades: another hugely influential production was Tycho Brahe's *Instruments for the Restoration of Astronomy*, some of whose plates are shown on pages 50–1. Tycho originally ordered a small number of copies to be printed for private circulation, but after his death the publisher Levinus

Hulsius (c. 1546–1606) brought it to a much wider audience through a new edition in 1602. Meanwhile, after a somewhat rocky start, the *Aegidianum* grew in stature following its relocation in 1575 to Altdorf, around 24 km (15 miles) east of the city. By 1622, it had been elevated to the status of university by imperial decree, with teaching split between a lower 'grammar school' and the university proper. When the grammar school returned to its old home in the city a few years later, it retained the name *Aegidianum* to distinguish it from the University of Altdorf. Both institutions played important roles in Doppelmayr's life and work.

From the early 1600s, Nuremberg began to suffer a loss of wealth, population and political influence amid changes to trade routes and the ravages of the Thirty Years' War (1618–48). However, throughout this time – and into Doppelmayr's own lifetime and beyond – it retained its reputation as a centre of learning, publishing, and artistic and scientific endeavour.

* * *

Doppelmayr's family had its roots around Oberreichental in the Grafschaft of Waldenfels, in present-day northwest Austria. His great-grandfather, a shoemaker called Sigmund, had emigrated to Nuremberg in 1562, probably in search of work. Sigmund made a success of his new life, gaining the rights of a citizen in 1593 and becoming a member of the city's Greater Council in 1626. When he died in 1632, his son Johannes (1603–61) was soon named in his place on the council. Although the 'Greater Council' was simply an advisory body to the patrician 'Lesser Council', where the real power lay, the election was a sign that the family was now well established among the city's merchant class.

It was with the third generation of Nuremberg Doppelmayrs – and in particular with Johann Gabriel's father Johann Sigmund (c. 1641–86) –

that an interest in natural philosophy first seems to have surfaced. While staying in the town of Öttingen, Johann Sigmund not only got married, but also came into contact with a figure who would influence both his own life and that of his son. The philosopher Johann Christoph Sturm (1635–1703) was a mathematician with a growing reputation as a skilled experimentalist. Johann Sigmund had shown early academic talent of his own (before the need to learn and, ultimately, take over the family business had limited his opportunities for scholarship), and when Sturm was called to the University of Altdorf in 1669, he took the opportunity to renew their acquaintance.

Thereafter, Johann Sigmund seems to have become something of an unofficial pupil of Sturm, studying mechanics and optics, and learning to make his own scientific instruments, including at least one for his 'mentor' – an air pump that may have been used in demonstrations at Sturm's *Collegium Experimentale sive Curiosum*, a private experimental club he established modelled on the scientific academies being set up elsewhere in Europe at the time. Johann Sigmund also corresponded with the astronomer Gottfried

Kirch (1639–1710) – later the first director of the Berlin Observatory – and from 1679 seems to have spent time at an observatory recently established by the artist, engraver and astronomer Georg Christoph Eimmart (1638–1705) on the Vestnertor bastion of Nuremberg Castle. By this point, he had two sons: Benedikt Jakob (b. 1667) and Johann Gabriel. The older brother, as heir, was groomed to take over the family business (although, ultimately, he took a different path and seems mostly to have lived off his inheritance). His junior by a decade, Johann Gabriel largely had to find his own way.

When Johann Sigmund died in 1686, his younger son was just eight years old. Legal statutes limited his mother Maria Catharina's power to decide her son's future, so guardians were appointed by the authorities. One of these was Johann Fabricius (1644–1729), a lecturer in theology at Altdorf (unrelated, as far as can be established, to his early 17th-century namesake famous for his early observation of spots on the Sun). Until 1689, Johann Gabriel was tutored at home, but from the age of twelve he attended the *Aegidianum*, where he received instruction in topics ranging from mathematics and theology to logic and the natural sciences. Latin was a matter of course for any educated young scholar of the time – and the lingua franca of Enlightenment Europe – but Doppelmayr also learned Greek. From the age of fifteen, he was promoted *ad lectiones publicas*, meaning that he could continue to study by attending lectures given by the gymnasium's most renowned professors.

At some point after his mother's remarriage in 1688, Doppelmayr became a permanent house guest of his guardian Fabricius, which doubtless eased the transition to studies at the University of Altdorf in 1696. Initially, he studied law, but some of his father's interest in natural philosophy had clearly rubbed off, and he also joined lectures by the renowned Sturm. Doppelmayr soon fell under his spell and began to concentrate on mathematics and physics, leading to dissertations on the Sun, and on vision and the camera obscura, a popular optical novelty of the time, under Sturm's guidance.

In 1699, Doppelmayr left Altdorf with his former university room-mate Moritz Wilhelm Böhmer (1676–1712). Their first destination was the recently founded University of Halle, where they continued their studies. Despite at first reapplying himself to law, it was here that Doppelmayr reached the crucial decision to devote himself to mathematics and natural philosophy. He decided to travel first to Holland, and then on to England – both countries at the heart of a growing discourse around natural

FIG. 8.
In this contemporary line engraving, Doppelmayr is depicted as an archetypal Enlightenment natural philosopher and man of letters, worthy of correspondence with some of the leading thinkers of his time.

IOHANN GABRIEL DOPPELMAIR.
*Mathem: Prof. Publ. Noriberg Acad.
Imperial. Leopoldino-Carolinæ Na:
turæ curiof. ut et fociet Reg. Borufs. fcient.
Sodalir.*

FIG. 8.

philosophy where he hoped to both learn new skills and discover the latest ideas.

Of these two destinations, Holland might have seemed the more immediately attractive for a young scholar at the turn of the 18th century. The Dutch Golden Age of the previous decades had seen huge innovations in art, commerce and science - not least inventions such as the telescope, simple microscope and pendulum clock, which were to bring new insight and precision to natural philosophy. The tolerance of the Dutch Republic also made it a haven for scholars seeking refuge from the religious turbulence that still periodically wracked other parts of Europe. When anti-Protestant repression in France triggered the arrival of a wave of Huguenot craftspeople after 1685, skilled printers transformed Holland into a thriving hub for the exchange of ideas that would fuel the coming Enlightenment.

Doppelmayr and Böhmer's 'Grand Tour' took them to Utrecht and then Leiden, where Doppelmayr stayed for some weeks with the German mathematician and astronomer Lothar Zumbach von Koesfeld (1661–1727). Zumbach was involved in the manufacture of globes and a variety of other devices for charting the sky, and it seems likely that it was here that Doppelmayr learned useful techniques for converting the various different forms of celestial coordinate used to pinpoint objects in the sky into positions on maps and charts of different projections. He also seems to have found the time to take instruction from a local optician in the art of grinding glass for optical instruments.

In early May 1701, the companions sailed from Rotterdam to London, where Doppelmayr sought out the company of mathematicians and natural philosophers. England's reputation for fresh thinking about the natural world had been fostered by the disparate scholars, physicians and experimentalists that had crystallized to form the Royal Society (officially the Royal Society of London for Improving Natural Knowledge) under a 1662 charter from Charles II (1630–85). Although not the first such scientific academy (both France and Italy had produced similar, though private, institutions), the society's innovative programme of investigations - and from 1665 its publication of the first solely scientific journal, the *Philosophical Transactions* - made it the first among equals of its day. Despite this, English natural philosophers were frequently somewhat at odds with their continental contemporaries.

In London and Oxford, Doppelmayr made the acquaintance of important figures, including Savilian Professor of Astronomy David Gregory (1659–1708), venerable scholar John Wallis (1616–

FIG. 9.

1703) and probably also John Flamsteed (1646–1719), the first Astronomer Royal. Subsequently, he was invited to attend lectures and discourses at the Royal Society, establishing relationships that would last for the rest of his life. Sadly, there is no evidence of a direct meeting with the famously spiky genius Isaac Newton, who was shortly to become president of the Royal Society. Late in the year, Doppelmayr returned to Holland (he had previously had a lucky escape when a friend persuaded him to give up his berth on another ship that subsequently foundered in a storm), before returning to Nuremberg by August 1702.

*   *   *

Doppelmayr's newly honed skills and his exposure to fresh ideas clearly made an impression on the scholars of Nuremberg, for by 1704, at the tender age of twenty-six, he was appointed as the new professor of mathematics at his old school, the *Aegidianum*. It was here that he would remain for the rest of his life, devoting himself to research, teaching and, most significantly for posterity, the popularization of the latest scientific ideas. In 1716, he married Susanna Maria Kellner (1697–1728), the daughter of a prominent local apothecary. They had four sons, but only the second survived infancy. (This surviving son, Johann Sigmund, showed his own early aptitude for mathematics and was taught by his father at the gymnasium, but later followed his mother's side of the family to become an apothecary.) By 1728, Doppelmayr was widowed; he passed the remainder of his life without remarrying, preferring to devote himself to scholarly studies, learned conversation, his garden and his publishing endeavours.

Doppelmayr may have no great scientific discovery of his own as a claim to fame, but he brought a sharp intellect and curious mind

FIG. 10.

to his work in interpreting mathematical and astronomical ideas for a wider audience, and in translating the theories of others. If a single thread can be seen to string together the majority of his publications, it is the assembly and dissemination of arguments for the correctness of the Copernican model of the universe.

Educators at the University of Altdorf and the gymnasium were free to teach the view of the universe that they personally preferred, and as late as 1669, Sturm's forerunner, Abdias Treu (1597–1669), had defended the ancient ideas of Aristotle with little modification. Sturm himself weighed the evidence and agreed that the planets orbited the Sun, but he remained acutely aware of the fact that the motion of Earth could not be proved.

The smouldering German debate about the structure of the universe finally flared back into life in the early 18th century after Johann Philipp von Wurzelbauer (1651–1725) – a prominent Nuremberg merchant who also acted as assistant to Eimmart at his observatory – translated a curious but influential work from the Low Countries called the *Cosmotheoros*. Written by the celebrated astronomer and instrument maker Christiaan Huygens (1629–95) and published posthumously at his own behest, the book was predominantly an extended speculation about life on other worlds, but was deeply rooted in a Copernican interpretation of the universe. When Wurzelbauer's

translation was published in 1703, it sparked a renewed wave of interest in what became known in Germany as the 'Copernican-Huygenian system'.

As a fellow member of the Nuremberg astronomical community, Doppelmayr must have read the new translation and was clearly intrigued. He also spotted an opportunity to answer the renewed demand for explications of the Copernican model of the universe. What may be his earliest published work, from 1705, shows an obvious intent to promote the Copernican theory's practical use among the German scientific community. This was a translation into Latin of the *Astronomia Carolina* (1664) by English astronomer Thomas Streete (1621–89). Streete's work combined a description of the Copernican model, modified by the laws of planetary motion of Johannes Kepler, with a series of tables for the accurate prediction of planetary positions. It offered a startling demonstration of the power of the Sun-centred approach and had already been used by Newton in developing his *Principia*. However, Doppelmayr's translation made it far more accessible to astronomers in continental Europe.

From 1706 onwards, Doppelmayr joined forces with Johann Baptist Homann (1664–1724), a former Dominican monk who had worked as a map engraver for other Nuremberg publishers before establishing his own firm in 1702. This alliance was the beginning of a partnership that would continue beyond Homann's lifetime and culminate in the spectacular *Atlas Coelestis*. Their first collaboration was an engraved plate illustrating the track of the solar eclipse that swept across Europe on 12 May 1706. Homann published this alongside a plate explaining current knowledge of the solar system (ultimately Plate 2 of the *Atlas*), and commissioned Doppelmayr to write a small accompanying book that explained the theory and mathematics behind the plates in greater detail. It is notable that the text on the plates was engraved in Latin, whereas the commentary was printed in German; this shrewd move broadened Homann's market beyond the Latin-reading elite, and allowed Doppelmayr to bring the ideas of Copernicanism to the masses. When they went on sale in 1707, both the plates and book proved highly popular. Clearly convinced that they had struck a chord, the pair collaborated again to publish Doppelmayr's *Short Introduction to Astronomy* the following year.

The professor of mathematics was now increasingly in demand among scientifically minded Nuremberg publishers. In 1708, he expanded an existing treatise on the marking of sundials, and in 1712, he adapted an encyclopaedic work by the French instrument maker Nicolas Bion (1652–1733) on the construction of mathematical instruments. In later editions, Doppelmayr added useful tables, instructions to make tools for surveying and an expanded section of astronomical instruments (including some ingenious inventions of his old instructor Zumbach).

Doppelmayr returned to theoretical matters – and his fruitful promotion of Copernicanism – in 1713, with a translation that combined a pair of works by English philosopher John Wilkins (1614–72). *The Discovery of a World in the Moone* (1638) was Wilkins's discussion of the Moon's motions, surface features and the possibility that it might be inhabited, while *A Discourse Concerning a New Planet* (1640) was an argument for the ideas of Copernicus and Galileo and the concept that Earth was itself a planet. Doppelmayr dropped the somewhat confusing titles for his volume, naming it instead *John Wilkins' Defence of Copernicus.*

Throughout this productive period, Doppelmayr's reputation grew steadily. Although his travels seem to have been limited, he corresponded widely. In 1715, he was elected to both the German National Academy of Sciences in Halle and the Berlin Academy. In 1724, he was offered a professorship at the new Russian capital of St Petersburg. Doppelmayr declined the position, fearing the climate, but he was later admitted to the Imperial Russian Academy of Sciences anyway. In 1733, he received a visit from Martin Folkes (1690–1754), vice president of London's Royal Society, who subsequently recommended him for fellowship.

Meanwhile, in 1730, Doppelmayr produced an invaluable reference work for later historians in the form of the *Historical account of the Nuremberg mathematicians and artists* – an index and potted biography of the city's mathematicians, instrument makers and artists. The following year's *Illustrated Experimental Physics* catalogued some 700 experiments conducted by and for Sturm's *Collegium Experimentale*, many of which Doppelmayr

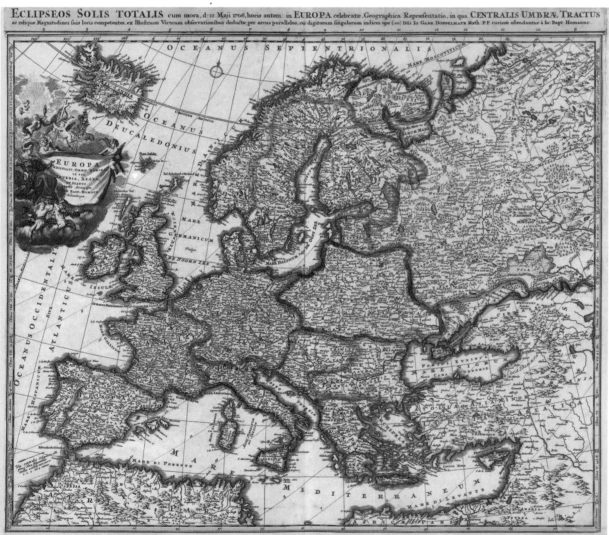

FIG. 11.

FIG. 11.
Doppelmayr and Homann's original map of the eclipse seen across Europe in May 1706. Here, the path of totality – where the Moon blocked the Sun entirely from view – is shaded blue. The 'penumbral' regions of the eclipse track – where more and more of the Sun was visible – are marked by dotted lines. The map was the most accurate produced at that time, and it was popular enough to be pirated (with some rather inelegant modifications) by Homann's rival Peter Schenk the Elder. Doppelmayr revisits the event in Plate 13 of the *Atlas Coelestis.*

FIGS. 12–13.
These two Doppelmayr celestial globes are held in the Dresden State Art Collections. At near right is a large 32-cm (12-in.) globe of 1728, engraved with the forty-eight classical constellations of Ptolemy, as well as some thirty later additions, with stellar coordinates calculated for 1730. At far right is a 20-cm (8-in.) globe of 1730.

FIG. 12.

FIG. 13.

himself had reproduced and elaborated upon in lectures that attracted not only scholars and artisans from Nuremberg, but also distinguished patrons from further afield.

Around this time, Doppelmayr became fascinated by the emerging science of electricity. He conducted many experiments and built devices of his own. However, some claim that a shock sustained during these experiments led to the stroke that paralysed the right side of his body and was ultimately responsible for his decline and death, at the age of seventy-two, in 1750.

From 1728, Doppelmayr also successfully branched out into the manufacture of globes. Nuremberg had a prestigious record in this area, dating back to the world's oldest surviving terrestrial globe (and, as far as we know, the first to be made in western Europe since antiquity).

While early large globes could be used for a variety of scientific and navigational purposes, if properly set up and equipped with a variety of additional rings and measuring devices, improvements to map making in the 16th century saw flat charts take their place for these practical purposes. By Doppelmayr's time, they survived as educational tools, but also as icons of knowledge itself. Just as today, the ostentatious display of such items conveyed a number of signals about the interests and status of the owner.

The construction of globes was a highly specialized craft for all involved: the cartographer who had to plot locations on Earth (or the positions of the stars) onto the distorted lens-shaped gores required to wrap around a globe, the artist who had to embellish these features, and the craftsman who had to put it all together. Doppelmayr partnered with the workshop of copperplate engraver Johann Georg Puschner (1680–1749) to make both terrestrial and celestial globes, which were sold in Nuremberg and Vienna. They were hugely

successful and became the most widely sold globes across German-speaking Europe at the time, continuing to be made (with some updates and amendments) into the early 19th century.

Alongside these endeavours, Doppelmayr continued a productive relationship with Homann and his successors across almost forty years, the fruits of which fill the following pages. Homann himself rapidly built a reputation as one of the foremost cartographers of the age, and in 1715 he was appointed Imperial Geographer by the Holy Roman Emperor Charles VI (1685– 1740). The honorific brought with it significant business advantages – not only in terms of prestige, but also through the grant of imperial printing privileges (an early form of copyright that gave Homann's firm some protection from illegal reproductions). Doppelmayr's skill as a mathematician would have been invaluable to Homann in determining the best means of rendering the spherical surfaces of Earth and the imaginary celestial sphere onto the flat surface of the printed page.

Doppelmayr also began supplementing Homann's terrestrial atlases with additional plates illustrating various aspects of astronomy. For the *New Atlas of the Whole World* (1707), the Copernican plate that had originally accompanied the eclipse chart was included, alongside a plate showing the surface features of the Moon. For the *Atlas of 100 Charts* (1712), meanwhile, he added a plate illustrating the rival Tychonic interpretation of the universe, alongside charts plotting the motions of the planets around the sky. These early charts appear in the final *Atlas Coelestis* as Plates 2 and 11, and Plates 3 and 7 through 10 respectively.

It is unclear when Homann and Doppelmayr first had the notion of collating and supplementing their existing plates to produce a grand atlas of the heavens, but work was clearly underway on the project in one form or another

FIG. 14.

FIG. 15.

FIGS. 14–15.
Two further celestial globes designed by Doppelmayr and manufactured by Johann Georg Puschner. At far left is a 20-cm (8-in.) globe of 1730. At near left is a miniature 10-cm (4-in.) globe of 1736. For this smallest size of globe, Doppelmayr used a more limited selection of constellations. He removed the faintest of his six magnitudes of stars and made several other changes.

prior to 1724, since several other plates (including those depicting the constellations) were listed among Homann's property at the time of his death that year. Experts have spotted the influence of the Polish astronomer Johannes Hevelius (1611–87) in these plates, while finding no trace of Flamsteed's authoritative catalogue and atlas of the late 1720s. Doppelmayr's own comments in the *Historical Account* indicate that the project was conceived in its final form by 1730, but more plates certainly remained to be completed in the late 1730s and early 1740s. These later dates are confirmed by their attribution to Homann Heirs (as the company became known after 1730) and also by the astronomical data incorporated within them.

Published in 1742, the *Atlas Coelestis* was widely sold and became a respectable bestseller for its time – certainly making enough impact for an expanded reissue (the *Atlas Novus Coelestis*) in 1748 or shortly thereafter, which added a somewhat scattershot selection of additional plates from other designers. Researcher Robert Harry van Gent of the University of Utrecht has identified eighteen copies of the original version held in specialist libraries around the world, and some seventeen additional copies from 1748 or after. However, entire copies and individual plates appear fairly frequently for private sale.

Doppelmayr's *Atlas* marks a substantial shift from earlier celestial cartography; unlike the previous atlases and charts of astronomers such as Johann Bayer (1572–1625), Hevelius and Flamsteed, it did not provide new observational data, nor did it attempt to introduce new constellations. Instead, it aimed to summarize the state of knowledge about the universe at the time, illustrate many of the calculations and theories behind it, and demonstrate the power of the Copernican model to explain celestial phenomena. It is undeniably a work of what

we would today call 'popular science', and with its wealth of illustrations, captions and annotations arguably exemplifies the ancestry of modern illustrated publishing.

The *Atlas* represents a marvellous synthesis of art and science, from a period when the use of elaborate allegories and constellation figures had not yet begun to dwindle in the face of a more 'nuts and bolts' approach. Certain plates had a long and significant afterlife: for example, Plate 15's world map, determined from the latest astronomical observations, was sold in large numbers as a standalone print and – although not without its errors – was considered one of the finest of its time and a basis for later and more accurate work. German amateur astronomer Johann Georg Palitzsch (1723–88), meanwhile, had Doppelmayr's constellation charts at his side to plot the location of a faint comet he discovered in Pisces on Christmas Day 1758. By confirming Edmond Halley's (1656–1742) prediction that the comet seen in 1682 would return seventy-six years later, Palitzsch's discovery put the theories that Doppelmayr had promoted throughout his career beyond doubt.

Finally, from today's perspective – at the far end of a telescoping series of astronomical revolutions that have repeatedly reduced the significance of Earth's place in the universe – the *Atlas* remains a salutary reminder that revolutions in the history of science and ideas are always more gradual, more tangled and more intriguing than the 'Just So' stories that we tell ourselves with the benefit of hindsight. In highlighting how long the Copernican debate continued, and revealing the valid doubts that led many to prefer rival systems of the universe, the *Atlas Coelestis* provides a fascinating insight into the ways in which knowledge advances by degrees – as well as a monumental summary of the universe as Doppelmayr and his contemporaries understood it.

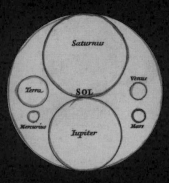

CELESTIAL BODIES
(TYCHO).

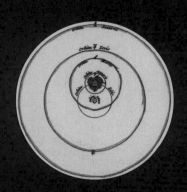

SYSTEM OF
RICCIOLI.

CELESTIAL BODIES
(RICCIOLI).

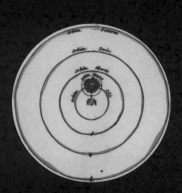

EGYPTIAN
SYSTEM.

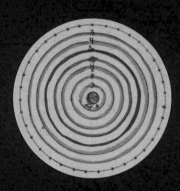

SYSTEM OF
PLUTARCH.

SYSTEM OF
JOHANNES COCCEIUS.

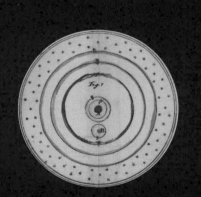

SYSTEM OF
WILLIAM GILBERT.

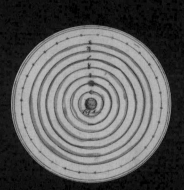

SYSTEM OF
PORPHYRY.

PTOLEMAIC AND
ALFONSINE SYSTEM.

SYSTEM OF
SÉBASTIAN LE CLERC.

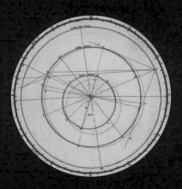

ORBITAL RELATIONSHIPS
(INNER PLANETS).

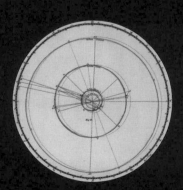

ORBITAL RELATIONSHIPS
(OUTER PLANETS).

T	H	E
A	T	L
A	S	·
C	O	E
L	E	S
T	I	S

ATLAS COELESTIS
studio et labore
JOH. GABRIELIS DOPPELMAIERI. Math. P.P.
impensis
Heredum Homannianorum.
Noribergæ. A. MDCCXLII.

ATLAS
COELESTIS

IN QVO

MVNDVS SPECTABILIS

ET IN EODEM
STELLARVM OMNIVM

PHOENOMENA NOTABILIA,

CIRCA IPSARVM LVMEN, FIGVRAM, FACIÉM, MOTVM, ECLI-
PSES, OCCVLTATIONES TRANSITVS, MAGNITVDINES DISTAN-
TIAS, ALIAQVE
SECVNDVM

NIC. COPERNICI

ET EX PARTE

TYCHONIS DE BRAHE

HIPOTHESIN.

NOSTRI INTUITU, SPECIALITER, RESPECTU VERO AD AP-
PARENTIAS PLANETARVM INDAGATV POSSIBILES E PLANETIS PRI-
MARIIS, ET E LUNA HABITO, GENERALITER

E CELEBERRIMORUM ASTRONOMORUM OBSERVATIONIBUS
GRAPHICE DESCRIPTA EXHIBENTVR

A

IOH. GABRIELE DOPPELMAIERO,

ACADEMIARVM IMPP. LEOPOLDINO CAROLINAE ET PETRO-
POLITANE, SOCIETATVMQVE REGG. SCIENTIARVM BRITANNICAE ET
BORVSSICAE, SODALI, NEC NON PROFESSORE PVBL. MATHEMA-
TVM NORIMB.

NORIMBERGAE

Sumptibus Heredum Homannianorum. A. 1742.

BEFORE COPERNICUS.

*Models of the heavens based on an Earth-centred universe were initially simple,
but became inevitably more complex in order to reconcile them with
increasingly accurate measurements.*

The *Atlas Coelestis* opens with a decorative frontispiece depicting four of the great figures in astronomy up to Doppelmayr's time. They are standing in a fanciful landscape with Nuremberg, Germany, on the horizon, beneath a sky filled with countless solar systems, mirroring the cosmology of French philosopher René Decartes (1596–1650). From left to right, these key figures are Claudius Ptolemaeus (better known as Ptolemy, *c.* 100–170 CE), Nicolaus Copernicus (1473–1543), Johannes Kepler (1571–1630) and Tycho Brahe (1546–1601). Copernicus, Kepler and Brahe are all figures who we shall encounter many times amid the pages of the *Atlas*, but it was Ptolemy who established the model of the cosmos that they all, to a greater or lesser extent, reacted against. To fully understand the unfolding Scientific Revolution that forms the stage for Johann Gabriel Doppelmayr's work, it is helpful to know something of the centuries of Greek 'natural philosophy' regarding the heavens that Ptolemy himself inherited, distilled and refined.

For thousands of years, the most intuitive interpretation of the universe was one that placed Earth at the centre of everything, with the Sun, Moon, planets and stars all circling it. How else would one explain the daily wheeling of the stars across the sky, the rising and setting of the Sun, and the phases of the Moon? As far as we can tell, this Earth-centred, or geocentric, view of the cosmos was shared by all ancient civilizations before the time of classical Greece.

Placing Earth centre stage understandably gave rise to the view that events in the heavens might influence the earthly realm, and, therefore, the many traditions of astrology that underlie most astronomy prior to the 17th century. Such traditions varied from the ancient Mesopotamian idea that major events on Earth and in the heavens followed similar cycles (hence, carefully watching the skies for repetitions of past celestial events could provide omens of momentous terrestrial happenings), to the Egyptian-influenced horoscopic astrology of the classical Mediterranean, in which the positions of celestial bodies at certain times allowed more direct divinations of personal fates. In these cases and many others, astrology made the understanding of celestial motions a matter of direct concern for everyday life.

It was only around the 5th century BCE that some Pythagorean philosophers – followers of the mystic philosopher Pythagoras (*c.* 570–495 BCE) who believed that numbers lay at the heart of nature – began to consider more complex theories. Philolaus of Croton (*c.* 470–385 BCE), for example, proposed that Earth and the Sun both orbited an unseen central fire that lay forever out of sight from the Mediterranean hemisphere.

Most philosophers, however, retained their belief in a static central Earth, reinforced by what seemed like sound physical and philosophical arguments. Aristotle (384–322 BCE) argued that Earth could not be in motion because otherwise we would see the effects of that movement – for example, if Earth ploughed through space, why did this not create a wind? If it was spinning on

PAGES 24–25.
The frontispiece and title page of the first edition of Johann Gabriel Doppelmayr's *Atlas Coelestis*, published in 1742.

FIG. 1.
In this plate from a 1708 edition of Andreas Cellarius's *Harmonia Macrocosmica*, celestial bodies – including personifications of the Sun, Moon and planets – are carried on chariots around their circular orbits of the fixed Earth.

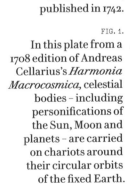

FIG. 1.

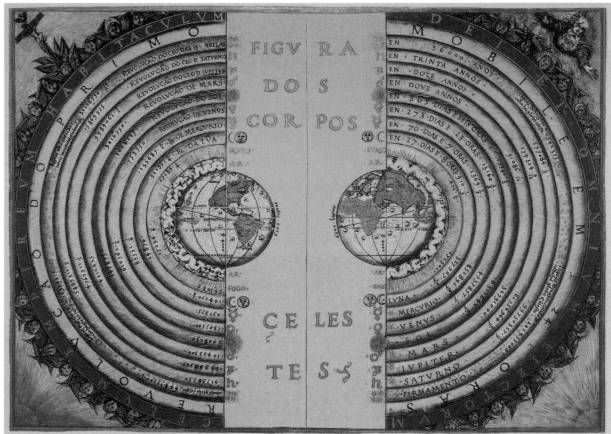

FIG. 2.

FIG. 2.
An illustration
of the geocentric
universe from the
Cosmographia
(1568) of Portuguese
cartographer
Bartolomeu Velho.
The figures given
around the perimeter
of each sphere show
the impressive level
of sophistication to
which the Ptolemaic
system was developed.
They not only
incorporate a 36,000-
year rotation period
for the stellar sphere
(accounting for the
phenomenon known
as 'precession'), but
also permit estimates
of each sphere's
distance from Earth
and its circumference.

its axis, why did objects flung vertically into the air not land behind the thrower? And if it was circling the Sun, why did *that* not show itself in the course of a year as a change in the apparent directions of the other celestial objects? Above all, Aristotle argued that Earth could not be in motion without some perpetual force being applied to keep it moving. This idea had roots in a broader Greek view of the universe that Aristotle supported – one in which the changeable, messy province of the everyday world was composed of four elements – earth, air, fire and water – in constant flux. Celestial objects, in contrast, inhabited a higher realm of perfection; composed of a fifth element, aether, they could move perpetually, following circular paths rather than the straight lines and arcs of motion on Earth.

The Greek system of Aristotle's time saw the universe as a series of nested spheres, arranged in concentric shells around Earth. The idea had its origins among the Pythagoreans, but was crystallized by Aristotle's older contemporary Plato (*c.* 428–348 BCE). Plato's interest in geometry brought him to believe that the universe should form a perfect spherical shape around a spherical Earth, with the motions of bodies within it themselves limited to spheres and circles. The Moon inhabited the innermost of these, with its changeable appearance blamed on proximity

to the terrestrial realm. Beyond it lay the Sun, and then the planets Venus, Mercury, Mars, Jupiter and Saturn – bright lights that appeared to follow intricate paths as they wandered against the outermost sphere of stars.

It was the motion of the planets that caused the greatest headache for classical astronomers. Venus appeared to execute large loops around the sky but was never seen far from the Sun; Mercury followed the same principle with even tighter loops. Mars, Jupiter and Saturn, meanwhile, wandered slowly through the constellations in a band of sky called the zodiac, moving gradually from east to west. Periodically, however, they would stop and reverse their motion, tracing retrograde loops across the sky before renewing their westward advance.

Plato challenged his disciples at the well-known Academy in Athens, Greece, to formulate a model that described these complex paths through the circular motion of multiple spheres or circles. All of these shells remained centred on Earth, and each rotated on its own axis at its own constant rate, but the motions combined as each one was driven by the next, resulting in the somewhat erratic wanderings of the planets across the sky. Eudoxus of Cnidus (*c.* 408–355 BCE) appears to have been the first to rise to the challenge – his original writings are lost forever, but references to his works survive in later Greek texts. In one of these, *On Speeds*, he developed a system that invoked a total of twenty-seven spheres to describe celestial motions: three each to drive the Moon and Sun, four for each of the known planets, and an all-encompassing outer sphere to carry the stars.

Neat though Eudoxus's model might have seemed, it was little use as a predictive tool; as astronomers continued to track the skies, they found that year upon year, the planets drifted further and further from their expected positions. Callippus (*c.* 370–300 BCE), a student of Eudoxus, added seven more spheres in order to correct the motions of Mercury, Venus, Mars and the Moon, and to take account of his discovery that the Sun did not move steadily through the stars over the course of the year.

Aristotle inherited this system; his chief innovation was to add further spheres between the sets influencing each celestial body, cancelling out the influence of the ones directly beyond it so that each planet could effectively be considered in isolation. Yet complexities continued to pile up – not least when Autolycus of Pitane (*c.* 360–290 BCE) questioned how the changing brightness of the planets might be accounted for when the nested spheres permitted no variation in their distances.

Astronomers and philosophers argued long and hard over how this creaking system (which in some interpretations now laboured under the weight of fifty-seven distinct spheres) might be fixed. It was in this environment that the possibility of a Sun-centred system was apparently first discussed by Aristarchus of Samos (*c.* 310–230 BCE). Again, his original work is lost, although it may have been influenced by his separate realization that the Sun was

significantly larger than Earth, and concern that it made little sense for a larger body to circle a smaller one (see page 103). These tentative ideas seem to have met with little interest – the persuasive physical arguments against a moving Earth still remained, and would do so until Galileo Galilei (1564–1642) discovered the principle of inertia in the early 17th century.

As early as the 3rd century BCE, however, Apollonius of Perga (*c.* 240–190 BCE) suggested an alternative approach that could do away with the vast majority of Aristotle's nested shells. What if, instead of sitting directly on a sphere around Earth, each planet was carried on a smaller sphere (an 'epicycle') that was itself rotating around a point carried on a larger geocentric sphere (the 'deferent')? The approach had immediate appeal. It seemed to offer an intuitive explanation for the retrograde looping motions of the outer planets, and it was popularized in the following decades by Hipparchus of Nicaea (*c.* 190–*c.* 120 BCE). However, it too failed the ultimate test of offering long-term accurate predictions for planetary motion. The problem was that deferents and epicycles were still expected to rotate at uniform rates in order to fit in with Platonic notions of perfection – yet it proved impossible for any combination of uniform rates to produce the appearance of changing planetary speeds measured in the heavens. The situation was ultimately saved – and the useful life of deferents and epicycles prolonged by a further thousand years and more – by Ptolemy in the 2nd century CE: a feat whose lasting influence explains his prominence on the *Atlas* frontispiece.

Ptolemy was the pre-eminent astronomer and geographer of his day, and his great book on the

FIG. 3.
In this illumination from a 15th-century French manuscript, Christ supports a globe containing the five layers of elements in the Aristotelean universe: aether, fire, air, water and earth.

FIG. 4.
This illumination from a 15th-century Greek edition of the *Cosmographia* (a work better known today as the *Geography*) depicts Ptolemy measuring the heavens and Earth.

FIG. 3.

FIG. 4.

cosmos – blessed with the ungainly original title of *Mathematical Syntaxis*, but known to history as the *Almagest* (150 CE) – summarized knowledge of the heavens up until his time. Ptolemy's most important original addition was a concept known as the 'equant point'. This, he suggested, was a location in space, removed from the centre of Earth, that provided a vantage point from which a particular deferent would appear to rotate at a uniform rate. Thus, while the pivot point of a planet's epicycle might appear to vary its speed through the heavens when seen from Earth, its motion was actually steady when seen from that planet's equant. The idea of uniform circular motion was, thus, effectively saved by splitting it in two – for each celestial body, the rotation of the deferent would be circular in relation to Earth at its centre, and uniform in relation to its equant.

To modern eyes, Ptolemy's solution might seem to be driven by little more than a dogmatic adherence to the Platonic ideal. However, it proved to be an effective mathematical tool, finally allowing planetary positions to be predicted with accuracy over fairly long periods (within the tolerance of the instruments of the day). Thus, the approach outlined in the *Almagest* became the standard tool of Western astronomy throughout the late classical world. Practical stargazers, tasked with charting the motions of the planets for astrological or religious reasons, could concern themselves with refining the knowledge and accuracy of past and present positions of the planets in order to improve predictions of the future.

With the fall of the Western Roman Empire, Ptolemy's ideas and writings were preserved only in the East, at the Greek-speaking court of Byzantium. It was from here that they were transmitted to the Islamic world, and soon adopted – and adapted – by scholars from a tradition unhampered by allegiance to ancient Greek philosophy.

The astronomers of the Islamic Golden Age, driven by both the pursuit of knowledge and the time-keeping and calendrical needs of their young religion, developed sophisticated new instruments for measuring and predicting celestial motions. Hence, while they still held the Ptolemaic model in high regard (even the *Almagest*'s modern title derives from *al-Majisti*, Arabic for 'the greatest'), they could not help but discover its shortcomings. Around 1027, mathematician Hasan ibn al-Haytham (*c.* 965–1040) summarized his own concerns in a work titled *Doubts on Ptolemy*, which inspired many further attempts at refinement and reconciliation. In 1247, for example, Persian scholar Nasir al-Din al-Tusi (1201–74) devised an entirely new

FIG. 5.

FIG. 5.
This depiction of the Moon in the constellation Cancer is from the *Kitab al-Bulhan,* or *Book of Wonders*, a 14th-century Persian manuscript on astronomy, astrology and other practices, compiled by Abd al-Hasan al-Isfahani.

mechanism to replace Ptolemy's equant in describing the apparent back-and-forth motion of the inner planets. Intriguingly, this system, known today as a Tusi couple, was later used by Copernicus himself, raising questions about whether he was aware of al-Tusi's ideas. Many other Islamic astronomers began to suggest that the world might at least rotate, if it did not actually move through space.

As classical Greek texts began to re-enter medieval Europe (frequently through cultural melting pots such as Spain and Sicily) and were translated into the more widely read Latin, new ideas from the Islamic world arrived alongside them. In astronomy, one of the most obvious signs of this mingling was a wealth of new names for individual stars. More significant, however, was the *zij*, or book of astronomical tables. Originating in Persia, these books provided the parameters for calculations of future planetary motions. After the tradition migrated into Europe through the sponsorship of Alfonso X of Castile (1221–84), it would be the stubborn refusal of celestial bodies to conform to the predictions of such tables, despite countless refinements, that ultimately triggered a revolution in Western astronomy.

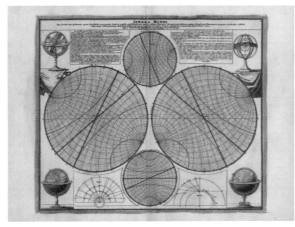

PLATE 1. | 34–35 |

PLATE 2. | 40–41 |

PLATE 3. | 46–47 |

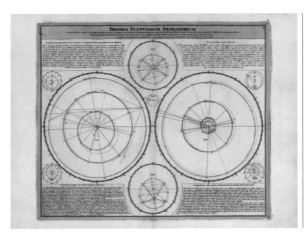

PLATE 4. | 56–57 |

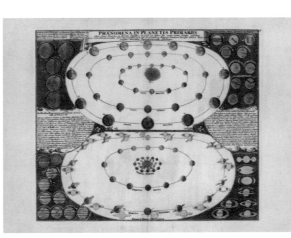

PLATE 5. | 64–65 |

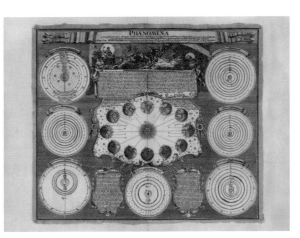

PLATE 6. | 70–71 |

PLATE 7. | 78–79 |

PLATE 8. | 84–85 |

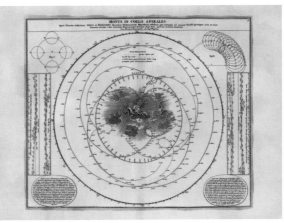

PLATE 9. | 90–91 |

PLATE 10. | 96–97 |

PLATE 11. | 104–105 |

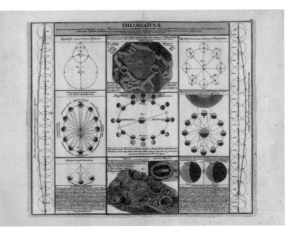

PLATE 12. | 112–113 |

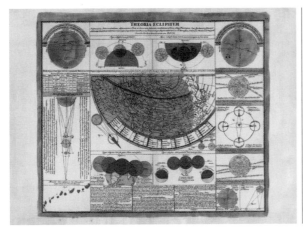

PLATE 13. | 120–121 |

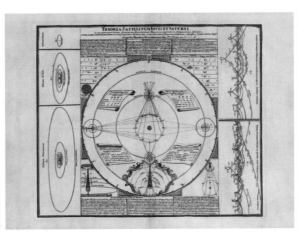

PLATE 14. | 128–129 |

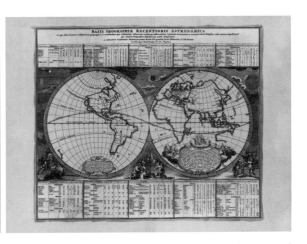

PLATE 15. | 134–135 |

PLATE 16. | 140–141 |

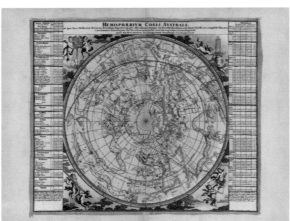

PLATE 17. | 148–149 |

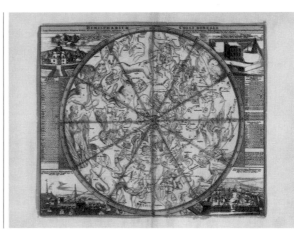

PLATE 18. | 156–157 |

PLATE 19. | 162–163 |

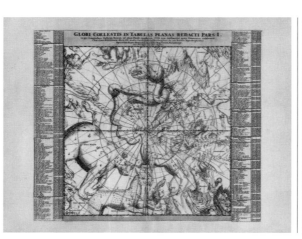

PLATE 20. | 168–169 |

PLATE 21. | 174–175 |

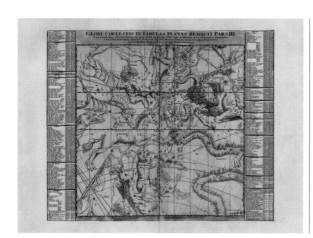

PLATE 22. | 180–181 |

PLATE 23. | 186–187 |

PLATE 24. | 192–193 |

PLATE 25. | 198–199 |

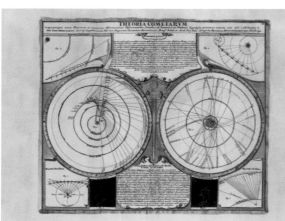

PLATE 26. | 206–207 |

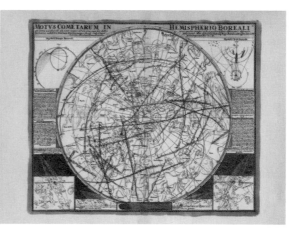

PLATE 27. | 212–213 |

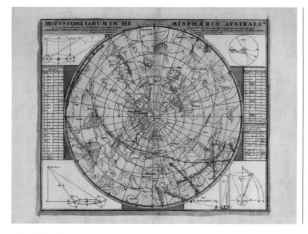

PLATE 28. | 220–221 |

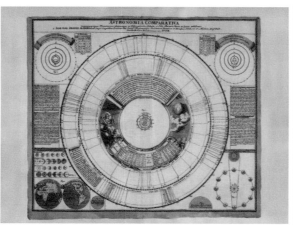

PLATE 29. | 226–227 |

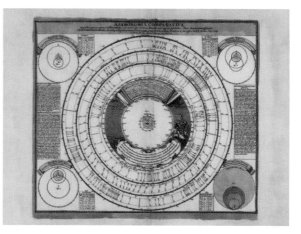

PLATE 30. | 230–231 |

PLATE 1.

THE SPHERE OF THE EARTH.

(SPHÆRA MUNDI)

*Doppelmayr describes the principles that relate an observer's
location on Earth to the visibility of the celestial sphere.*

The first plate of Doppelmayr's *Atlas* describes a principle that astronomers have used to interpret the sky since ancient times: the relationship between Earth's own surface and the so-called 'celestial sphere'. With no cues to the relative distance of celestial objects, even today many people find it natural to interpret the night sky as a dome of vast and uniform distance. The philosophers of the classical world took this assumption as the basis for an entire system of cosmology, assuming that objects beyond Earth were moving in front of a remote sphere of 'fixed stars', which they usually assumed to be revolving on its axis once every twenty-four hours, while the Earth stood still at the centre of the cosmos. Centuries of pre-telescopic stargazing were largely devoted to reconciling this model with the actual motions revealed in the heavens. While 400 years of astronomy, with the aid of optical instruments, have revealed clues that show the reality to be very different, the concept of a celestial sphere – on which angles can be measured

and the motions of celestial bodies projected – is so useful that it is still used today.

The Greeks, like most ancient peoples, were well aware that Earth was a sphere. They recognized that different stars came into view, and celestial objects reached different altitudes above the horizon, as one moved north or south, and that the Sun rose earlier in the east and later in the west – both phenomena that could not be explained by any sort of model in which Earth was a flat plane. Other evidence came from the appearance of Earth's circular shadow cast on the Moon during lunar eclipses, and the way in which the tops of tall buildings came into view as a ship approached harbour while lower ones were still out of sight.

The celestial sphere concept that astronomers have used since classical times, and that Doppelmayr illustrates, is, therefore, an extension of Earth's own geometry into the heavens. It is mounted on an axis that extends from Earth's geographical north and south poles through the north and south celestial poles, about which it appears to rotate westwards once a day (in reality, of course, Earth spins eastwards). For an observer on Earth's surface, the bulk of the planet blocks half of the celestial sphere from view at any particular moment, but the visible half is constantly changing as Earth's rotation brings new constellations into view and others sink out of sight. Only one point remains fixed in the sky – the celestial pole that is visible from a particular hemisphere. In northern latitudes, the moderately bright star Polaris (in the constellation of Ursa Minor) lies very close to the north celestial pole and acts as a handy marker in the sky, but its current position is pure coincidence; the southern celestial pole has no such obvious marker. In the course of twenty-four hours, stars circle around the celestial pole. Some are close enough to the pole to never sink below the horizon and are termed 'circumpolar', but the majority rise from the east, reach their greatest height above the horizon as they cross the north–south meridian line that

FIG. 1.
This stunning pocket armillary produced by Johann Baptist Homann prior to 1715 demonstrates how our view of the heavens can be projected as a sphere encompassing the sky, matching Earth's own geometry. Inside the celestial and terrestrial globes, a mere 6.4 cm (2½ in.) in diameter, sits a compact movable armillary sphere.

FIG. 1.

FIG. 2.

FIG. 2.
These illustrations of an armillary sphere and an orrery were published by Tobias Conrad Lotter (1717–77) at Augsburg in 1774. The armillary is an ancient instrument that extends key terrestrial features, such as the equator, tropics and ecliptic, onto a sphere around Earth. This allows the effects of Earth's true rotation or the apparent motion of objects on the celestial sphere to be charted. The orrery, in contrast, is a more modern device (invented in 1704) that models and predicts the positions of the planets in space, usually using a Sun-centred model.

runs across the sky through the zenith point directly overhead, and then set in the west.

Midway between the two celestial poles, naturally enough, the sky is divided into two hemispheres by a celestial equator. Because these features are all extensions of those on Earth, the poles lie higher in the sky at higher latitudes, and lower on the horizon closer to the equator. Conversely, the celestial equator reaches its highest point in an observer's sky directly opposite their celestial pole, and is separated from it by an angle of 90 degrees so that it skims the horizon at polar latitudes and rises high overhead in equatorial ones. The height of the celestial equator in their sky naturally determines just how much of the 'opposite' celestial hemisphere a particular observer can see.

Another important line shown on Doppelmayr's plates is the ecliptic. Marking the path apparently taken by the Sun against the stars over a year, it is, in fact, a projection of the plane of Earth's orbit around the Sun onto the sky (the Sun appears amid different background stars in the famous zodiac constellations as we see it from different directions over the course of Earth's annual circuit around it). Because Earth's axis of rotation is tilted relative to its orbit, the ecliptic is tilted at an angle of 23.5 degrees to the celestial equator, so that half its

track lies in the southern hemisphere of the sky and half in the northern. Doppelmayr's chart highlights the fact that the ecliptic has poles of its own, and addresses how to determine the positions of celestial objects in various different possible systems of coordinates (analogous to Earth's latitude and longitude). These systems arise depending on whether one measures angles relative to the celestial poles and equator, relative to the ecliptic, or relative to one's local horizon. The latter system, which involves measuring altitude (angle above the horizon) and azimuth (direction relative to due north or south) is the most intuitive, but also the least practical for cataloguing celestial positions because it changes from moment to moment and is unique to a particular observer's location.

Doppelmayr depicts these complexities not only through the central series of spherical diagrams but also using globes and astronomical models known as 'armillaries' in the corners of the plate. At the bottom, he also addresses two other important points – the difference in an object's measured position at the surface of Earth compared to its true centre, and the deceptive nature of refraction in Earth's atmosphere, which deflects the images of celestial objects from their true positions (although, fortunately, in a way that can be calculated and taken into account).

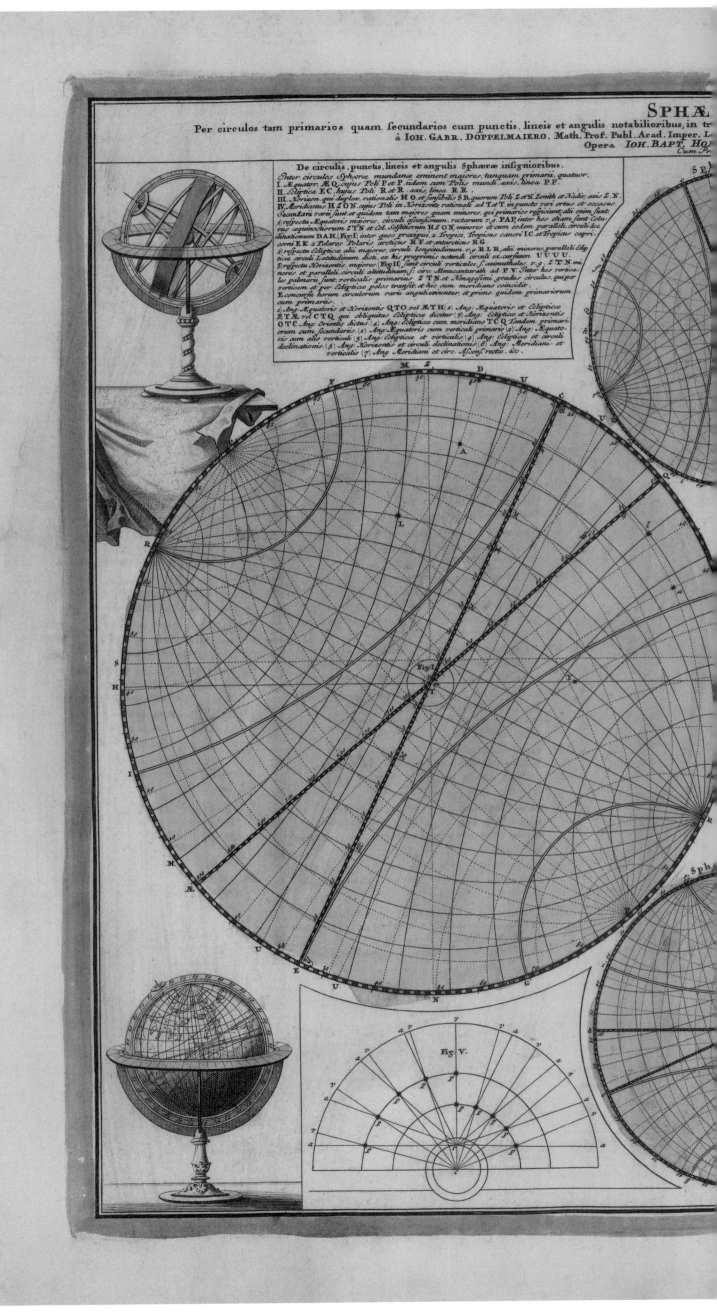

SPHÆ

Per circulos tam primarios quam secundarios cum punctis, lineis et angulis notabilioribus, in tr
à IOH. GABR. DOPPELMAIERO, Math. Prof. Publ. Acad. Imper. L
Opera IOH. BAPT. HO
Cum Pr

De circulis, punctis, lineis et angulis Sphæræ insignioribus.

Inter circulos Sphæræ mundanæ eminent majores, tanquam primarii, quatuor,
I. Æquator ÆQ, cujus Poli P et P. iidem cum Polis mundi, axis, linea PP.
II. Ecliptica EC, hujus Poli R et R, axis, linea RR.
III. Horizon, qui duplex, rationalis HO, et sensibilis SD, quorum Poli Z et N, Zenith et Nadir; axis Z N.
IV. Meridianus H Z O N, cujus Poli in Horizonte rationali ad T et T, in puncto vi ortus et occasus.
Secundarii varii sunt, et quidem tam majores quam minores, qui primarios respiciunt; alii enim sunt:
1) respectu Æquatoris majores, circuli ascensionum, recturam v. g. PAP, inter hos etiam sunt Colurus æquinoctiorum Z T N et Col. Solstitiorum H Z O N, minores et cum eodem paralleli, circuli de-
clinationum DAH, Fig. T, inter quos præcipui, a Tropico, Tropicus cancri I C et Tropicus capricorni R K, a Polare Polario arcticus R Y et antarcticus R G.
2) respectu Eclipticæ alii majores, circuli longitudinum v. g. R I R, alii minores, paralleli Eclipticæ circuli Latitudinum dicti, ex his præcipuis notandi circuli ex-cursuum U U U U.
3) respectu Horizontis, majores, Fig. II, sunt circuli verticales f. azimuthales, v. g. Z T N minores et paralleli, circuli altitudinum f. circ. Almicantarath ad P V. Inter hos verticales palmarii sunt verticalis primarius Z T N et Nonagesimi gradus circulus, qui per
verticem et per Eclipticæ polos transit, et hic cum meridiano coincidit.
E concursu horum circulorum varii anguli oriuntur, et primo quidem primariorum
cum primariis.
1) Ang. Æquatoris et Horizontis Q T O, vel Æ T H (2) Ang. Æquatoris et Eclipticæ
Æ T A, vel C T Q qui obliquitas Eclipticæ dicitur (3) Ang. Eclipticæ et Horizontis
O T C Ang. Orientis dictus. (4) Ang. Eclipticæ cum meridiano T C Q Tandem primariorum cum secundariis. (1) Ang. Æquatoris cum verticali primario (2) Ang. Æquatoris cum alio verticali (3) Ang. Eclipticæ et verticalis. (4) Ang. Eclipticæ et circuli
declinationis. (5) Ang. Horizontis et circuli declinationis (6) Ang. Meridiani et
verticalis (7) Ang. Meridiani et circ. Ascensi rectæ. &c.

Sp

Sph

Fig. I.

Fig. V.

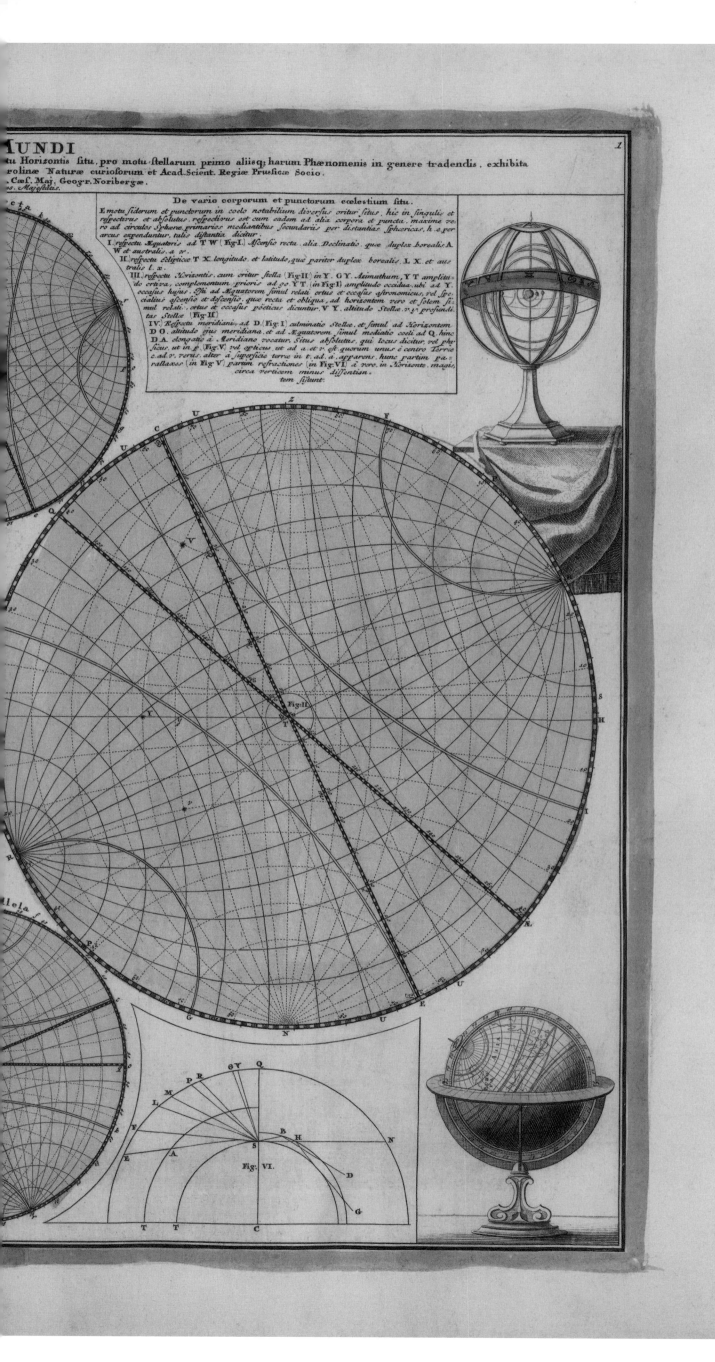

Doppelmayr uses a large sphere,
split in two and viewed as if seen
from the inside, to describe the
key points of the celestial sphere
and the coordinate systems
arising from them. Earth (*Terra*)
lies at the centre, described by
a small circle marked *T*, with
two lines bisecting the celestial
sphere marked by letter labels
around the edges: *Æ–Q* for the
celestial equator, and *E–C* for
the ecliptic line followed by the
Sun. The celestial poles (direct
extensions of Earth's own poles)
are denoted *P*, and the poles of
the ecliptic *R*. Two coordinate
systems can then be constructed
based on these fixed points and
lines. First, ecliptic latitude
or equatorial declination is
described along lines parallel
to the ecliptic or equator from
o to 90 degrees. Second, ecliptic
longitude or equatorial right
ascension is determined by great
circles through the poles of each
system, using the intersection
of the celestial equator and the
northbound ecliptic (the Sun's
position at the northern spring
equinox) as a starting point
for measuring the angles.

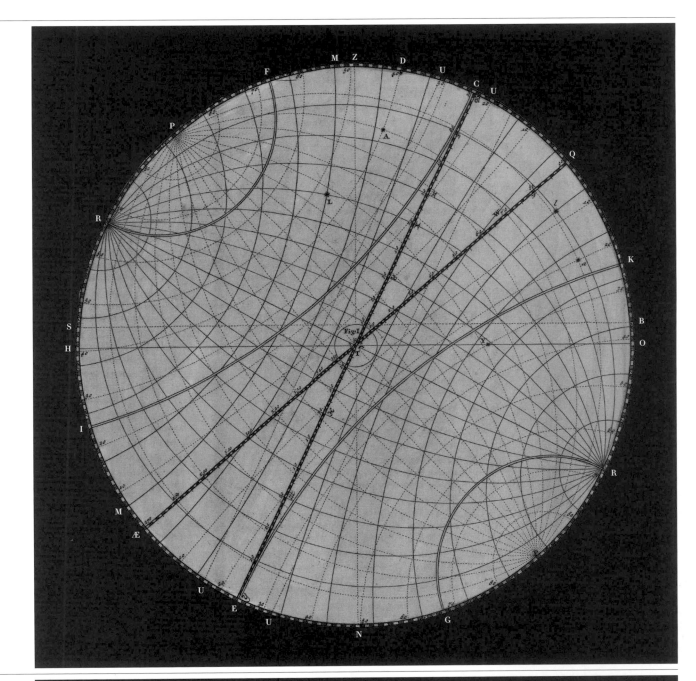

The right half of the large celestial
sphere presents a mirror image
of the left, plotting the coordinates
on the ecliptic and celestial
equator as they continue around
the sky. In addition to the fixed
points in the sky, Doppelmayr
marks points of significance to
the observer at a location on top
of the small central Earth: *Z* is the
zenith, or point directly overhead,
and *N* the nadir, directly beneath.
The *H–O* line marks the 'rational
horizon' – an extension of a line
through the centre of Earth
onto the celestial sphere, and
the *S–B* line denotes the 'sensible
horizon' – an extension of a plane
just touching Earth's surface at
the observer's location, taking
into account the bulk of Earth
blocking the celestial sphere
from view. A third coordinate
system can be constructed for
any observing location, based
simply on a celestial object's
altitude (the angle above the
horizon) and azimuth (the angle
made between due north [or
south] on the horizon and the
point where a line from the
zenith passing through the
object meets the horizon) at
a particular moment.

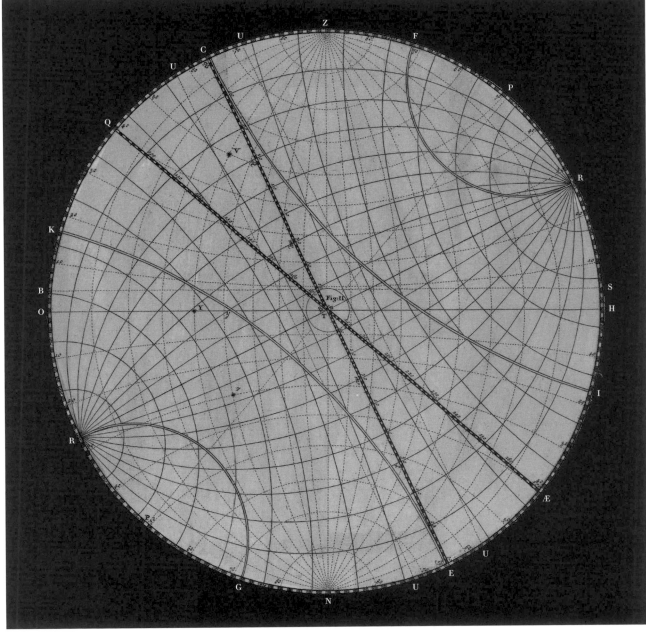

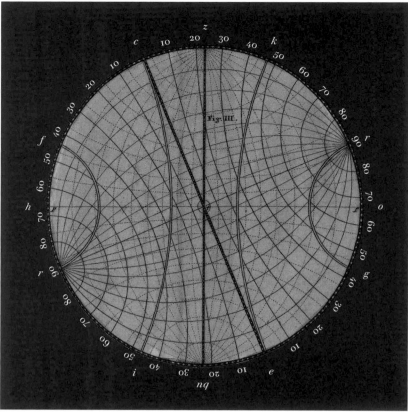

While the two main diagrams show the celestial sphere tilted at an arbitrary angle, Doppelmayr accompanies them with a pair of illustrations of the sphaera recta (Fig. 3, the situation where the celestial equator is perpendicular to the horizon, as seen at Earth's equator) and the sphaera parallel (Fig. 4, where the celestial equator is parallel to the horizon, as seen at the poles).

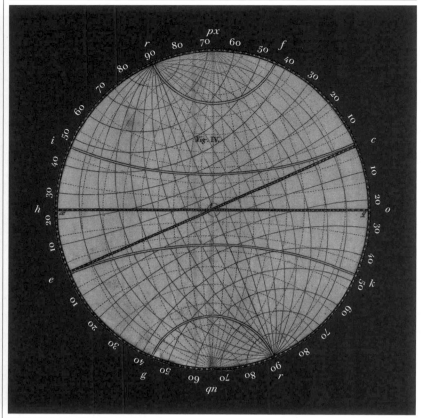

THE POLAR VIEW. (FIG. 4.)

Notably, in all Doppelmayr's sphere illustrations, the thin solid lines of the ecliptic coordinates give that system more prominence than the equatorial one. Ecliptic coordinates were especially suitable for charting the motions of planets, whereas equatorial ones only became more popular once telescope mounts were developed that allowed a telescope to move directly on lines of declination and right ascension.

ARMILLARY SPHERES.

A pair of armillaries act as corner decorations at the top of the plate. One is the more familiar design, with a central Earth, parallel rings marking the celestial equator and tropics, and a tilted band to denote the ecliptic. The other supports a geocentric model of the solar system, with planets supported on concentric rings.

CELESTIAL AND TERRESTRIAL GLOBES.

At the bottom of the plate, a pair of celestial and terrestrial globes provide decorative embellishment. They also emphasize the similarity of the coordinate systems used to measure locations on Earth and in the heavens.

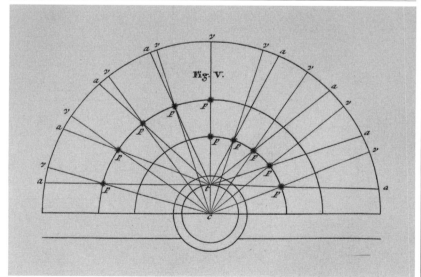

THE EFFECT OF PARALLAX. (FIG. 5.)

In this diagram, Doppelmayr illustrates the difference in the apparent positions of nearby celestial objects when they are measured from the surface of Earth (t), rather than from its centre (c). The same effect arises when Earth's daily rotation carries an observer along with it (so-called 'diurnal parallax'), but in practice it only affects the very closest objects.

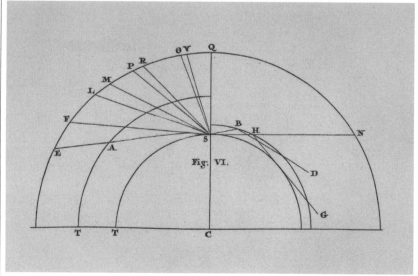

THE EFFECT OF REFRACTION. (FIG. 6.)

The phenomenon of atmospheric refraction (bending of light in Earth's atmosphere) altering the apparent positions of astronomical objects had been recognized since classical times. Although it was not possible to fully correct for this in measurements, Doppelmayr provides a diagram showing that its effect is greater near the horizon (E) and effectively non-existent at the zenith (Q).

PLATE 2.

SYSTEM OF THE SUN AND PLANETS.

*Doppelmayr summarizes the Copernican model
of the solar system and demonstrates its usefulness
for explaining celestial phenomena.*

Plate 2 of the *Atlas* presents a vision of the Enlightenment universe. Originally compiled for Johann Baptist Homann's (1664–1724) *Neuer Atlas* (1707), at its heart lies a model of the solar system, centred on the Sun according to the theories of Nicolaus Copernicus (1473–1543), and elaborated with the discoveries made over some 130 years of telescopic observations.

What we now think of as the Copernican Revolution was a long time coming – and followed a prolonged and tortuous path to acceptance. The practice of 'positional astronomy' – the idea that the ancient system of epicycles and equant points could deliver accurate predictions if only it was provided with sufficiently precise initial measurements of planetary positions and movements – reached its peak in medieval Spain during the late 13th century, where Islamic, Jewish and European ideas and scholarship mixed freely. Here, Alfonso X of Castile (1221–84) sponsored the compilation of astronomical tables that drew on a wide variety of earlier sources and fresh observations to deliver unprecedented accuracy. The resulting *Alfonsine Tables* were used to create ephemerides – charts of the heavenly bodies that could be used in casting horoscopes.

This process began to reveal shortcomings in the complex model of the universe proposed by the Greek astronomer Ptolemy (*c.* 100–170 CE). More accurate measurements made the tables themselves more precise, but also revealed errors in their predictions that might have been overlooked in previous centuries. Thus, Ptolemy's model, like that of Aristotle (384–322 BCE) before him, began to accrue awkward elaborations: epicycles within epicycles just to keep the cosmic clockwork in line with observation.

The first rumblings of a revolution came in 1377, when French philosopher Nicolas d'Oresme (*c.* 1320–82) published his *Book of the Heavens and the Earth*. In it, he demonstrated that the daily motion of the stars, at least, could be explained as well by a rotating Earth as by a rotating outer celestial sphere. He foreshadowed Galileo Galilei's (1564–1642) later concept of inertia by arguing that the elements would share Earth's motion and so we should not expect a perpetual wind from the east. He suggested that spinning the relatively small Earth about its axis might prove more economical to the scheme of the universe than causing a vast starry sphere to rotate in a matter of twenty-four hours. Finally, he directly addressed a thorny issue that would come back to haunt Galileo in particular: the fact that several biblical accounts mention the Sun, and one (in the Book of Joshua) even has it briefly stopped on its path. D'Oresme suggested this was just the Bible speaking to the language and common experience of its audience, and should not be taken as a statement on the true construction of the universe. Nevertheless, he ultimately held back from reaching any radical conclusions, insisting that he, like all right-thinking people, believed the heavens, rather than Earth, stood still.

A century and a half later, Copernicus launched his theory in a very different climate. The transformations unleashed by the Renaissance and the Protestant Reformation saw many long-accepted dogmas being openly questioned, while the invention of the printing press allowed new ideas to spread more quickly than ever before. Copernicus was particularly inspired by the *Epitome of the Almagest* (1496) by Regiomontanus (1436–76) and Georg von Peuerbach (1423–61), which drew attention to some of the problems in Ptolemy's theory of lunar motion. After confirming these for himself through observation, he began to read more widely and develop his own ideas. By 1514, he had summarized these in a small book usually referred to as the *Little Commentary*, which he circulated among friends and fellow astronomers in manuscript copies.

Although chiefly famous for placing the Sun, rather than Earth, at its centre, the Copernican vision of the universe was not as simple or as comprehensive as later depictions (including Doppelmayr's) imply. While the planets were now placed on their familiar paths around the Sun with only the Moon orbiting Earth, Copernicus was forced to retain the smaller epicycles that caused them to wander back and forth even as they generally drifted westwards across the sky. The main reason for this was that he still believed in the necessity of uniform circular motion. Changes to the apparent speed and direction of the other planets could not be entirely explained by our shifting point of view on Earth, and so a further mechanism was required. Even with this unwanted complication, Copernicus's system offered a powerful alternative to Ptolemy's, and word began to spread through academic circles across Europe. Legend has it that the first copies of his work *On the Revolutions of the Heavenly Spheres* were brought to Copernicus as he lay dying from a stroke in May 1543. Whether he knew about the printer's addition of a preface by

theologian Andreas Osiander (1498–1552), insisting that the book's hypothesis should be treated merely as a mathematical tool rather than a description of the true nature of the universe, we will never know.

Modern research has challenged the long-standing view that the complex work made little impact at the time. In fact, a census of all known surviving copies from its early printings suggest that the book was read (and annotated) by many astronomers keen to make use of its mathematical tools. It seems true, however, that many turned a blind eye to its implications for cosmology – perhaps unsurprising given the fervent religious debates of the time and the fact that its ideas were derided by Protestants. Somewhat ironically, given their later clash with Galileo, parts of the Catholic Church offered the theory a warmer welcome, at least while it remained firmly in the realm of mathematical hypothesis. It was only from around 1609 that the invention and development of the telescope revealed new phenomena in the sky for which a Sun-centred, rather than Earth-centred, universe seemed the only plausible explanation.

FIG. 1.

FIG. 1.
Perhaps the best-known plate from Andreas Cellarius's *Harmonia Macrocosmica* (1660), this illustration depicts the Sun at the centre of the solar system. It demonstrates how our planet's tilted axis of rotation can tip the northern hemisphere towards and away from the Sun at different times of year, giving rise to the familiar pattern of seasons. Earth is shown at four different points in the year. Anti-clockwise from top, these are the winter solstice, vernal or spring equinox, summer solstice and autumnal equinox.

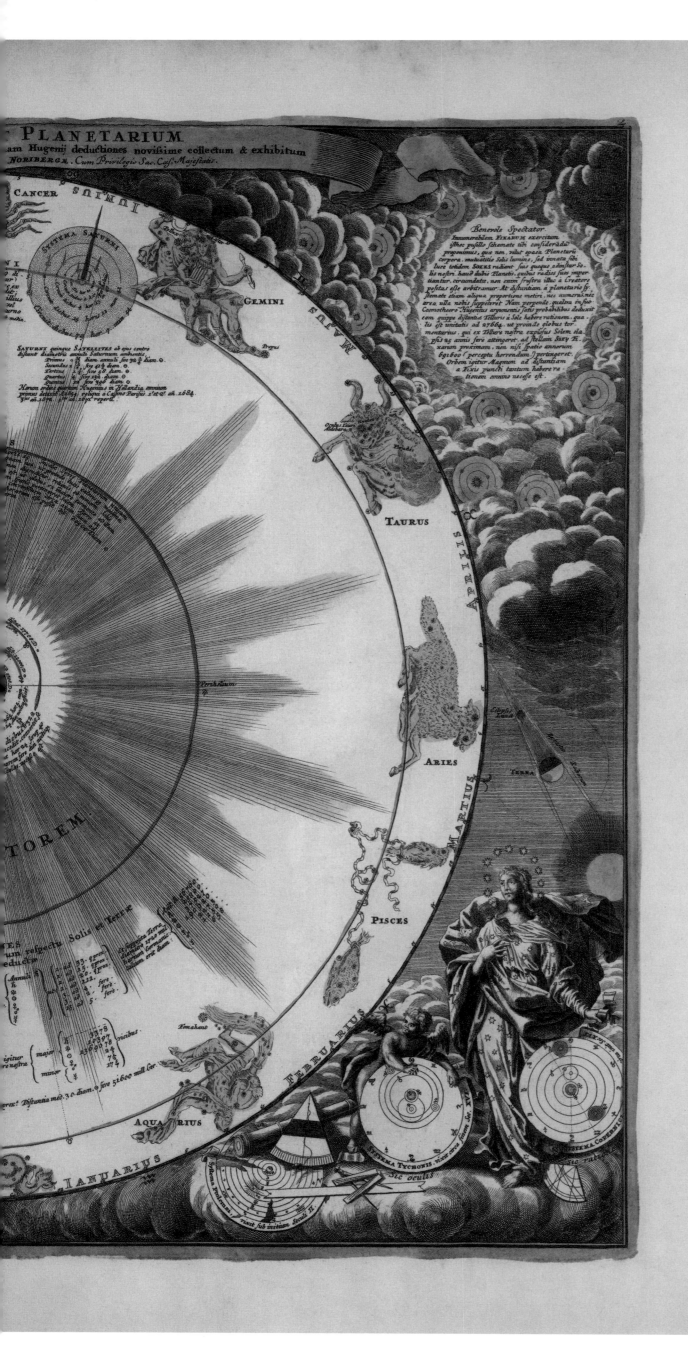

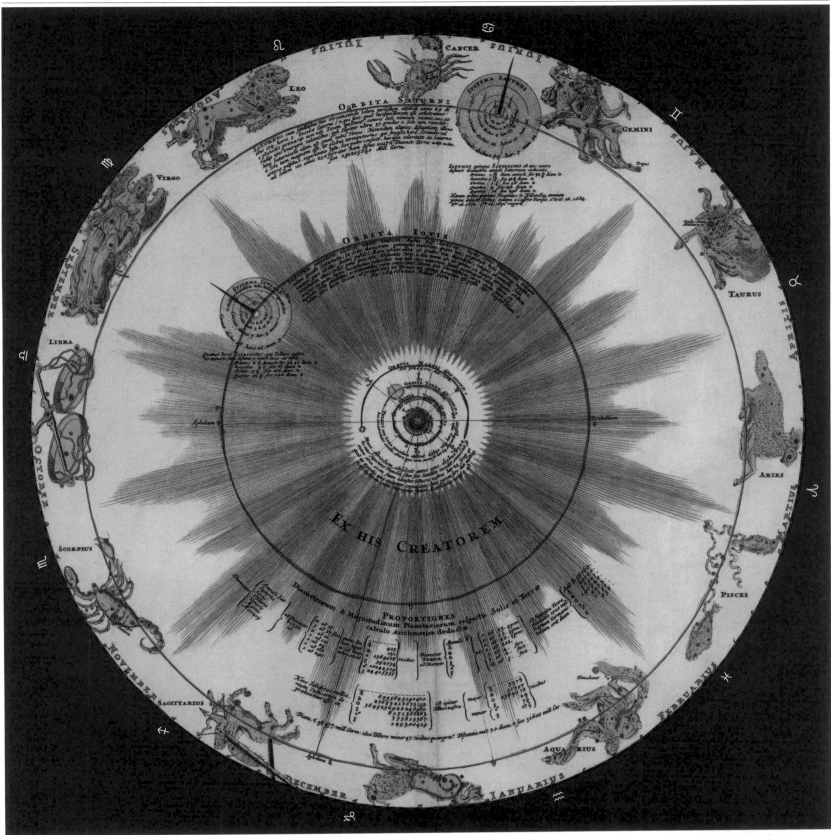

THE COPERNICAN SOLAR SYSTEM.

The plate's central illustration makes an attempt to depict the relative scales of planetary orbits, showing that the four inner planets cluster relatively close to the Sun, while those of Jupiter and Saturn are much further out. Rays surrounding the Sun also hint at its dwindling influence at greater distances.

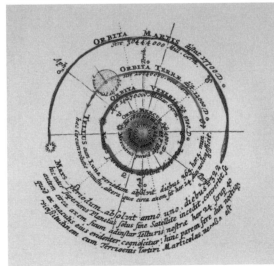

THE INNER PLANETS.

The region around Mercury, Venus, Earth and Mars is crowded with information about their orbits. Distance from the Sun is given in Earth diameters, and orbital periods in days and hours. The current directions of perihelion (each orbit's closest point to the Sun) and aphelion (its greatest distance) are shown.

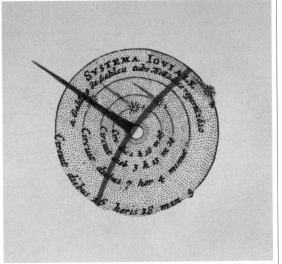

THE JOVIAN SYSTEM.

Jupiter is depicted with its four major satellites, today known as the 'Galilean moons'. The satellites, numbered 1 to 4 moving outwards, are shown with their orbital periods. The familiar names Io, Europa, Ganymede and Callisto were not widely adopted until the 20th century.

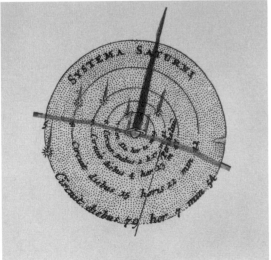

THE SATURNIAN SYSTEM.

Saturn – the outermost planet known in Doppelmayr's time – is depicted with its surrounding ring system and five known moons: Tethys, Dione, Rhea, Titan and Iapetus (numbered 1 to 5). Titan was discovered by Christiaan Huygens in 1655, and the other four by Giovanni Domenico Cassini between 1671 and 1684.

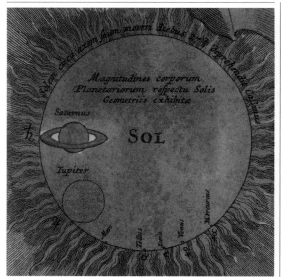

THE SCALE OF THE BODIES OF THE PLANETS WITH RESPECT TO THE SUN.

At upper left, the plate reproduces an illustration from Huygens's *Cosmotheoros*, depicting the relative sizes of the Sun and planets. The table embedded below centre in the main map of the Copernican system further shows how diameters and volumes relative to Earth can be calculated.

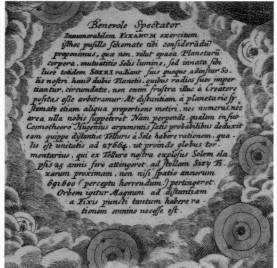

THE PLURALITY OF WORLDS.

At upper right, there is a lengthy description of the fixed stars as Suns in their own right, each surrounded by its own solar system. After discussing Huygens's calculation that Sirius (the sky's brightest star) is some 27,664 times further away than the Sun, it goes on to describe the vast time it would take to travel to these other stellar systems.

MODELLING THE UNIVERSE.

Allegorical figures and astronomical instruments accompany the three great historical models of the solar system. These are the Ptolemaic system (depicted in a state of decay), the Tychonic system (subtitled *sic oculis*, meaning 'so with the eyes') and the Copernican view (subtitled *sic ratione*, meaning 'so with reason').

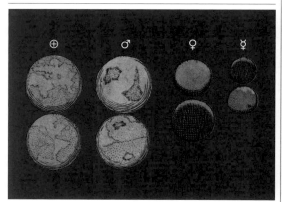

MARKINGS AND APPEARANCES OF THE PLANETS.

Here, the plate shows the typical surface features and appearances of the four inner planets. They include the seas and continents of Earth, the dark markings on the face of Mars and the changing phases of Venus and Mercury. All are explored in more detail on Plate 5.

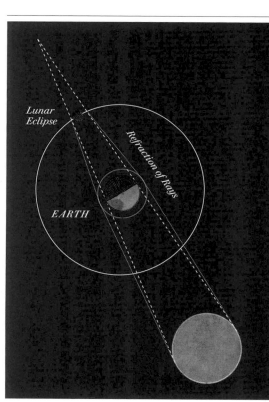

LUNAR ECLIPSES.

A companion diagram to the solar eclipse depicts the geometry of lunar eclipses, in which the Full Moon passes through the long cone of shadow cast by Earth. Because Earth is larger than the Moon, the required alignment is far less precise, and lunar eclipses can be seen across Earth's entire night-time hemisphere.

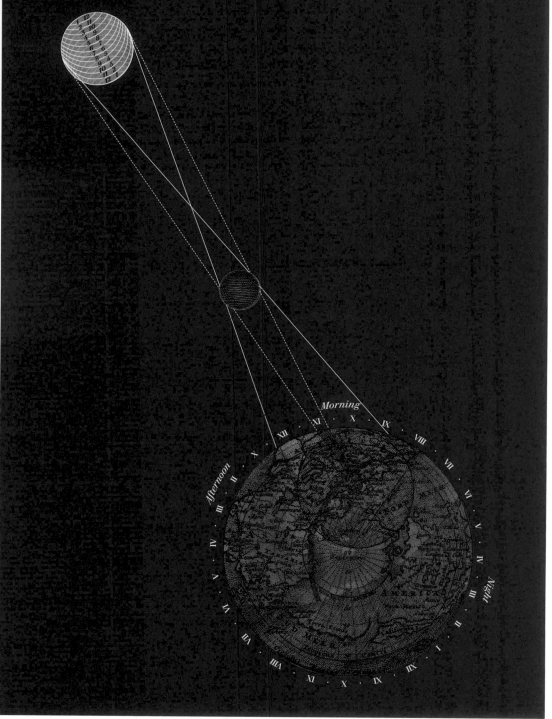

THE SOLAR ECLIPSE OF 12 MAY 1706.

First appearing in Homann's *New Atlas of the Whole World* (1707), the most recent solar eclipse seen over Europe at the time is depicted on Plate 2. The eclipse was the first to have its path across the Earth's surface accurately predicted in advance, along with the extent of the umbral (total) and penumbral (partial) shadows cast by the Moon.

PLATE 3.

THE TYCHONIC SYSTEM OF THE UNIVERSE.

Doppelmayr depicts the principal alternatives to the Copernican system in the early Enlightenment period.

For Johann Baptist Homann's (1664–1724) *Atlas of 100 Charts* (1712), Doppelmayr designed a plate depicting the main alternative to Nicolaus Copernicus's (1473–1543) Earth-centred solar system in the late 1500s and early 1600s, known today as the Tychonic model. The plate is included as Plate 3 in the *Atlas Coelestis*.

The system is named after Danish nobleman Tycho Brahe (1546–1601), the most renowned astronomer of his age. Born with connections to several noble families and heir to a considerable fortune, Tycho had an early interest in the stars and was particularly inspired by the sight of a solar eclipse in August 1560. The eclipse in question had been predicted, although its eventual timing was off by a day, and this sparked Tycho's realization that more precise measurements were the key to more accurate predictions.

Techniques for measuring the positions of objects in the sky had been essentially unchanged since classical times. Since our view of the sky is constantly rotating and carrying the stars and planets with it, the precise direction and elevation (angle above the horizon) of any celestial object changes from moment to moment. However, positions on the rotating 'celestial sphere' (see Plate 1) could still be measured. For instance, an object's declination (its position measured in degrees north or south of the celestial equator) could be tracked by measuring its elevation above the horizon as it crossed the meridian (a north–south line across the sky where it is at its highest). Meanwhile, relative right ascension (east–west separation) in the heavens could be measured in terms of the difference in time when objects crossed the meridian line.

Angular measurements were taken with a variety of ingenious instruments. For example, the mural quadrant was a quarter-circle mounted on a wall that was precisely north-south aligned. The arc of the quadrant was divided into degrees, while a pivoted sighting bar allowed the observer to precisely align with an object crossing the meridian and then read off the elevation. The cross-staff, meanwhile, had a crosspiece or transom with a sight on either end, sliding up and down a graduated pole with an eyepiece. With the transom positioned so that two objects of interest appeared in either sight, the marks on the pole revealed the angle between them.

Supported by Frederick II of Denmark (1534–88), Tycho built two observatories on the island of Hven in the Øresund strait, where he installed a variety of instruments, including many of his own design. The increased scale of these devices enabled him to measure angles with far greater precision, allowing him to begin compiling tables of planetary positions that would ultimately – in the hands of his later assistant and successor, Johannes Kepler (1571–1630) – prove key to explaining the true nature of the solar system.

Late in his lifetime, Tycho became renowned for his own model of the universe, the legacy of which would linger well into the 17th century. In part, this was motivated by practical concerns: as early as 1563, he had observed a close conjunction of Jupiter and Saturn and noted that tables using both the Copernican theory and the venerable Ptolemaic model failed to describe it accurately. However, it was two major celestial events of the 1570s – the new star or nova that lingered in the heavens for more than a year after its brilliant eruption in November 1572, and the Great Comet that swept across the skies in 1577–78 – that convinced him that the old Aristotelean cosmology must be wrong. Collaboration with observers elsewhere in Europe allowed Tycho to determine that neither of these objects showed parallax (the shift in direction of nearby objects relative to more distant ones when they are viewed from different locations). He concluded that they must, therefore, be happening at a great distance, and, furthermore, that the comet followed a path that would cross over the supposed spheres of the other celestial bodies.

While these discoveries undermined some of Tycho's belief in the reality of the spheres, they did not shake his faith in the Bible or the evidence of his own senses. The Book of Joshua spoke of God making the Sun stand still for a period of time (a keystone of religious support for a universe with the Sun in motion), and there was no evidence for the physical effects that might reveal a rotating and moving Earth. He also had underlying doubts about how a massive body such as Earth could move through space.

Tycho, therefore, developed an ingenious compromise, which he revealed in 1588. The system swept away the solid spheres of the Aristotelean cosmos, but kept Earth in its central place. Unlike previous models, it allowed for two centres of motion: the Moon and Sun followed circular paths around a stationary Earth, while the other planets orbited the Sun. To explain their different motions, Mercury and Venus's orbits were assumed to be smaller than the Sun's separation from Earth, so they could never stray far from the Sun in Earth's sky. Those of Mars, Jupiter and Saturn encompassed Earth, thereby allowing them to appear on the opposite side of the sky from the Sun. In order for the paths of the planets (and comets) to cross each other, Tycho abandoned spheres in favour of a fluid cosmos, made of an airy material similar to the pneuma of the ancient Stoic philosophers (the principal rivals to Aristoteleanism in late antiquity). The planets swam in this airy material thanks to an animating force or soul that was inherent within them.

In the decades after Tycho proposed his system, several other 'geo-heliocentric' models were suggested, two of which are illustrated at the bottom of Plate 3. Jesuit priest and astronomer Giovanni Battista Riccioli (1598–1671) proposed a variant in which Jupiter and Saturn were restored to an outer orbit around Earth, while the somewhat mysterious 'Egyptian system' did the same for Mars, leaving just Mercury and Venus to orbit the Sun. These alternative models of the solar system would remain a subject for debate until well into Doppelmayr's career.

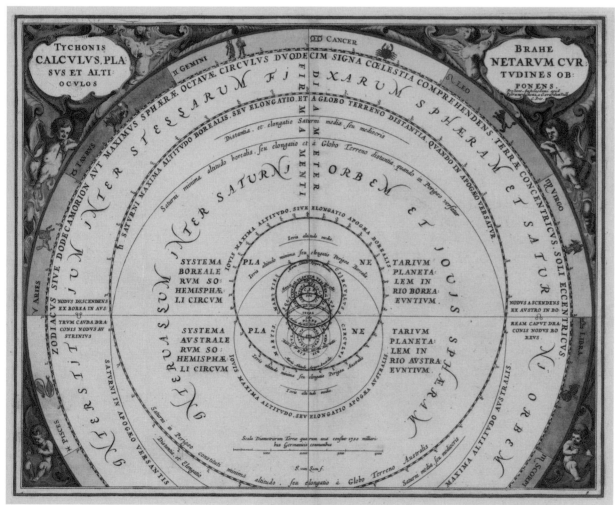

FIG. 1.

FIG. 1.
This diagram of the Tychonic system from Andreas Cellarius's *Harmonia Macrocosmica* (1660) shows the orbits of the planets around the Sun in two positions relative to Earth, with their maximum, minimum and median distances indicated. By allowing substantial changes in the distance of the planets from Earth, Tycho's model – like the Copernican system – could explain changes in their apparent brightness, which the Ptolemaic view could not.

SYSTEMA MU

Secundum celeberrimorum Astronomorum Tychonis de Brahe et Io. B

operâ Ioh. Bar

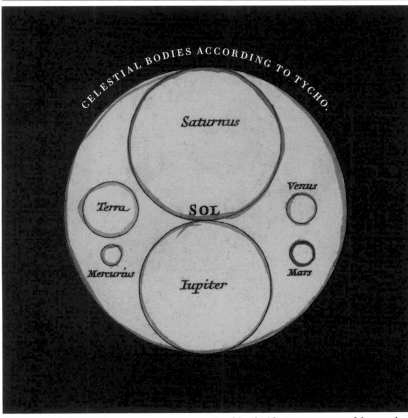

CELESTIAL BODIES ACCORDING TO TYCHO.

Combined with measurements of the angular size of the planets in Earth's sky, Tycho's model predicted relative scales of the planets that were broadly accurate. However, they were all far too large compared to the Sun.

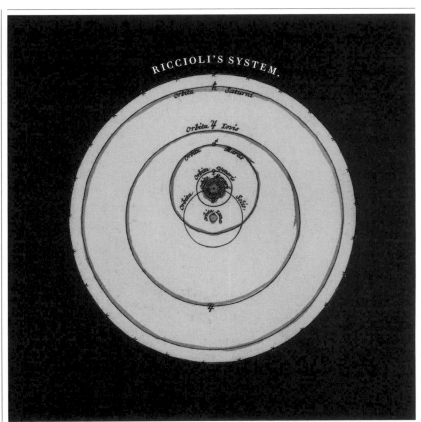

RICCIOLI'S SYSTEM.

In Riccioli's geo-heliocentric system, Mercury, Venus and Mars follow circular orbits around the Sun, while the Moon, the Sun, Jupiter and Saturn orbit Earth. Mars's orbit is arranged such that it encloses Earth, allowing it to make a full circuit in Earth's skies.

CELESTIAL BODIES ACCORDING TO RICCIOLI.

Riccioli's model of the solar system suggested that the Sun was far larger than even the biggest planets. However, it still underestimated the difference in scale.

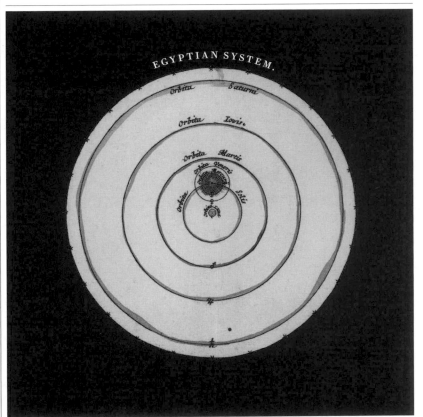

EGYPTIAN SYSTEM.

In the relatively simplistic Egyptian system, Mercury and Venus are satellites of the Sun (explaining why their orbits are anchored to the Sun's position in the sky), while the Moon, the Sun and the outer planets all circle Earth.

TYCHO BRAHE.

The greatest of the pre-telescopic astronomers is shown in the finery of a Danish nobleman. The depiction draws on a late 16th-century portrait by Dutch artist Jacob de Gheyn II.

GIOVANNI BATTISTA RICCIOLI.

Here, the influential priest and astronomer is depicted in the dress of the Jesuit order, displaying a copy of his book *New Almagest* (1651).

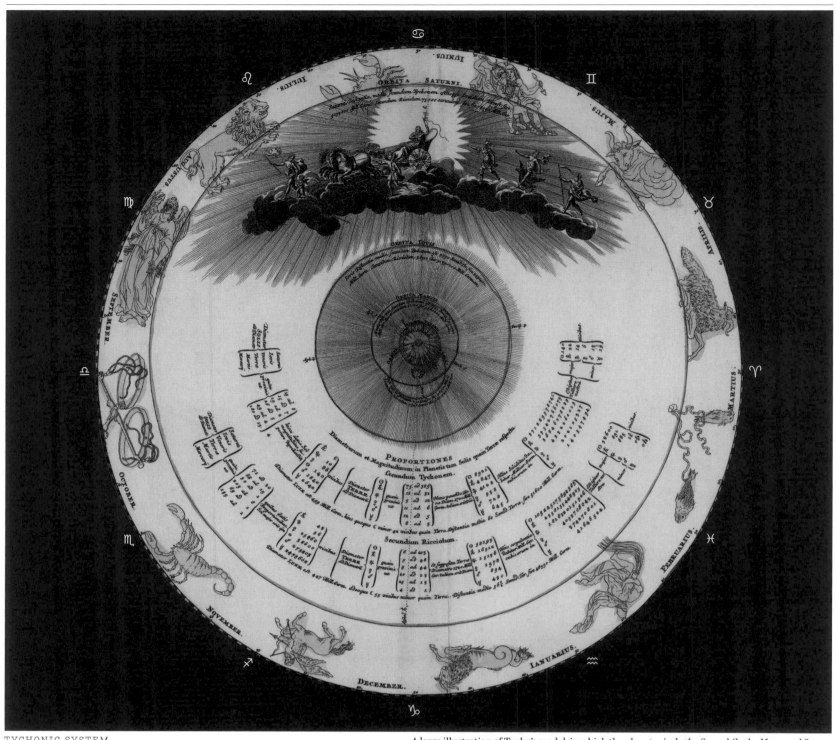

TYCHONIC SYSTEM.

A large illustration of Tycho's model, in which the planets circle the Sun while the Moon and Sun orbit the fixed Earth, dominates the plate. Depictions of the zodiac constellations around the edge are shown alongside the months when the Sun aligns with them. Above centre, the classical gods associated with each planet are shown flanking Helios, god of the Sun.

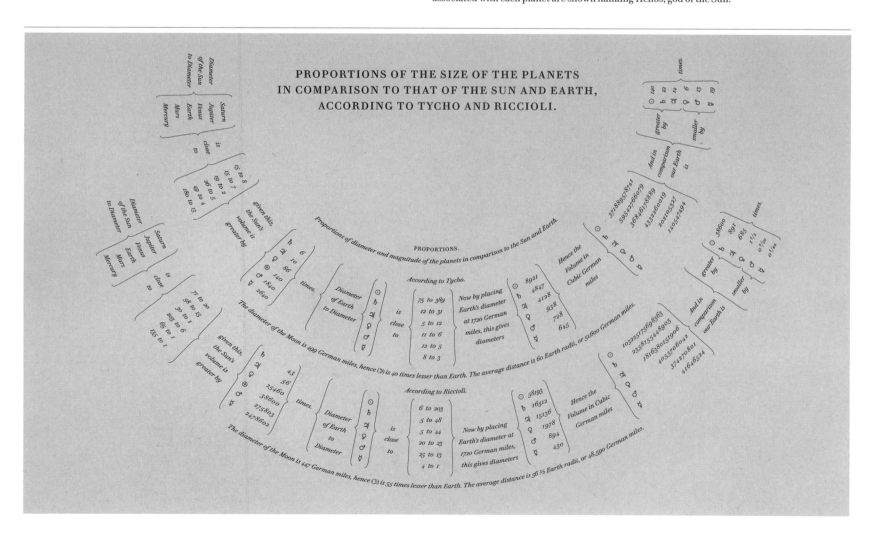

PROPORTIONS OF THE SIZE OF THE PLANETS
IN COMPARISON TO THAT OF THE SUN AND EARTH,
ACCORDING TO TYCHO AND RICCIOLI.

PLATE 3.

QVADRANS MINOR
ORICHALCICVS INAVRATVS.

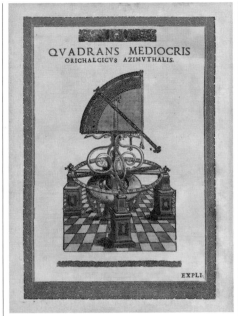

QVADRANS MEDIOCRIS
ORICHALCICVS AZIMVTHALIS.

EXPLI-

QVADRANS ALIVS ORICHAL.
CICVS ETIAM AZIMVTHALIS·

EXPLI-

QVADRANS MAGNVS CHA,
LIBEVS, IN QVADRATO ETIAM CHA-
libeo comprehensus, utrinque Azimuthalis.

EXPLI-

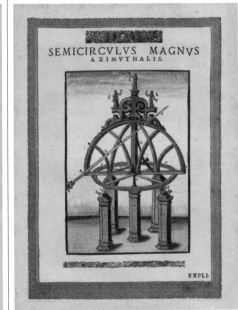

SEMICIRCVLVS MAGNVS
AZIMVTHALIS.

EXPLI-

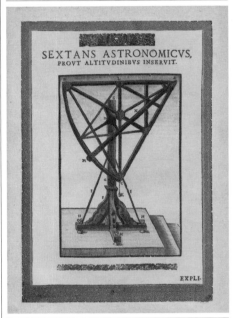

SEXTANS ASTRONOMICVS,
PROVT ALTITVDINIBVS INSERVIT.

EXPLI-

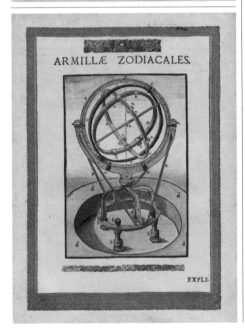

ARMILLÆ ZODIACALES.

EXPLI-

ARMILLÆ ÆQVATORIÆ

EXPLI-

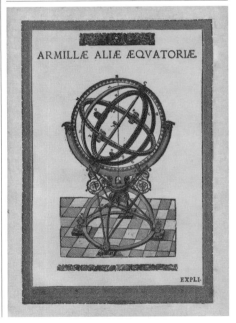

ARMILLÆ ALIÆ ÆQVATORIÆ

EXPLI-

ASTRONOMIÆ INSTAURATÆ MECHANICA (1602).

Tycho Brahe's *Astronomiæ instauratæ mechanica* (*Instruments for the Restoration of Astronomy*) describes the cutting edge of astronomical technology on the eve of the telescopic revolution. Written in 1598, it outlines the instruments used in Tycho's great observatories on the Danish island of Hven, with which he recorded the positions of objects in the sky to an unprecedented degree of accuracy. This selection of plates illustrates various quadrants and sextants used for measuring positions and separations, alongside armillary spheres used to model both the zodiacal and equatorial

ARMILLÆ ÆQVATORIÆ MAXIMÆ,
SESQVIALTERO CONSTANTES
circulo.

EXPLI-

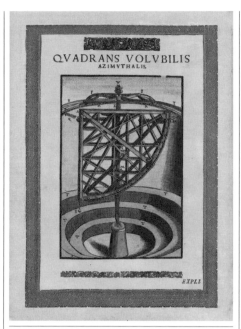

QVADRANS VOLVBILIS
AZIMVTHALIS.

EXPLI-

GLOBVS MAGNVS
ORICHALCICVS.

EXPLI-

ARCVS BIPARTITVS
MINORIBVS SIDERVM
distantiis inserviens.

EXPLI-

SEXTANS CHALYBEVS PRO
DISTANTIIS PER VNICVM OBSERVATOREM
dimetiendis.

EXPLI-

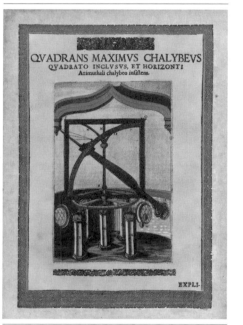

QVADRANS MAXIMVS CHALYBEVS
QVADRATO INCLVSVS, ET HORIZONTI
Azimuthali chalybeo insistens.

EXPLI-

SEXTANS ASTRONOMICVS
TRIGONICVS PRO DISTANTIIS
rimandis.

EXPLI-

PARALLATICVM ALIVD, SIVE
REGVLAE TAM ALTITVDINES QVAM
Azimutha expedientes.

EXPLI-

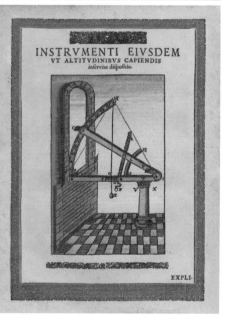

INSTRVMENTI EIVSDEM
VT ALTITVDINIBVS CAPIENDIS
inserviat dispositio.

EXPLI-

coordinate systems of the heavens. Further plates include,
at top right of this page, the Great Globe – perhaps Tycho's greatest
achievement. This 1.6-m (5-ft) hollow wooden sphere took a decade
to build to the required accuracy, after which it was covered in brass
plates onto which the positions of stars and other objects could be
precisely etched. Pivoting auxiliary circles permitted rapid conversion
between the unchanging coordinate systems of the sphere and the
localized altitude and azimuth coordinates unique to a particular
time and location.

PLATE 4.

THEORY OF THE PRIMARY PLANETS.

(THEORIA PLANETARUM PRIMARIORUM)

Doppelmayr demonstrates how the Copernican system, modified by Kepler's elliptical orbits, accounts for the motions of the planets.

Plate 4 demonstrates how Nicolaus Copernicus's (1473–1543) Sun-centred model of the solar system can make intuitive sense of some of the most obvious phenomena in the motions of the planets. It also describes the revolutionary ideas of Johannes Kepler (1571–1630), as well as the adornments other astronomers added to Kepler's simplicity in an attempt to explain what drove the planets in their orbits.

Long before the planets were recognized as balls of rock and gas distinct from stars, they had first drawn attention to themselves through their eccentric motions in the sky. While the stars wheeled around the heavens in fixed patterns that never seemed to change, five bright lights wandered among them on varying paths. Two of these lights rarely strayed far from the Sun. One, the brightest of all, traced large loops and often appeared as a brilliant beacon in the dark sky, lingering after sunset or heralding the new day – small wonder that ancient civilizations frequently associated it with their goddess of beauty (the Roman Venus). The other – much fainter, faster moving and harder to spot – made only brief appearances in the sky at dawn or dusk; the Romans named it Mercury, after the fleet-footed messenger of the gods.

The remaining three planets moved differently. Largely unshackled from the Sun, they would circle westwards around the entire sky along the band of stars known as the zodiac. Once in each cycle, they would approach the Sun's own position and disappear into the sunset sky, before re-emerging weeks or months later to be visible before sunrise. What was more, their general westward track was frequently interrupted by periods of 'retrograde' motion, in which they tracked east across the sky for weeks or months before resuming their general westward trend. The least predictable of these three wanderers, which had a baleful red colour and could vary significantly in brightness, became associated with gods of war, such as the Greek Ares and Roman Mars. Steadier in its motion and more predictable in its brilliance was the planet frequently associated with Zeus, the king of the gods, known from Roman times as Jupiter. Finally, the system was completed by the fainter and more sedate planet associated by both the Greeks and Romans with the king's father: Cronus, or Saturn.

Tracking the motions of these planets and predicting various events in their passage around the sky became the key concern of ancient astronomy. Such events could include the timing of their conjunctions – or comings-together – with the Sun, Moon, stars and each other; the greatest distance, or 'elongation', from the Sun achieved by the so-called 'inferior planets' Venus and Mercury; and the timing of

FIG. 1.
One of the most common applications of astrology was to medieval medicine. The dominance of certain planets and constellations was seen as influencing the motions of the four classical elements and (in a model originated by Greek philosopher Empedocles) the four humours or fluids within the body. These, in turn, were linked to bodily organs, physical illnesses and psychological states in a complex model encapsulated by illustrations such as the *Anatomy of Man and Woman* from the *Très Riches Heures du Duc de Berry* (1415).

FIG. 1.

FIG. 2.

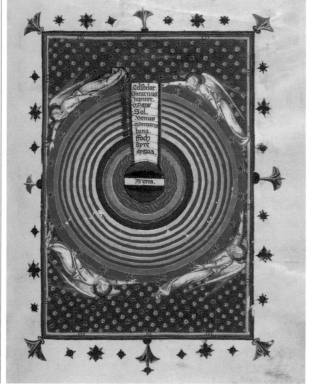

FIG. 3.

FIGS. 2–3.
Two depictions of the Ptolemaic system that dominated medieval astrology. Far left is an illumination from a 1464 edition of Gautier de Metz's *L'image du Mond*; near left is a late 14th-century edition of the *Breviari d'Amor* by Occitan poet Matfre Ermengau (d. 1322). Both show not only the spheres of the planets and outermost fixed stars, but also the inner sublunary spheres that were considered the rightful positions of the elements fire, air, water and earth.

'oppositions', when the free-roaming 'superior' planets lay directly opposite the Sun in the sky and were, therefore, visible throughout the night. Above all, there was the question of how and when the Sun, Moon and planets moved in and out of the various constellations.

All of these questions had an immense practical importance because, until well into the 17th century, what we think of today as the science of astronomy was inextricably linked to astrology – the forecasting of events on Earth based on those in the heavens. Although modern astrology is widely regarded as a fairly harmless superstition, the classical and medieval form was part of a sophisticated worldview that encompassed everything from the organization of states to the treatment of illness. Few scholars believed that the celestial bodies themselves were affecting events and people on Earth, but they did hold to a widespread view that events on Earth and in the heavens followed pre-ordained cycles; history might not repeat itself, but it certainly rhymed. If, for example, a great king died unexpectedly while hunting during a conjunction of certain planets, then a shrewd ruler might well wish to know when the next such conjunction was due, and modify their plans accordingly. The ability to predict such events, which seem little more than curiosities to modern life, was the driving force of astronomy for two millennia or more.

Thus, astronomer/astrologers took a great interest in Copernicus's system long before they properly digested its true implications. While geocentric models of the universe had been forced to employ complex epicycles and other mechanisms to keep the inferior planets anchored to the shifting location of the Sun as it circled Earth, the Copernican model had a far simpler explanation: with Earth as a third planet, the orbits of each of the two Sunward planets cover only a limited angle in our skies. All planets orbit in the same direction (counter-clockwise as seen from above) and circle the Sun at different speeds and in different periods, so that the distance and direction from one to another changes. An inferior planet has an orbit smaller than Earth's and reaches its greatest elongation east or west of the Sun as it rounds the outer edge of its orbit seen from our point of view. It comes closest to Earth at a point called 'inferior conjunction', when it lies in exactly the same direction as the Sun (although because the orbits are slightly tilted in respect to each other, it does not usually pass across the face of the Sun itself). At its furthest from Earth, meanwhile, it lies on the opposite side of the Sun at 'superior conjunction'. As it moves from superior back to inferior conjunction via its greatest eastern elongation, the planet is visible in the evening sky after sunset (since it lies east of the Sun and sets after it). After inferior conjunction, it appears in the morning sky and loops through its western elongation before returning to the next superior conjunction.

PLATE 4.

In addition, the Sun-centred system offers an easy explanation for the most obvious aspects of motion among the three superior planets. Mars, Jupiter and Saturn have just a single conjunction with the Sun, when they lie directly on the far side of the Sun from Earth. The combined motions of Earth and the planets then causes them to steadily drift west from the Sun, appearing in the pre-dawn sky and gradually increasing their elongation so that they rise earlier and earlier. At opposition, they lie on the point in their orbit that is in the opposite direction to the Sun as seen from Earth. Thereafter, they continue westward and slowly close in on the Sun's location in the sky, approaching it from the east and disappearing into the sunset glow as they near the next conjunction.

For all its superficially attractive simplicity, however, Copernicus's system alone had inherent problems that delayed its widespread adoption. Unable to break free from the shackles of circular motion at a uniform rate – inherited by Western philosophers from Aristotle (384–322 BCE) – he was forced to fall back on a system of epicycles (smaller orbits upon the larger ones) almost as complex as that of Ptolemy (c. 100–170 CE) in order to keep his planetary motions roughly in line with the observed reality.

For this reason, the initial response to Copernicus's ideas was to treat them as a useful calculating tool, but not necessarily an alternative model that overthrew long-held notions of physical reality. This may have helped soften the initial shock of his proposals, making it easier for European astronomers to begin thinking in Copernican terms, using them as a means of predicting planetary motions, and weighing up their benefits and shortcomings. Within a generation, studies of the 'new star' of 1572 and the Great Comet of 1577 would undermine the old Aristotelean belief in the unchanging heavens (see page 218). These events inspired the great Danish astronomer Tycho Brahe (1546–1601) to develop his own hybrid system as a rival to Copernicus (see Plate 3), but the undeniable reality of widely seen events in the sky also had a more profound effect on scholarly understandings of the universe. If the heavens could change, then perhaps they were not made from a perfect aether, but from the same base and changeable elements found on Earth? What was more, in Aristotle's model, objects below the heavenly spheres found their place in the universe depending on their affinity with the four classical elements of earth, air, fire and water, but if the stars and planets were made of similar materials, then some other force must be at work – perhaps a force similar to those experienced on Earth itself?

Into this debate, in the late 16th century, stepped Kepler, a German astronomer and astrologer who had become a committed Copernican while studying at the University of Tübingen in Germany. Kepler's beliefs were rooted partly in observation, but equally in mathematical speculation and a theological model of the universe. Believing that the Sun could be equated with God the Father, he argued that it must be the source of all motive power in the solar system (turning on its head the traditional notion that the motion of the outer stellar sphere drove the inner planetary ones at their different rates). In *The Cosmographic Mystery* (1596), he outlined an elegant and somewhat mystical theory that linked the diameters of the planetary spheres to the spacing created when the regular 'platonic solids' of geometry are nested within each other.

Kepler's early theories seem strange to modern eyes, but he combined them with a talent for mathematical calculation and a tendency towards what we would now think of as a scientific approach. Pursuing the best astronomical data to aid development of his theory led him into correspondence with Tycho, and in 1600 he travelled to work with the Dane at his new observatory in Bohemia. Tycho was

FIG. 4.

The frontispiece of the *Rudolphine Tables*, published by Kepler in 1627, acknowledges a debt to ancient Greek astronomers, such as Hipparchus (near left) and Ptolemy (far right), but reserves its places of honour for a seated Nicolaus Copernicus and the standing figure of Tycho Brahe. Kepler himself is shown in the plinth at lower left, with a printing press on the right and Brahe's observatory island of Hven in the centre.

FIG. 4.

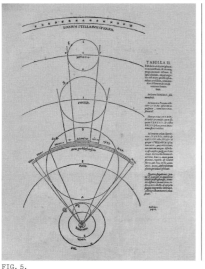

FIG. 5.

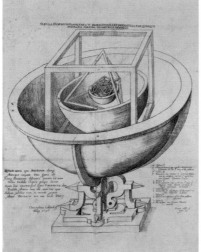

FIG. 6.

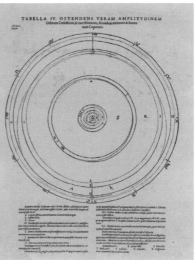

FIG. 7.

FIGS. 5–7.
Kepler's *The Cosmographic Mystery* (1596) is a complex and, at times, obscure work that bridges medieval and Enlightenment views of the universe. Its most famous suggestion was that the size of the orbits of the six known planets could be explained if each orbit defined a sphere, with a series of nested platonic solids (shapes comprised entirely of identical regular faces, of which just five exist) governing the gaps between them.

by now employed as court astronomer to the capricious Holy Roman Emperor Rudolf II (1552–1612), and Kepler was keen to gain access to his unrivalled measurements of planetary motion (particularly those of Mars) in order to refine his model. Although the older astronomer guarded these records closely, Kepler's mathematical prowess soon saw him taken into Tycho's confidence, and following Brahe's unexpected death in 1601, he took over the role of Imperial Mathematician, with a remit to complete the *Rudolphine Tables* – the most ambitious star catalogue and table of planetary motions yet attempted.

Over the following years, Kepler's access to his mentor's precise measurements led him to key discoveries that supported the Copernican theory over all others. He became convinced that the motions of Mars could not be accurately described by a circular orbit with an epicycle, and began to investigate whether the paths taken by the planets might be ovoid, or egg-shaped. This, of course, was only plausible if the idea of heavenly spheres was discarded entirely – a step that Kepler's Lutheran religion and unique cosmology prepared him to take. In Kepler's view, the Sun radiated motive force that was imparted to the planets, and that decreased with distance from the Sun. Hence, planets in closer orbits moved faster than those further out, and individual worlds sped up and slowed down as their paths took them closer to, or further away from, the central orb.

Careful calculations soon allowed Kepler to make estimates of just how much the Earth–Sun and Earth–Mars distances varied as they moved around their orbits. By 1602, he had established that a line drawn between the Sun and a planet 'sweeps out' equal areas in equal times as the planet moves along its orbit (that is, the area enclosed by the orbit itself and lines from the Sun to the planet at, for instance, thirty-day intervals are the same regardless of the length of the line). By late 1604, he had also reduced the shape of his ovoid planetary paths to simple ellipses (a circle stretched along one axis), with the Sun at one of the two focal points on either side of the centre. These two discoveries formed the basis of Kepler's first two laws of planetary motion, published in his *New Astronomy* (1609). They were joined by a third law, linking a planet's orbital period to its mean distance from the Sun, in his *Harmony of the World* (1619).

Although the rules of elliptical motion would, at last, provide astronomers with the model they needed to forecast the movements of the planets, they were adopted only gradually. Scholars such as Galileo Galilei (1564–1642) and René Descartes (1596–1650) stuck with their preconceptions of circular motion, but – largely through the spread of his three-volume textbook, the *Epitome of Copernican Astronomy* (1615–21) – Kepler eventually won an audience for his ideas. Lingering respect for Aristotle combined with Kepler's own mystical leanings and the vagueness of his ideas about the central motive force led several leading astronomers – including Ismaël Bullialdus (1605–94) in France, Seth Ward (1617–89) in England and Nicolaus Mercator (1620–87) in Germany – to propose alternative systems that adopted Kepler's ideas in part, while still applying special pleading to others. Nevertheless, when combined with successful practical demonstrations of Kepler's laws, these all made important contributions to the wider acceptance of elliptical orbits in the mid- to late 17th century. In doing so, they set the stage for the ideas of Isaac Newton (1643–1727), whose wider laws of motion and of universal gravitation would not only explain the special case of Kepler's laws, but also reveal the true nature of the force behind them.

THEORIA PLANET

In qua ipsorum motus in Copernicano Systemate tam ex Kepleri et recentiorum Astronomoru[m]

IOH. GABR. DOPPELMAYERO, Mathem. Prof. Publ. Acad. Cæs. Leop

Sumptibus *IOH. BAPTISTÆ HOMA*

De planetarum primariorum motu in genere, ex Kepleri recentiorumque Astronomorum Hypothesi.

Terra cæterique primarij in Orbitis circa solem, quoad situm et magnitudinem, diversis figuris Ellipticæ, à circulari tamen non multum recedentis, tam ex Kepleri quam aliorum assertionibus per observationes innumeras hactenus firmatis, hac perpetuo moventur lege, ut, radio (Fig. 3) ex S Solis centro et communi omnium foco ad locum Planetæ in T r g. ducto, quilibet aream A S T, Tempori, quo planeta ab ejus Aphelio A ad T movetur, proportionatam, hoc est, æqualibus temporibus, æquales areas sive portiones ellipticas describat et ut tempus integræ cujusque periodi ad tempus, quo planeta ab Aphelio A ad T r g. provehitur sic area totius orbitæ ellipticæ ad aream A S T se habeat. His positis sequitur quod planeta quivis circa Aphelium tardius, circa Perihelium celerius, prout à sole remotior aut illi propior est, et sic motu semper inæquali circa Solem volvatur, hæc inæqualitas planetis propria et vera prima, ad differentiam secundæ, quæ optica et è motu terræ resultans a nobis deprehenditur, Astronomis dicta, pro hac indaganda Keplerus circulum cuivis orbitæ circumscripsit, ut designet arcus hujus, exhibeatque mensuram temporis, quod planeta in illis insumit, quorum subsidio Anomalias harum motuum exhibuit, et quod quærebatur ex voto exsolvit.

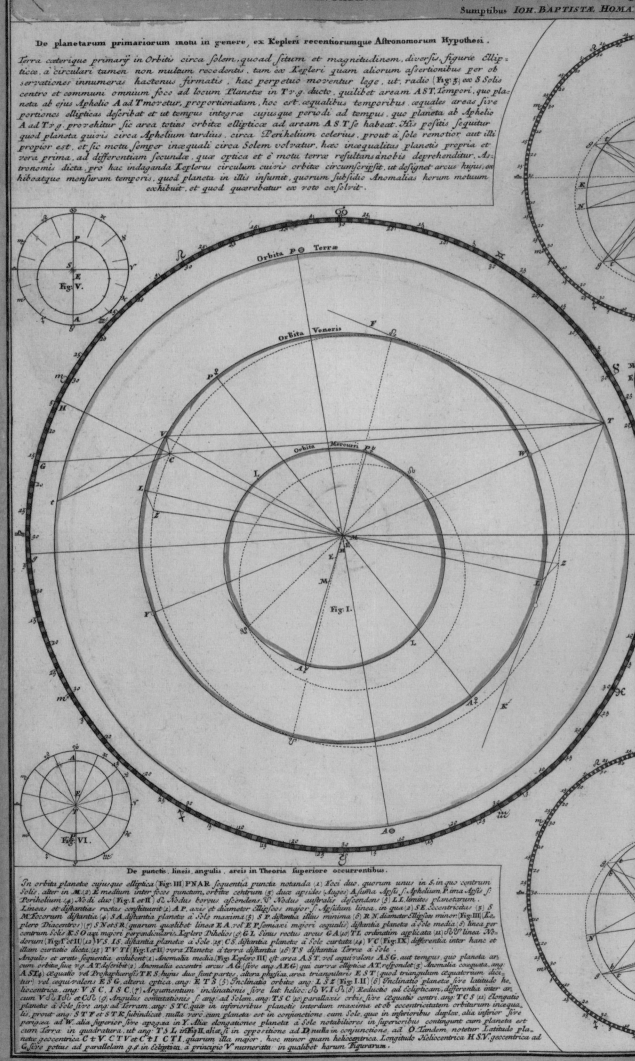

Orbita P ☉ Terræ

Orbita Veneris

Orbita Mercurii

Fig. I.

Fig. V.

Fig. VI.

De punctis, lineis, angulis, areis in Theoria superiore occurrentibus.

In orbita planetæ cujusque elliptica (Fig. III) P N A R sequentia puncta notanda (1) Foci duo, quorum unus in S, in quo centrum Solis, alter in M, (2) E medium inter foces punctum, orbitæ centrum (3) duæ apsides (Auges) A summa Apsis s. Aphelium. P ima Apsis s. Perihelium, (4) Nodi duo (Fig. I et II) d. Nodus boreus ascendens, ♃ Nodus australis descendens (5) L L limites planetarum. Lineæ et distantiæ rectæ constituunt (1) A P, axes et diameter Ellipseos major s. Apsidum linea, in qua (2) S E Eccentricitas (3) S M Eccorum distantia (4) S A distantia planetæ à Sole maxima (5) S P distantia illius minima (6) R N diameter Ellipsis minor (Fig. III) Keplero Diacentros (7) S N et S R, quarum quælibet linea E A, vel E P semiaxis major aequalis distantia planetæ à Sole media (8) linea per centrum Solis K S O axi majori perpendicularis Keplero Dihelios (9) G L Sinus rectus arcus G A, (10) T L ordinatim applicata (11) S ♃ linea Nodorum (Fig. I et II) (12) V S. IS. distantia planetæ à Sole (13) C S. distantia planetæ à Sole curtata (14) V C (Fig. IX) differentia inter hanc et illam curtatio dicta (15) T V TI (Fig. I et II) vera Planetæ à terra distantia (16) T S distantia Terræ à Sole. Anguli, et areæ sequentia exhibent (1) Anomalia medias (Fig. Kepleri III) est area A S T vel aequivalens A S G, aut tempus, quo planeta arc cum orbita suæ v. g. A T describit (2) Anomalia eccentri arcus A G (sive ang. A E G) qui curvæ ellipticæ A T, respondet (3) Anomalia coæquata, ang. A S T, Œquatio vel Prosthaphæresis T E S, hujus duæ sunt partes, altera physica, area triangularis E S T (quod triangulum Œquatorium dicitur) vel aequivalens E S G, altera optica. ang. E T S (5) Inclinatio orbitæ ang. L S Z (Fig. I II) (6) Inclinatio planetæ sive latitudo heliocentrica, ang. V S C, I S C, (7) Argumentum inclinationis sive lat. heliocentrica ♌ V I ♌ (8) Reductio ad ecliticam, differentia inter ar. cum V ♌ I ♌ et C ♌ (9) Angulus comutationis s. ang. ad Solem, ang. T S C (10) parallaxis orbis sive aequatio centri ang. T C S (11) Elongatio planetæ à Sole, sive ang. ad Terram, ang. S T C quæ in inferioribus planetis interdum maxima et ob eccentricitatem orbitarum inæqualis, prout ang. S T F et S T K, subindicat, nulla vero cum planetæ est in conjunctione cum Sole, quæ in inferioribus duplex, alia inferior sive perigæa ad W. alia superior sive apogæa in Y. Aliæ elongationes planeta à Sole notabiliores in superioribus contingunt cum planeta est cum Terra in quadratura, ut ang. T S L in Fig. II. alia s. in oppositione ad D nulla in conjunctione, ad ♀ Tandem notetur Latitudo planetæ geocentrica C ♉ V. C T V et C ♉ I. C T I. quarum illa major, hæc minor quam heliocentrica. Longitudo Heliocentrica H S V, geocentrica ad G sive potius ad parallelam g s. in Ecliptica, à principio V numeratu in qualibet harum Figurarum.

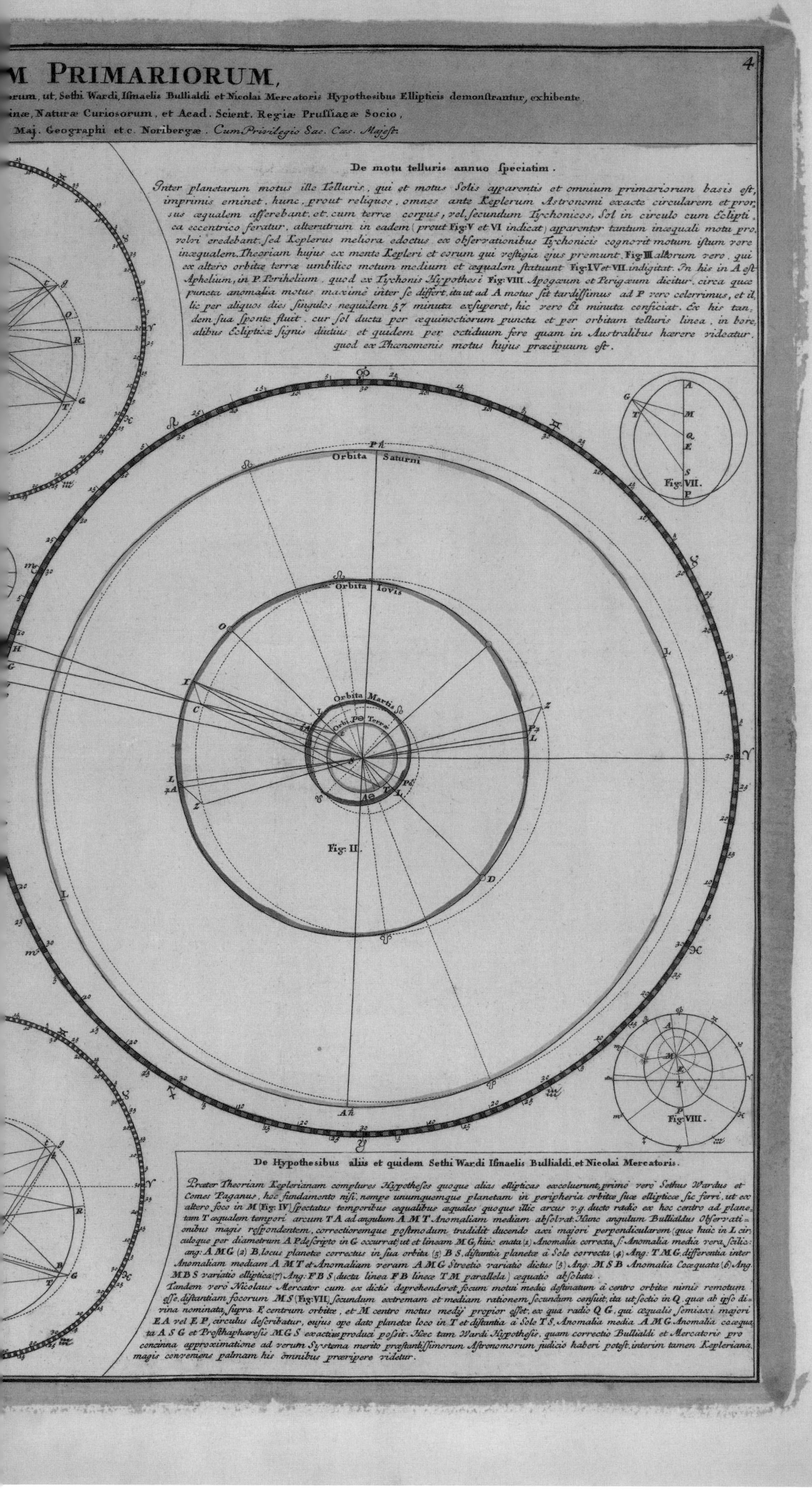

M PRIMARIORUM,

rum, ut, Sethi Wardi, Ismaelis Bullialdi et Nicolai Mercatoris Hypothesibus Ellipticis demonstrantur, exhibente.

inæ, Naturæ Curiosorum, et Acad. Scient. Regiæ Prussiacæ Socio,

Maj. Geographi etc. Noribergæ. Cum Privilegio Sac. Cæs. Majest.

De motu telluris annuo speciatim.

Inter planetarum motus ille Telluris, qui et motus Solis apparentis et omnium primariorum basis est, imprimis eminet, hunc, prout reliquos, omnes ante Keplerum Astronomi exacte circularem et prorsus æqualem asserebant, et, cum terræ corpus, vel secundum Tychonicos, Sol in circulo cum Eclipti ca eccentrico feratur, alterutrum in eadem (prout Fig.V et VI indicat) apparenter tantum inæquali motu pro volvi credebant, sed Keplerus meliora edoctus, ex observationibus Tychonicis cognovit motum istum vere inæqualem. Theoriam hujus ex mente Kepleri et eorum qui vestigia ejus premunt, Fig.III aliorum vero, qui ex altero orbitæ terræ umbilico motum medium et æqualem statuunt Fig.IV et VII. indigitat. In his in A est Aphelium, in P. Perihelium, quod ex Tychonis Hypothesi Fig.VIII Apogæum et Perigæum dicitur, circa quæ puncta anomalia motus maxime inter se differt, ita ut ad A motus sit tardissimus ad P vero celerrimus, et il lic per aliquos dies singulos nequidem 57. minuta exsuperet, hic vero 61. minuta conficiat. Ex his tan dem sua sponte fluit, cur sol ducta per æquinoctiorum puncta et per orbitam telluris linea, in bore alibus Eclipticæ signis diutius et quidem per octiduum fere quam in Australibus hærere videatur. quod ex Phænomenis motus hujus præcipuum est.

Orbita Saturni

Orbita Jovis

Orbita Martis

Orbita Terræ

Fig: II.

Fig. VII.

Fig. VIII.

De Hypothesibus aliis et quidem Sethi Wardi Ismaelis Bullialdi et Nicolai Mercatoris.

Præter Theoriam Keplerianam complures Hypotheses quoque alias ellipticas excoluerunt, primo vero Sethus Wardus et Comes Taganus, hac fundamento nisi, nempe unumquemque planetam in peripheria orbitæ suæ ellipticæ sic ferri, ut ex altero foco in M (Fig: IV) spectatus temporibus æqualibus æquales quoque illic arcus v.g. ducto radio ex hoc centro ad plane tam T æqualem tempori arcum T A ad angulum A M T Anomaliam mediam absolvat. Hunc angulum Bullialdus Observati onibus magis respondentem, correctioremque postmodum, tradidit ducendo axi majori perpendicularem (quæ huic in L cir culoque per diametrum A P descripto in G occurrat) ut et lineam M G, hinc enata (1) Anomalia correcta, f. Anomalia media vera, scilic: ang. A M G (2) B. locus planetæ correctus in sua orbita (3) B S. distantia planetæ à Sole correcta (4) Ang: T M G. differentia inter Anomaliam mediam A M T et Anomaliam veram A M G Stroetio variatio dictus (5) Ang: M S B Anomalia Cœæquata (6) Ang: M B S variatio ellipticæ (7) Ang: F B S (ducta linea F B lineæ T M parallela) æquatio absoluta.
Tandem vero Nicolaus Mercator cum ex dictis deprehenderet, focum motui medio destinatum à centro orbitæ nimis remotum esse, distantiam focorum M S (Fig: VII) secundum extremam et mediam rationem secuit, ita ut sectio in Q, quæ ab ipso di vina nominata, figra E centrum orbitæ, et M centro motus medij propior esset, ex qua radio Q G, qui æqualis semiaxi majori E A vel E P, circulus describatur, cujus ope dato planetæ loco in T et distantia à Sole T S, Anomalia media A M G Anomaliæ cœæqua tæ A S G et Prosthaphæresis M G S exactius produci possit. Hæc tam Wardi Hypothesis, quam correctio Bullialdi et Mercatoris pro concinna approximatione ad verum Systema merito præstantissimorum Astronomorum judicio haberi potest, interim tamen Kepleriana magis conveniens palmam his omnibus præripere videtur.

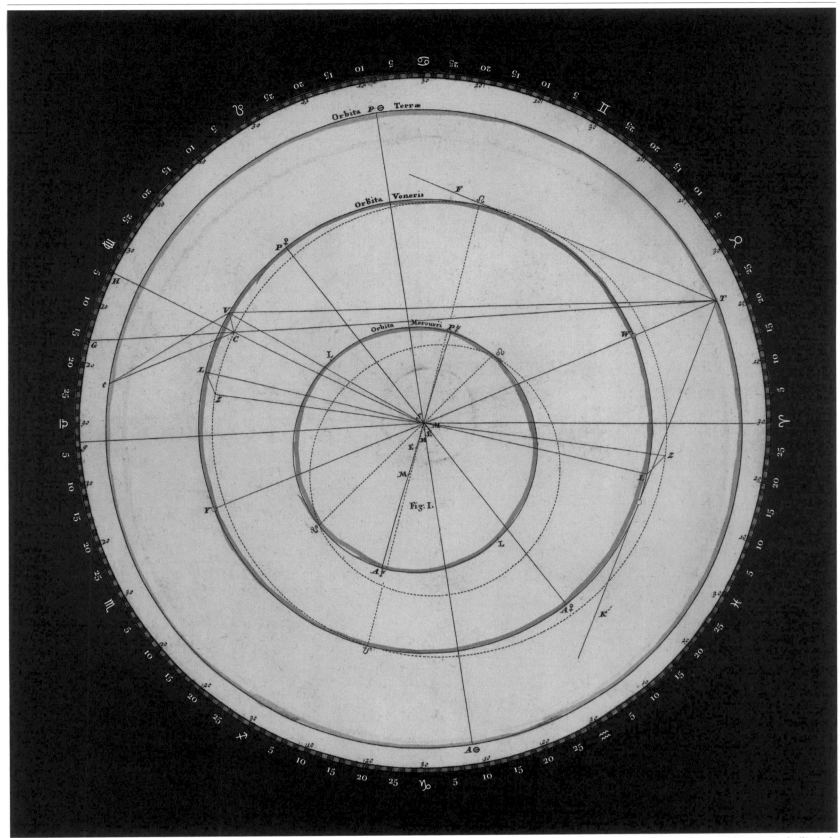

ORBITAL RELATIONSHIPS OF
THE INNER PLANETS. (FIG. 1.)

One of two large diagrams depicts the relationship between the orbits of Earth, Venus and Mercury. Each elliptical orbit is marked with its perihelion and aphelion points (closest to and furthest from the Sun). Also indicated are the locations of the nodes (marked ☊ and ☋), where the inclined orbits of the other planets pass north and south through the ecliptic.

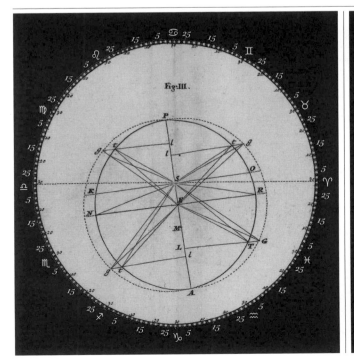

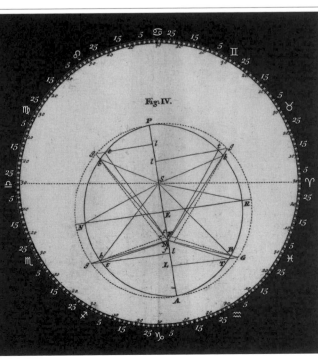

EXPLORING ELLIPTICAL
ORBITS. (FIGS. 3–4.)

Two diagrams describe key characteristics of elliptical orbits. First, the way in which a line from the Sun (*S*) to a planet sweeps out equal areas in equal periods (so the planet moves faster at perihelion and more slowly at aphelion). Second, the Sun's position at one of two focus points on either side of the orbit's geometric centre.

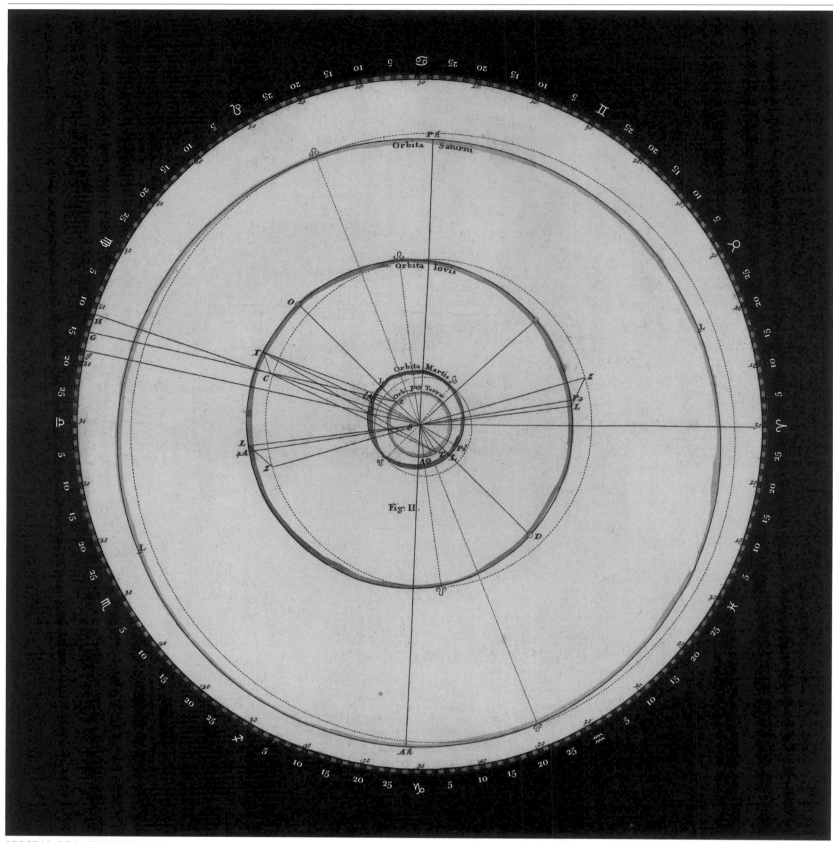

ORBITAL RELATIONSHIPS OF
THE OUTER PLANETS. (FIG. 2.)

A second large diagram explains the configuration of Mars, Jupiter and Saturn's orbits in relation to Earth's.
On both main diagrams, Doppelmayr constructs lines that explain how the varying motion of both Earth and
the other planets along their elliptical orbits – at slight inclinations to each other – gives rise to the observed
motions of the planets.

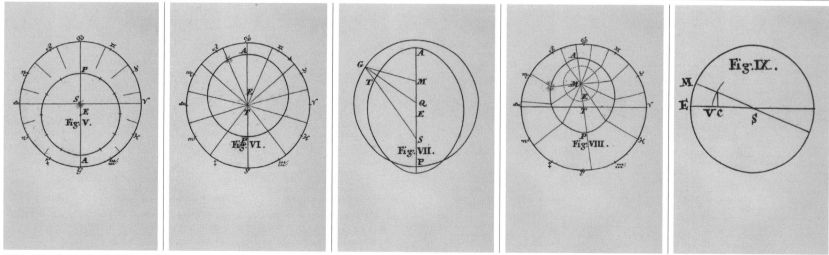

THE SUN'S PATH IN THE SKY.
(FIGS. 5-9.)

A series of supporting diagrams explore the explanations offered by different models for the Sun's varying speed
along the ecliptic, showing again how Kepler's theory is superior.

PLATE 4.

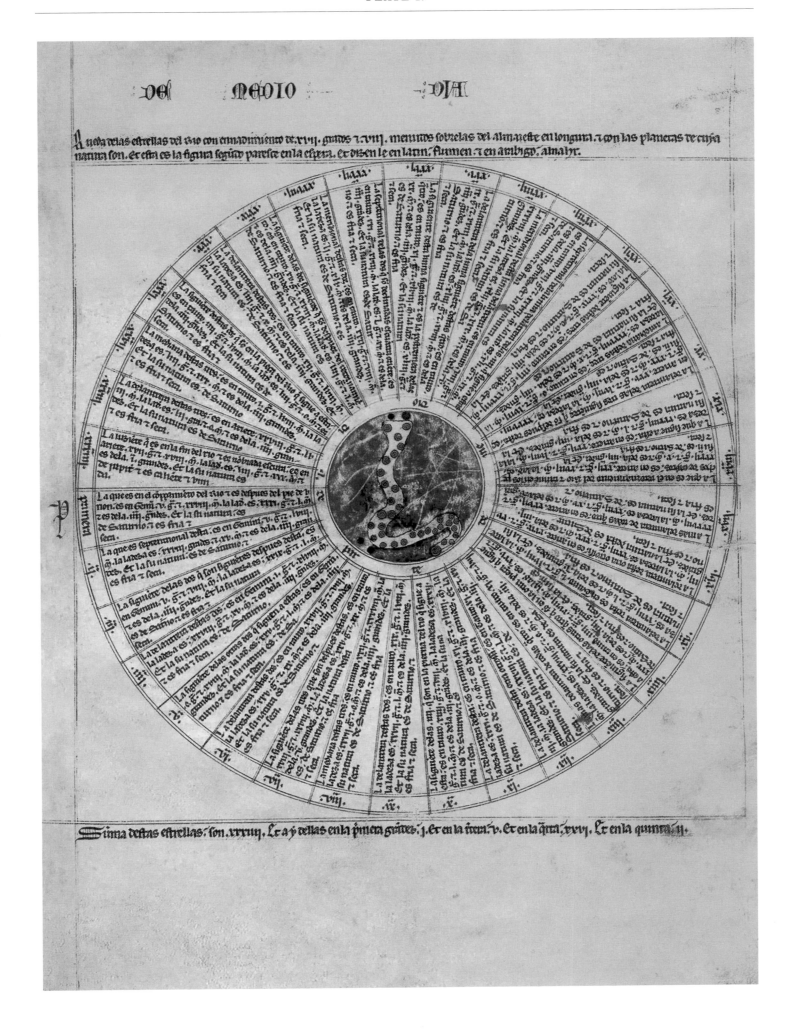

LIBROS DEL SABER DE ASTRONOMIÁ (12TH CENTURY).

The *Libros del Saber de Astronomiá* (*Books of the Wisdom of Astronomy*) was an extraordinary astronomical encyclopaedia commissioned in the late 12th century under Alfonso X of Castile. Compiled by Christian, Jewish and Muslim scholars at Toledo, it encompasses a vast range of knowledge, including detailed tables for use in astrological prediction (above). Alfonso also commissioned the *Alfonsine Tables* - an ephemeris of planetary positions that offered tools for predicting future movements with unprecedented precision - whose persistent inaccuracies did much to fuel doubts

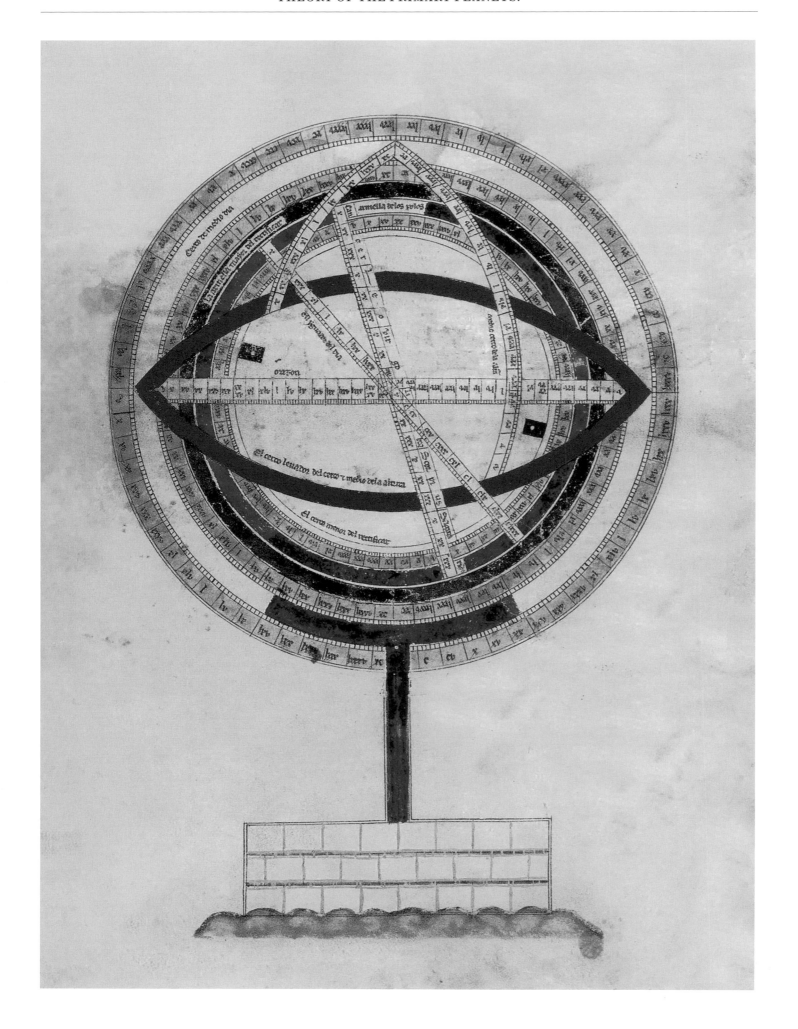

of the geocentric model. Alongside astrological tables, the *Books of the Wisdom of Astronomy* includes manuals for the use of instruments, such as the armillary sphere above and the astrolabe. As well as offering a tool for measuring inclinations of objects in the sky, disc-shaped astrolabes functioned as elaborate analogue computers, with sliding and rotating pointers to simplify various calculations, and a variety of useful data engraved on either side of the disc. They found uses not only in astronomy, but also as general surveying tools – for example, when calculating the height of distant objects.

PLATE 5.

PHENOMENA OF THE PRIMARY PLANETS.

(PHÆNOMENA IN PLANETIS PRIMARIIS)

*Doppelmayr shows how the orbits and orientations of the planets
in the Copernican system affect their appearance as seen from Earth.*

Plate 5 describes the appearance and phases of the planets, including markings observed erroneously on Venus and more accurately on Mars; the phases of Mercury; the shifting cloud bands of Jupiter; and early observations of Saturn's puzzling shape, eventually resolved by Christiaan Huygens (1629–95) as the planet's famous ring system. Observing these phenomena had only become possible thanks to a new device – the telescope.

The telescope's invention is often attributed to Dutch spectacle maker Hans Lippershey (*c.* 1570–1619). Lippershey discovered that a convex lens lined up behind a concave one, with a significant distance between them, could create a magnified image. He tried (and failed) to patent the device in 1608, and as reports of the new instrument circulated around Europe, many curious people attempted to build their own.

The most famous of these was Galileo Galilei (1564–1642), then professor of mathematics at the University of Padua in northern Italy. Galileo was already renowned as a successful inventor and a pioneer of scientific experimentation, and through his methodical approach he rapidly improved the basic telescope design, increasing magnification from around three times in his first attempt of early 1609, to around thirty times in the space of a few months. Late in 1610, this apparently allowed him to be the first person to record the Moon-like phases of Venus through a telescope, thus opening a new era in which the planets were transformed from mere lights in the sky into worlds with appreciable features of their own.

While Galileo's telescope allowed him to make several other key discoveries within the solar system – including spots on the surface of the Sun, Jupiter's four major moons, and the fact that there was something odd about the shape of Saturn – the optical arrangement outlined by Lippershey produced a sharp image only for objects in a very narrow 'field of view', and, thus, severely limited early telescopic observers. As early as 1611, however, Johannes Kepler (1571–1630) outlined an alternative arrangement in which both the front 'objective' lens and the rear 'eyepiece' were outward-curving, or convex. This produced a wider field of view and, theoretically, allowed higher

FIG. 1.

Johannes Hevelius's successful brewing business in Danzig (now Gdańsk, Poland) paid for him to construct an ambitious observatory straddling the roofs of three houses. This view from his *Machina Coelestis* (1673) highlights its centrepiece – the enormous 46-m (151-ft) Keplerian telescope that he used to map the Moon.

FIG. 1.

FIG. 2.

FIG. 3.

FIGS. 2–3.
In these two plates from *Machina Coelestis*, Hevelius describes the techniques and equipment used at his state-of-the-art observatory. Far left is an enclosed hut, with a hooded aperture for the eyepiece end of a telescope. In daylight, the telescope could be directed towards the Sun, projecting a bright image onto the screen. Near left shows the various tools that Hevelius used in the painstaking process of grinding, shaping and polishing precision lenses for his optical instruments.

magnifications, at the minor cost of flipping the image itself upside down. Perhaps surprisingly, no one seems to have built a 'Keplerian' telescope until Christoph Scheiner (1573–1650) – a Jesuit priest and scientific rival of Galileo – in 1630. Thereafter, however, Scheiner's account of the instrument's advantages led to its rapid adoption.

The underlying principle behind any telescope relies on the fact that rays of light from distant objects are effectively parallel to each other, and so a precisely ground glass lens (or later, a curved mirror) can redirect them into a tightening cone of rays that converge at a single point: the focus. The concave eyepiece of a Galilean telescope intercepts the converging rays before they reach the focus, and bends or 'refracts' them back onto diverging paths so that they reach the eye as if they were coming from a closer, or larger, magnified object. Meanwhile, the Keplerian design allows the rays to cross at a focus and then refracts them with a second convex lens to create the diverging light cone viewed by the observer.

The actual magnification achieved by any refracting (lens-based) telescope depends on the shape of the two lenses and on the distance between them. Unfortunately, curved glass lenses come with their own drawbacks. Principally, the light passing through the lens is bent by different amounts depending on its colour, resulting in a series of colourful 'fringes' known as 'chromatic aberration'. The stronger the lens's curvature, the greater the effect. While the challenges of chromatic aberration

would eventually be overcome in the late 18th century, early telescopic astronomers found an ingenious workaround: minimizing the curvature of the objective lens to create an extremely long cone of light that reached a focus far behind the lens, before being picked up by the eyepiece to create the magnified image. This reduced the problem of coloured fringes while permitting higher magnifications.

The final factor affecting telescopes from the mid-17th century until Doppelmayr's own time is today known as 'light grasp'. Because a telescope's objective lens has a larger light-collecting surface than a human pupil, it effectively delivers more of the light from distant objects into the eye, making faint objects appear brighter. The larger the lens, the more light can be delivered, but the longer the focal length (in fact, all else being equal, doubling the lens's diameter quadruples the focal length). As optical glassmakers improved their techniques for casting and polishing lenses of increasing size, telescopes had to become longer to accommodate them. The result was an era of bizarre-looking instruments – enormous tubes tens of metres long, supported on scaffolds, and even longer 'aerial telescopes' that abandoned tubes entirely in favour of mounting the objective on a distant mast and linking it to the observer at the eyepiece with strings, controlling wires and other mechanisms. Precarious though they often seemed, these devices, nevertheless, allowed the great astronomers of the 17th and early 18th centuries to begin observing the 'phenomena' of other planets.

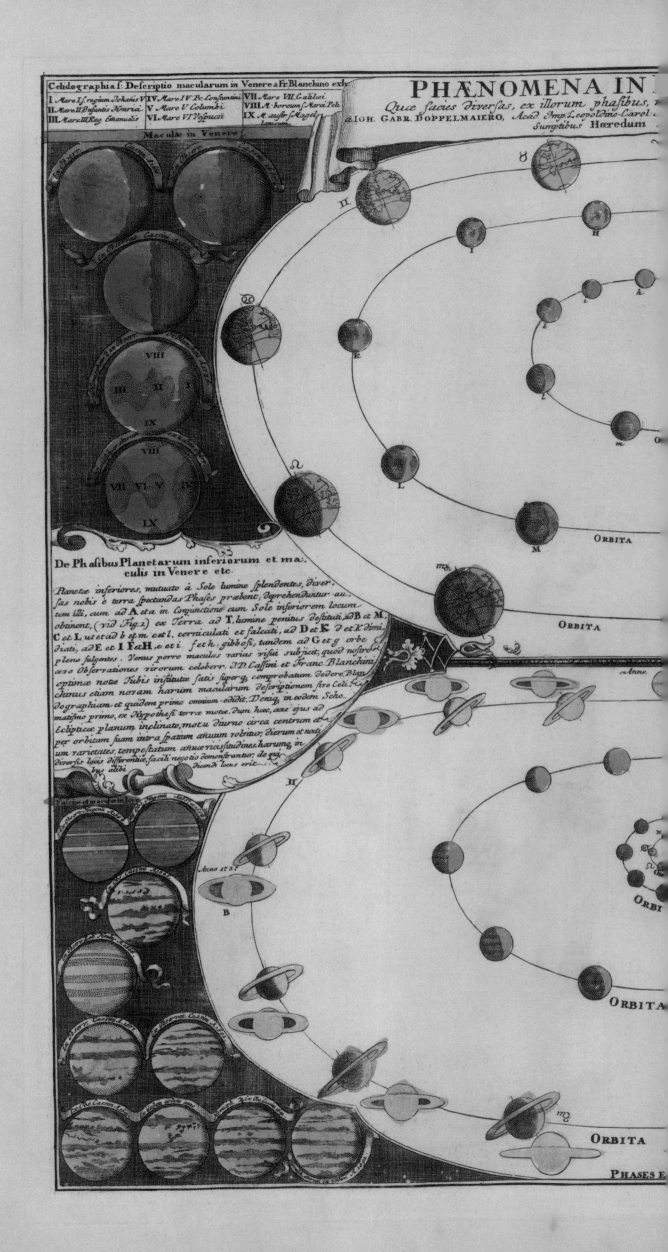

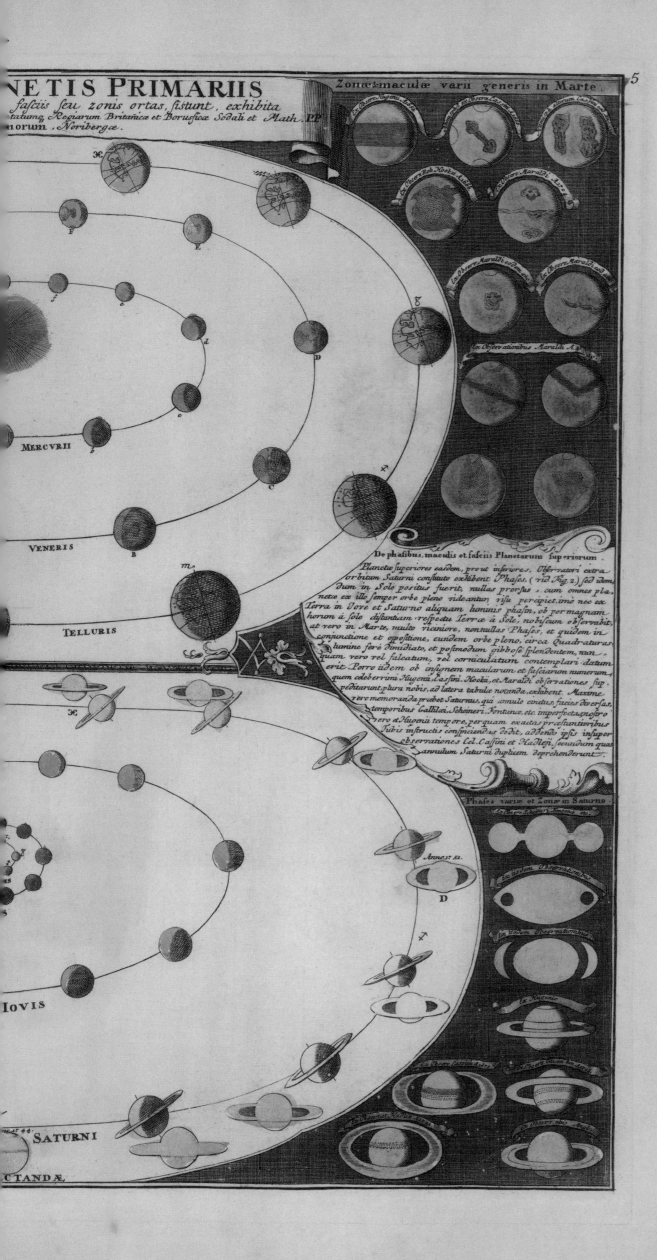

NETIS PRIMARIIS

fasciis seu zonis ortas, sistunt, exhibita
ntium, Regiarum Britañice et Borusficæ Sodali et Math PF
norum . Noribergæ.

Zonæ & maculæ varii generis in Marte

De phasibus, maculis et fasciis Planetarum superiorum.

Planetæ superiores eandem, prout inferiores, Observatori extra orbitam Saturni constituto exhibent Phases. (vid Fig. 2.) sed idem, dum in Sole positus fuerit, nullas prorsus , cum omnes planetæ ex illo semper orbe pleno videantur, risu percipiet. imo nec ex Terra in Jove et Saturno aliquam luminis phasin, ob per magnam horum à sole distantiam respectu Terræ à Sole, notabilem observabit, at vero in Marte, multo viciniore, nonnullas Phases, et quidem in conjunctione et oppositione, eundem orbe pleno, circa Quadraturas lumine feré dimidiato, et postmodum gibbose splendentem, nunquam vero vel falcatum, vel corniculatum contemplari datum erit. Porro iidem ob insignem macularum et fasciarum numerum, quem celeberrimi Hugenii, Cassini, Hookii, et Maraldi observationes suppeditarunt, plura nobis, ad latera tabulæ notanda, exhibent. Maxime vero memoranda præbet Saturnus, qui annulo cinctus, facies diversas, temporibus Galilæi, Scheineri, Fontanæ, etc. imperfectè nostro ævo vero et Hugenii tempore, perquam exactas præstantioribus tubis instructis conspiciendas dedit, addendo ipsis insuper observationes Cel. Cassini et Halleji, secundum quas annulum Saturni duplicem deprehenderunt.

Phases variæ et Zonæ in Saturno

Annus 32.

D

MERCVRII

VENERIS

TELLVRIS

IOVIS

SATVRNI

CTANDÆ

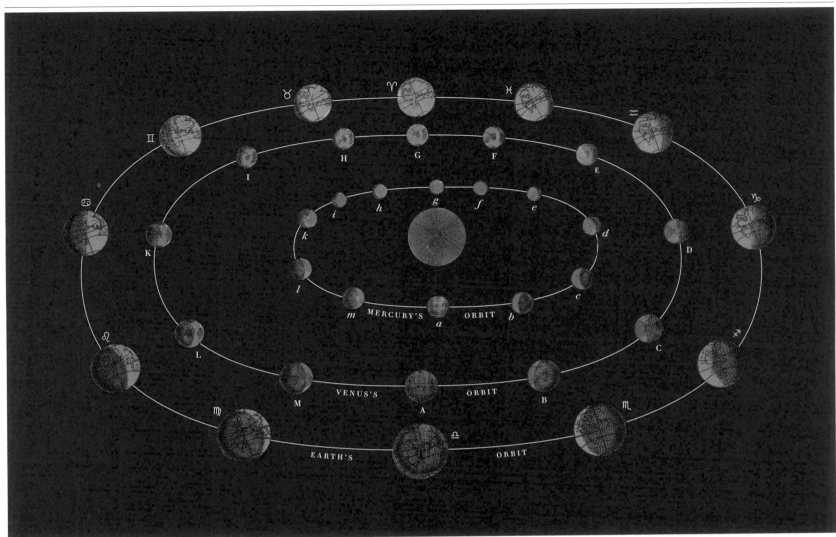

PHASES OF THE INFERIOR PLANETS
AND MARKINGS OF VENUS. (FIG. 1.)

The upper centre of the plate illustrates the orbits of Mercury, Venus and Earth, explaining why the inferior planets change their appearance as observers on Earth see differing amounts of their sunlit side. The illustration also hints at markings on Venus – a topic that remains controversial today.

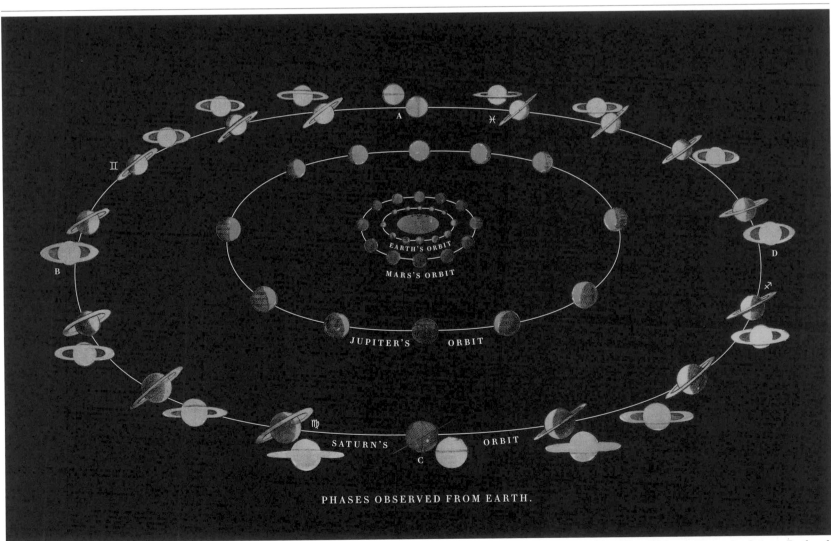

PHASES AND APPEARANCE OF
THE SUPERIOR PLANETS. (FIG. 2.)

The lower central illustration depicts the orbits of Mars, Jupiter and Saturn in relation to Earth and the Sun. These planets are shown to have a daylit and a dark hemisphere, but because we view them from the direction of the Sun, our view is mostly limited to the sunlit half. For Saturn, Doppelmayr incorporates a diagram originated by Huygens, showing how the rings change their appearance during each orbit.

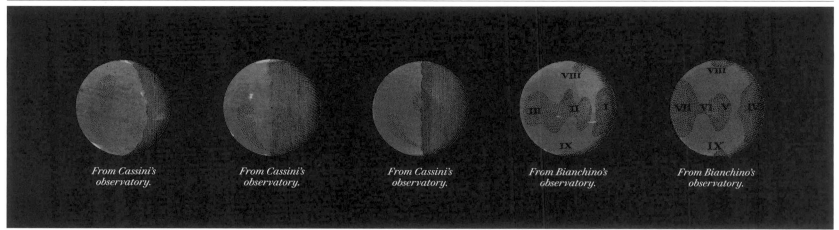

MARKINGS OF VENUS ACCORDING
TO BIANCHINO.

This section shows various dark markings on Venus reported by Francesco Bianchino in his 1728 book on the subject. Beneath each figure, Doppelmayr notes the reporting observatory. While Venus's brilliant white clouds generally appear uniform for optical observers, dark markings are still occasionally reported. These are, however, generally dismissed as illusions or telescopic artefacts.

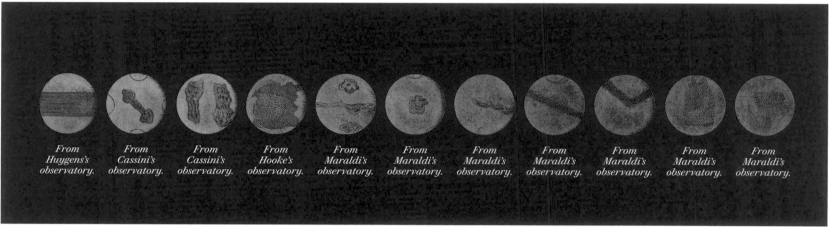

MARKINGS AND VARIATIONS
OF MARS.

This series of views reproduces observations of Mars by Christiaan Huygens, Giovanni Domenico Cassini, Robert Hooke and Giacomo Maraldi. While early interpretations of the Martian surface varied considerably, the triangle on Maraldi's final drawing is likely to be a representation of the region now known as Syrtis Major.

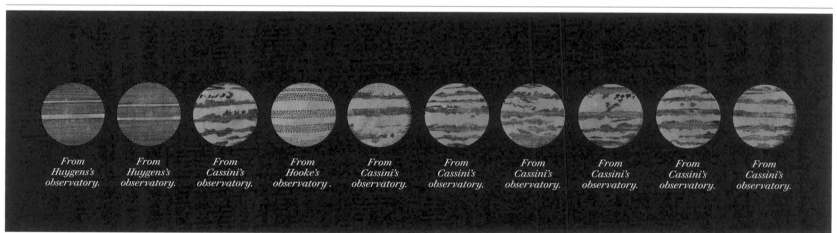

MARKINGS OF JUPITER.

Doppelmayr reproduces various sketches of Jupiter by observers including Christiaan Huygens, Giovanni Domenico Cassini and Robert Hooke. Cassini's sketches in particular show some understanding of the turbulent cloud bands that dominate the planet's appearance.

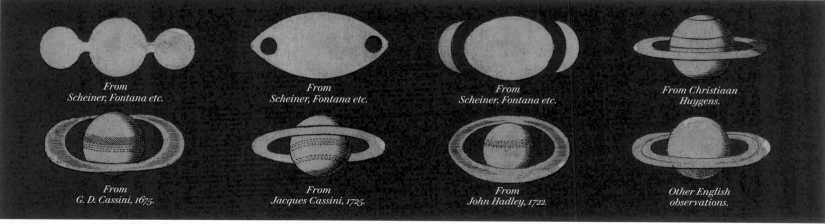

CHANGING INTERPRETATIONS
AND MARKINGS OF SATURN.

Doppelmayr presents a handful of interpretations of Saturn's strange appearance by early telescopic observers, culminating in Huygens's recognition of the rings. Four further images show differences in Saturn's appearance, owing to our changing views of the rings and the shadows that they cast upon the planet.

PLATE 6.

PHENOMENA.

(PHÆNOMENA)

Doppelmayr explains the phenomena of the seasons,
and depicts a variety of curious alternative cosmologies.

Concerned primarily with the changing length of days experienced on the surface of Earth, Plate 6 of Doppelmayr's *Atlas* is fittingly headed with a depiction of the god Helios riding the chariot of the Sun through the heavens. It is thought to be one of the later plates in the book, designed and engraved some time after 1735.

The annual pattern of seasons traditionally begins with the Sun at the northern hemisphere's vernal or spring equinox (when day and night are the same length). In today's calendar, this occurs on 20 or 21 March each year; the Sun rises due east, crosses (transits) the meridian due south at its highest point in the sky for the day, and then sets due west. Over the next few months, the Sun moves higher in the sky, edging closer to the north celestial pole. The points where it rises and sets creep further towards the northeast and northwest, and the days get longer until they reach a maximum at the summer solstice on 20 or 21 June. Thereafter, the Sun's motion reverses – it tracks slowly southwards, and the long days gradually diminish while the short nights grow. By 22 or 23 September, the Sun once again rises and sets due east and west, and the days and nights are of equal length at the northern autumnal equinox. The southward drift continues unabated until the winter solstice on 21 or 22 December, where the Sun rises in the southeast, transits at its lowest altitude over the southern horizon and slinks back below the southwestern horizon just a few hours later. Finally, the Sun's drift reverses again, until it returns to the next vernal equinox.

In terms of the celestial sphere, this annual pattern can be imagined in the same way, regardless of which cosmology one prefers – the Sun follows an eastward track against the background 'fixed stars' of the zodiac, along a line known as the ecliptic. This forms a 'great circle' around the sphere (one whose plane passes through its centre), but is tilted at an angle of 23.5 degrees to the celestial equator that divides the sky's northern and southern hemispheres. At (northern) winter solstice, the Sun is at its southernmost point on the ecliptic, from where it tracks northwards, crossing into the northern celestial hemisphere at the vernal equinox, reaching its northernmost point at the summer solstice, and then returning south via the autumnal equinox. For observers south of the equator, of course, the Sun passes higher in the sky when it is in the southern celestial hemisphere – hence the northern winter solstice is the southern summer one, and the pattern of seasons is reversed.

The tilt of the ecliptic forced geocentric astronomers to add yet another complication to their model of the universe – one that caused the solar and planetary spheres to slowly wobble to the north and south over the course of each year.

FIG. 1.
This stunning 13th-century illumination is from the *Book of Divine Works* originally composed by German mystic Hildegard of Bingen. It illustrates her vision of the connections between the human body and soul and the wider cosmos. Hildegard drew complex parallels between the four terrestrial elements of Aristotle, the four bodily humours of medieval medicine and the four seasons of the year, as well as between the idealized classical geometry of the human body and that of the cosmos.

FIG. 1.

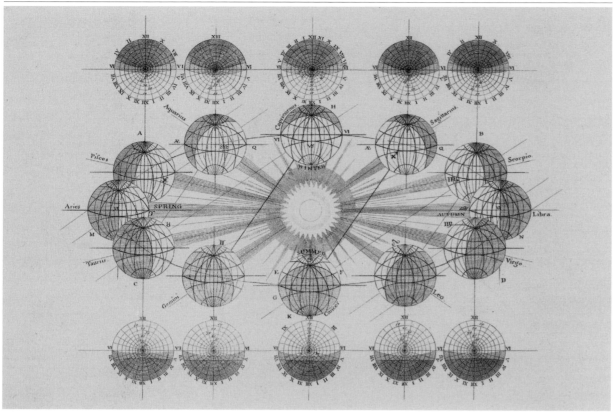

FIG. 2.

FIG. 2.

From *The Universal Vicissitude of Seasons* (1737) by Thomas Wright of Durham, this diagram exhaustively describes the cause of the seasons and the changes in the Sun's apparent path across the sky from different parts of Earth throughout the year.

They also faced a significant challenge from the growing realization that the lengths of the seasons were not the same; in fact, the intervals between the four marker points in the Sun's journey varied between less than eighty-nine days (from autumn equinox to winter solstice) to ninety-four days (from vernal equinox to summer solstice). In order to explain why even the Sun refused to travel at a uniform speed around the Earth, they were forced to fall back on the same epicycles and equant points that were needed to explain the motions of the planets.

The Copernican theory faced few such challenges. With Earth placed in orbit as a planet around the Sun, it was possible to imagine that its axis of daily rotation was simply tilted with respect to the plane of its orbit. In other words, while the North Pole points in a single constant direction in space (defining the poles and equator of the celestial sphere), that direction is not 'bolt upright' compared to Earth's orbit around the Sun. Instead, it is tilted at an angle of 23.5 degrees. As a result, sometimes each hemisphere will be angled towards the Sun (and see it rising higher in the sky) and sometimes it will be tipped away from it.

Doppelmayr illustrates the principle at the centre of the plate, showing how the Sun's altitude in the sky varies through the year as seen from latitudes including the Arctic and Antarctic circles as well as that of Nuremberg

itself. From the 17th century, the introduction of the elliptical model for planetary orbits provided a neat explanation for the differing lengths of the seasons: if Earth's own orbit is a slight ellipse, then it will move more slowly when further from the Sun (and the Sun will, therefore, slow its apparent path across the sky).

Doppelmayr accompanies his central artwork with a series of seven decorative charts showing various ancient and modern theories of the universe. At top right is the version of the Ptolemaic theory expounded in the *Alfonsine Tables* (notable for the insertion of two further spheres to 'drive' the fixed stars in accordance with a long-term astronomical cycle known as the 'precession of the equinoxes'). This is accompanied by the theories widely attributed to Plato (*c.* 428–348 BCE), Plutarch (46–119 CE) and Porphyry of Tyre (*c.* 234–305 CE), which differ in their ordering of the spheres of the Sun, Mercury and Venus. Along the bottom of the plate are curious cosmologies attributed to Johannes Cocceius (1603–69), William Gilbert (1544–1603) and Sébastian Le Clerc (1637–1714). Each of these marks an attempt to address the variation in seasons by somehow offsetting either Earth or the Sun from the centre of rotation. Gilbert's scheme is particularly notable for its early suggestion (in a broadly geocentric context) that Earth was rotating and the stars were scattered throughout space at varying distances.

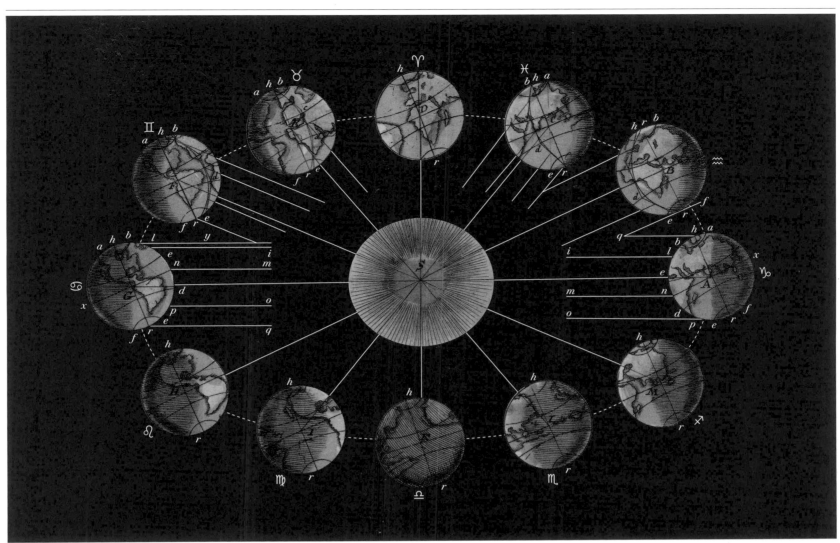

CAUSE OF THE SEASONS.
(FIG. 1.)

Doppelmayr's central diagram shows Earth at various stages in its annual orbit around the Sun, with the orientation of its axis of rotation remaining fixed in space. Rays emanating from the central Sun are used to show how illumination differs across Earth at different points in the year, with the solstices at extreme left and right, and the equinoxes at top and bottom.

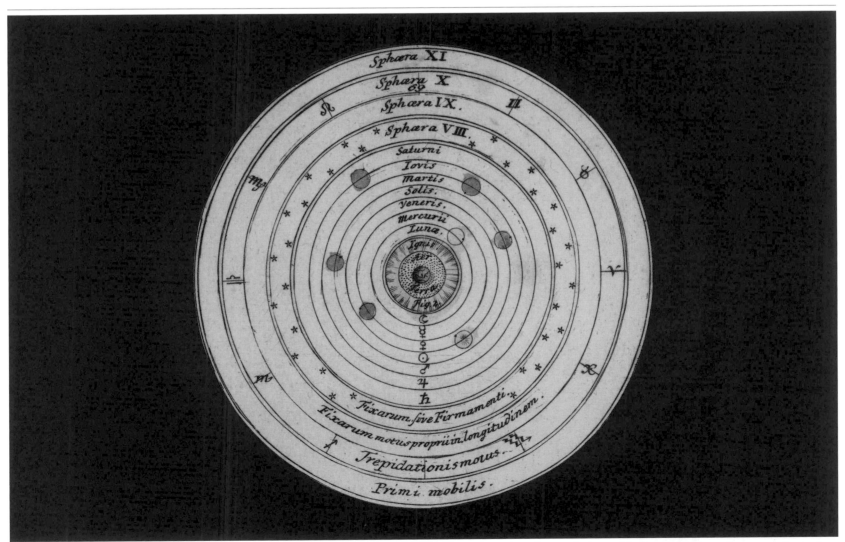

PTOLEMAIC AND ALFONSINE SYSTEM.
(FIG. 2.)

A detailed diagram shows the nested spheres of the Ptolemaic system as they were understood at the time of the 13th-century *Alfonsine Tables*. Within the orbit of the Moon lay the four spheres of the Aristotelean elements. Beyond the sphere of the fixed stars lie additional spheres to control the long, slow drift of precession and a now-redundant cyclical motion called trepidation. The outermost sphere of all – the *primum mobile* – drives motion throughout the entire system.

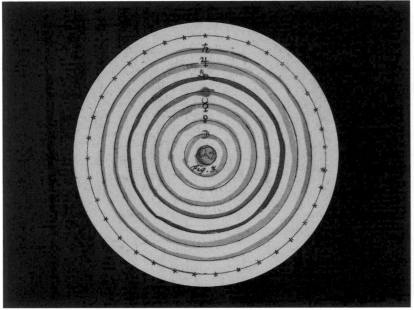

SYSTEM OF PLUTARCH. (FIG. 3.)

In his astronomical writings, the Greek biographer and philosopher Plutarch describes a geocentric system in which the spheres of Venus and Mercury lie between those of the Moon and Sun and are driven, in part, by the motion of the solar sphere, with the brilliant Venus next closest to Earth beyond the Moon.

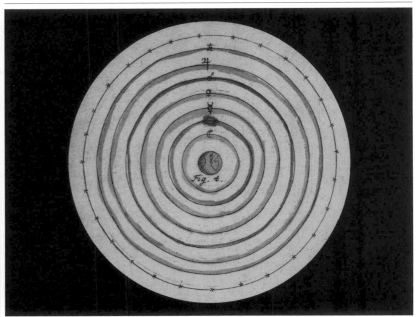

PLATONIC SYSTEM. (FIG. 4.)

The geocentric system here attributed to Plato (although properly the work of his student Eudoxus) places Earth at the centre of the universe, encircled by concentric spheres carrying the Moon, Sun, Mercury, Venus, Mars, Jupiter, Saturn and fixed stars. Doppelmayr's illustration simplifies the series of spheres driving each body's motion to a single circular pathway.

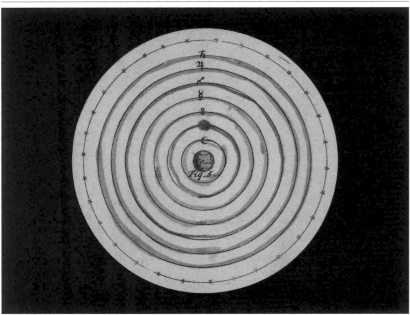

SYSTEM OF PORPHYRY. (FIG. 5.)

The Neoplatonic philosopher and astrologer Porphyry formulated a system similar to that of Plutarch. However, he placed Mercury's sphere beyond the Moon's, with that of the slower moving Venus's further away and closer to the Sun.

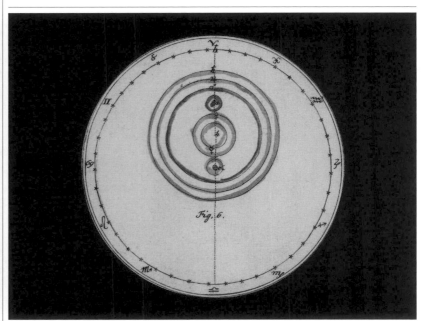

SYSTEM OF JOHANNES COCCEIUS. (FIG. 6.)

In this curious cosmology suggested by an obscure Dutch astronomer, the planets circle a point midway between the fixed Earth and the orbiting Sun. Mercury and Venus remain between Earth and the Sun at all times, while the orbits of the outer planets encircle both.

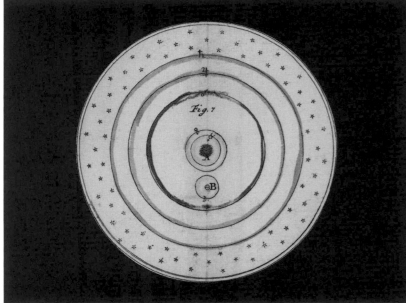

SYSTEM OF WILLIAM GILBERT. (FIG. 7.)

William Gilbert, a 16th-century pioneer of magnetism, was the first astronomer to argue that the stars lay at varying distances, rather than on a fixed sphere. He attributed most motion in the heavens to a daily rotation of Earth, but his depiction of the solar system notably avoided settling on a Copernican or Tychonic viewpoint.

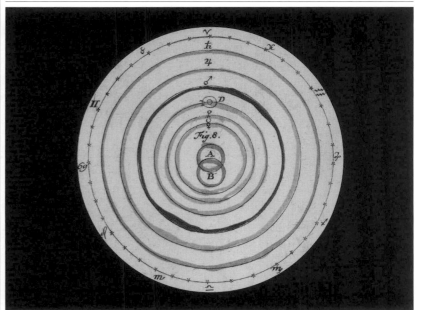

SYSTEM OF SÉBASTIAN LE CLERC. (FIG. 8.)

This ingenious but obscure variation on the Copernican system sets Earth and the other planets in circular orbits around the Sun (*A*). However, it has the Sun itself in a small orbit around the true central point (*B*) in an effort to explain problems that remained due to reliance on regular circular motion.

PLATE 6.

TRÈS RICHES HEURES DU DUC DE BERRY
(15TH CENTURY).

This stunning sequence of folios from perhaps the best-known illuminated manuscript of all time depicts the Labours of the Months: activities taking place among the nobility at court and the peasants working on the duke's lands throughout the year. Above each scene, an arch shows Phoebus (an alternative name for the Greek god Apollo) bearing the Sun in his golden chariot through the appropriate zodiac constellations. Developed from monastic and liturgical calendars, books of hours were a popular medieval form of devotional book,

containing cycles of prayers and readings appropriate to different times of the year. The *Très Riches Heures* was commissioned by French prince Jean, Duke of Berry, from Dutch miniaturists the Limbourg brothers. It contains some 206 illuminated folios on the finest quality calf-skin. About half of these are full-page illuminations, frequently using the duke's many castles as decorative backdrops. When all three brothers and their patron died in 1416, the book was left unfinished, but it continued to be worked on later in the 15th century by a variety of hands.

PLATE 7.

THE PHENOMENA OF IRREGULAR MOTION.

(PHÆNOMENA MOTUUM IRREGULARIUM)

Doppelmayr demonstrates how the Copernican model, coupled with elliptical orbits, can be used to explain the complex movements of the inner planets.

Plates 7 through to 10 of Doppelmayr's *Atlas* are among the earliest in their origins. Along with Plate 3, they were originally conceived for Johann Baptist Homann's (1664–1724) *Atlas of 100 Charts* published in 1712. Their early date is betrayed by the events shown in their complex charts (in this case, the motions of Mercury and Venus in the year 1710). However, even when republished a generation later, they provided a masterly explanation of intricate phenomena.

The motions of the inferior planets are some of the most complex in the sky; to describe them simply as loops around the present location of the Sun is to ignore an array of variations in the size, angle and speed of these movements, as well as the brightness of the objects involved. Venus, at least, has the benefit of being unmistakable: the brightest object in the sky after the Sun and Moon, and a more-or-less permanent resident of either the early morning or the evening sky. Mercury, in contrast, is the most elusive of the naked-eye planets known to the ancients. It is inherently much fainter than Venus and makes only fleeting appearances, where it can be hard to spot against the glow of dawn or dusk.

Although today the planets are named for the Roman goddess of beauty and the fleet-footed messenger god, it took ancient civilizations some time to recognize that each of these entities was a single body, making both dawn and dusk appearances. To the ancient Greeks, Venus had two names – Phosphorus in the morning and Hesperus in the evening. The realization that they were one and the same came around the 5th or 6th century BCE, but the tradition of giving alternate names to the morning and evening apparitions lingers to the present day (from Lucifer and Vesper in the Roman world of late antiquity, to today's morning star and evening star). The same dual nature applied at first to the innermost planet, known as Apollo on its morning apparitions and Hermes in the evening (in this case, of course, Hermes became the dominant name, and custody of the planet was transferred to that god's direct equivalent, Mercury, in the Roman era).

While Venus can reach a maximum angular separation from the Sun (known as its 'greatest elongation') of around 46 degrees, Mercury's greatest elongations are a mere 28 degrees. This means that Venus can linger in the sky until well after sunset, and rise well before dawn, to appear as a dazzling 'star' against the full darkness of night. Mercury, however, is trapped in perpetual twilight – the Sun drags the planet with it shortly after it sets during evening elongations, and rises hard on its heels in morning apparitions.

FIG. 1.

Italian artist Geronimo Frezza depicts Lucifer, the morning star, as a winged putto pouring light from a vase (1704). This engraving is a reproduction of a figure from the *Allegory of Time* (1611–12), a ceiling fresco by baroque artist Francesco Albani, painted for Rome's Palazzo Verospi.

LVCIFER

FIG. 1.

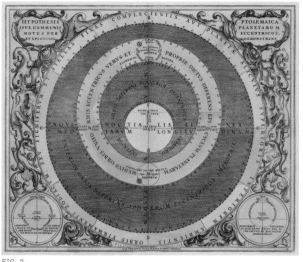

FIG. 2.

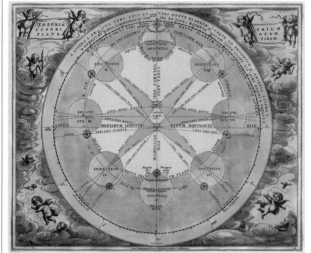

FIG. 3.

FIGS. 2–3.
These illustrations of the Ptolemaic system of epicycles and deferents are from Andreas Cellarius's *Harmonia Macrocosmica* (1660). They are a rare attempt to depict the physical implications of Ptolemy's epicycles, recognizing the need for a distinct 'depth' to each of the spheres within which the planets could move.

For ancient astronomers, the most significant puzzle in the behaviour of Venus was the stark variation in its speed against the background stars. Some 584 days separate one eastern (evening) elongation from the next, or one western (morning) elongation from its successor, but the planet can take as few as 141 days to pass from maximum eastern to maximum western elongation, before taking around 443 days to track laboriously back to its starting position.

This presented obvious problems for anyone trying to model a geocentric system with uniform circular motion around Earth, and inspired some Greek philosophers – most famously Aristarchus of Samos (*c.* 310–230 BCE) – to consider the idea of a heliocentric system, in which the difference in motions is an obvious consequence of the relative speeds and locations of Earth and Venus. Others proposed a semi-heliocentric system akin to the later ideas of Tycho Brahe (1546–1601), with Venus and Mercury circling the Sun even as the Sun circled Earth. We have seen previously how the theory of epicycles proposed by Ptolemy (*c.* 100–170 CE) – in which Venus moved on a smaller circle centred on its primary sphere – became the most widely accepted explanation for this sort of effect (see page 28). This also offered an intuitive explanation for Venus's changes in brightness over the course of its motions – an effect of its overall distance from observers on Earth.

Mercury's motions presented similar problems to those of Venus, as well as some additional challenges. Most notable is the fact that one elongation can differ significantly from the next – 28 degrees might be the planet's maximum separation from the Sun, but on some apparitions it reaches a mere 18 degrees. In line with this, while Mercury shares the Venusian pattern of a relatively slow eastward track followed by a fast return to the western

pre-dawn sky, the intervals between successive eastern or western elongations are prone to significant variations. Finally, Mercury's path can take it significantly north or south of the ecliptic, with major implications for its visibility at a specific apparition. In modern terms, most of these variations can be explained by the shape of Mercury's orbit – a distinctly elliptical eighty-eight-day path that varies between 46 and 70 million km (29–43 million miles) from the Sun, and is tilted relative to the ecliptic at an angle of 7 degrees. However, even once these characteristics were understood, the planet still showed some anomalies of motion that would not be resolved until early in the 20th century.

Doppelmayr's charts demonstrate how the motions of the inferior planets can be accurately explained in terms of a Copernican system with elliptical orbits. The plate is flanked by projections showing the movement of Venus and Mercury relative to the ecliptic over periods of several months in 1610, while at its heart lies a somewhat tangled diagram featuring the nested orbits of Earth, Venus and Mercury, surrounded by the outer ring of the ecliptic. The locations of the planets on their orbits are marked for significant days (accounting for their varying speed along their orbits), while a web of radiating lines demonstrates how the planets appear to move along the ecliptic and pass through various constellations as their alignment with Earth changes. Simplified diagrams at upper left and right demonstrate the principle, and explain how the size of the orbits of the inferior planets determines how close to the Sun they remain. At lower right is the November 1710 transit of Mercury (a relatively frequent event in which the planet passes across the face of the Sun as seen from Earth), while at lower left, Doppelmayr looks forward to the far rarer transit of Venus forecast for June 1761.

PHÆNOMENA M
quos Planetæ inferiores VENVS ET
Directionibus, Stationibus et Retrogradationibus fuis è Terra fpectandos pra
À Ioh. Gabriele Doppelmajero Mathem.

Fig. I.

Motus VENERIS per integram fuam revolutionem, à principio añi 1710 ad medium usque Augufti è Terra inæqualis fecundum Ecliptcæ ductum fpectandus.

PHÆNOMENON VENERIS,
ejusdem corporis fublucido Solis orbe,
maculæ inftar, tranfituro rariffimum.

Septentrio.

Conjunctio Solis cum Venere
Norimbergæ añ. Chrifti.
die 6. Iunij h. 13 56′ fecundum calcula
Kulleijani drca ned. defcen. conficienda.

Eclip tica.

ORBITA VENERIS

Fig. III.

Meridies.

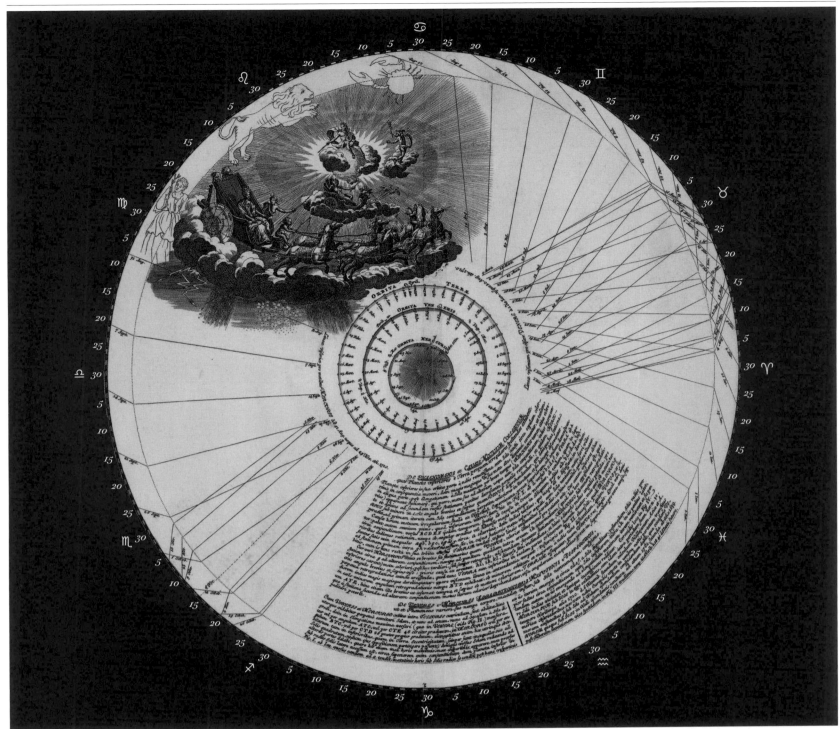

ORBITS OF MERCURY, VENUS AND EARTH.

The central diagram on Plate 7 plots the orbits of Mercury, Venus and Earth at its centre, with dates that show the locations of the planets at various times in 1710. The perihelion (closest point to the Sun) and aphelion (most distant point) of each orbit are clearly marked. Beyond the orbit of Earth, lines are constructed showing how the relative directions of the planets give rise to their wandering paths through Earth's skies. A central motif depicts the three planets in their respective circuits around the Sun.

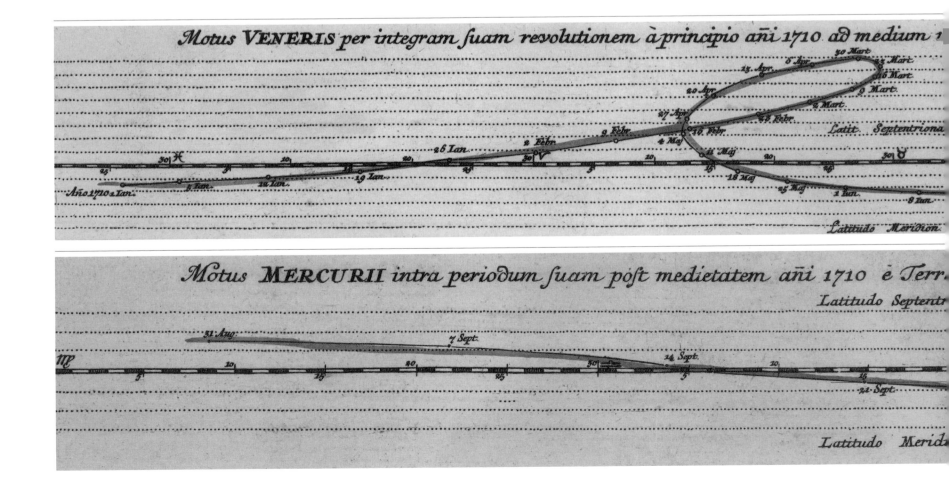

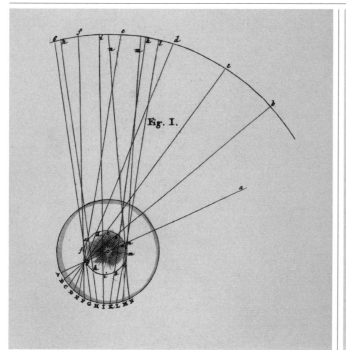

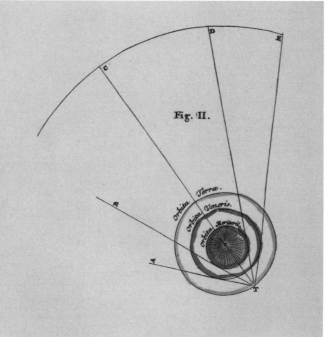

METHOD OF CONSTRUCTION. (FIG. 1.)

This simplified diagram to accompany Doppelmayr's explanatory text demonstrates how the calculations of direction from Earth to an inner planet (shown in the central diagram) are achieved.

ELONGATIONS AND CONJUNCTIONS. (FIG. 2.)

A small diagram illustrates key points on the orbits of the inferior planets as seen from Earth. These are elongations (maximum distance east or west from the Sun, when their motion briefly slows to a halt) and conjunctions (when they pass behind or in front of the Sun and their motion is at its fastest).

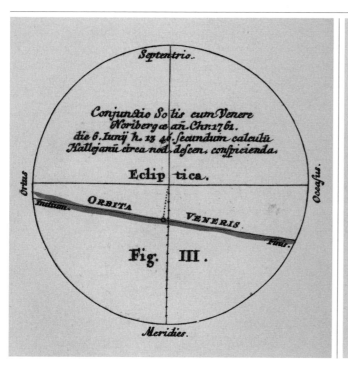

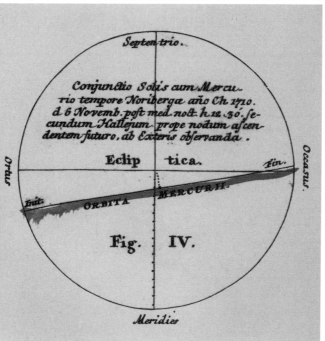

TRANSIT OF VENUS. (FIG. 3.)

Doppelmayr illustrates the path of the transit of Venus across the face of the Sun – predicted by Edmond Halley to occur in June 1761 – as seen from Nuremberg.

TRANSIT OF MERCURY. (FIG. 4.)

This illustration shows the observed transit of Mercury in 1710, also predicted by Halley. The track shows how Mercury's orbit is inclined to the ecliptic, explaining why transits do not occur every time the planets pass between Earth and the Sun.

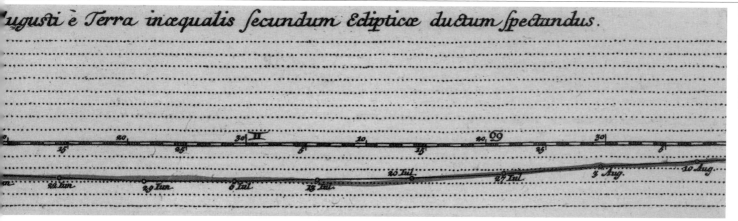

MOTION OF VENUS.

Doppelmayr charts the position of Venus relative to the ecliptic between January and August 1610, demonstrating how the plotted positions correspond to the lines on the central diagram.

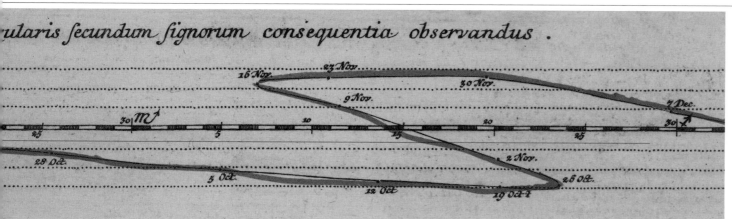

MOTION OF MERCURY.

The innermost planet follows a markedly elliptical orbit around the Sun in just eighty-eight days. Doppelmayr's accompanying chart shows how Mercury's observed location in the sky over a single orbit matches up with calculations based on the Copernican/Keplerian model.

PLATE 8.

EPHEMERIDES OF GEOMETRIC CELESTIAL MOTION.

(EPHEMERIDES MOTUUM COELESTIUM GEOMETRICÆ)

Doppelmayr describes the irregular motions of the planets recorded through 1708 and 1709, and explains how they arise in the Copernican model of the solar system.

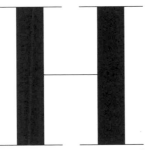aving previously examined the precise motion of the inferior planets, Doppelmayr applies a similar process to the entire solar system, depicted in Plate 8. He analyses the motions of the planets over a period of almost two years through a series of charts around a central diagram, concentrating especially on the superior planets: Mars, Jupiter and Saturn.

Although widely understood in modern terms as the planets whose orbits are larger than Earth's (as opposed to the inferior planets that have smaller orbits closer to the Sun), the concept of superior planets originated in Ptolemy's (*c.* 100–170 CE) geocentric cosmology. In this model, an inferior planet was one whose deferent (the large circular orbit upon which a planet travelled around Earth, while also circling its own epicycle) rotated at the same rate as the Sun's, so that the planet's movement around its epicycle formed loops around the Sun's position in the sky. A superior planet, meanwhile, was one whose deferent was independent of the Sun's, and rotated more slowly.

Philosophers differed slightly in their opinion of the arrangement of the two types of planet amid the nested spheres of the universe. For understandable reasons, they placed the slower moving superior planets in the outer spheres, closer to the fixed stars. However, while Ptolemy placed Mercury and Venus between the Moon and the Sun (with Mars, Jupiter and then Saturn beyond), Porphyry of Tyre (*c.* 234–305 CE) placed the Sun in the next position beyond the Moon, then Mercury and Venus before Mars.

The quest to understand the motions of the superior planets – and Mars in particular – came to dominate medieval astronomy. These planets begin each phase of visibility by emerging into the eastern sky before dawn, as they gain enough distance to the west of the Sun to not be drowned by its light. Over the following months, they extend their separation, becoming visible in the evening sky once they rise before midnight, and eventually reaching opposition (where they lie in exactly the opposite direction to the Sun and rise as it sets). Thereafter, they begin to 'close in' on the Sun from the east, becoming confined to evening skies and, eventually, disappearing into the afterglow of the sunset before beginning the cycle again.

Measuring the motion of the planets against the background stars, however, reveals that this superficial description of our everyday experience is somewhat misleading. In general, the superior planets are slowly moving around the zodiac in an eastward direction, but because the periods of their circuits are slower than the annual track of the Sun around the ecliptic (the apparent path of the Sun across the sky over the course of a year), they appear to move west in relation to the brightest object in the sky.

Complicating this story further are variations in the apparent speed and even direction of each planet's eastward-trending track. Approaching each opposition – the point at which the planet lies directly opposite the Sun and is brightest in the sky – their motion appears to slow and move westwards for a period, described as a backward, or 'retrograde', loop among the stars, before resuming its 'prograde' eastward course after opposition. Mars has the largest retrograde loops at the biggest intervals (of around two years), while Jupiter's and Saturn's are significantly smaller and more frequent (both recurring in a little over a year).

The interval between two successive retrograde loops, or two successive oppositions, is known as a planet's 'synodic period'. It is 780 days for Mars, 399 days for Jupiter and 378 days for Saturn. The period for each planet to return to its starting position among the stars – known as its 'sidereal period' – is very different, ranging from 687 days for Mars to 4,333 days for Jupiter and 10,759 days for Saturn.

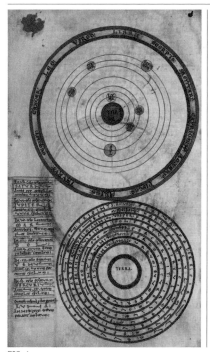

FIG. 1.

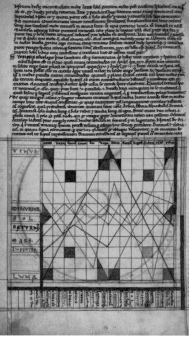

FIG. 2.

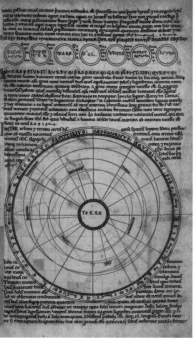

FIG. 3.

These pages from a rare English cosmography of the late 12th century put a variety of classical ideas in a Christian context for monastic use. At far left is an illustration of the Ptolemaic system, with a wheel below showing the periodic cycles of the celestial bodies. In the centre is a chart plotting the courses of the planets through the zodiac. At near left are diagrams relating to cosmic harmony: the idea that proportions of cyclical motion in the heavens are related to musical notes and intervals.

Explaining the relationship between these patterns – and, in particular, retrograde motion – was one of the main motivations for the system of epicycles used in the geocentric system of cosmology. Adding smaller circles to the main deferents allowed the apparent motions as observed from Earth to combine in ways that, at least notionally, could explain these occasional reversals. In the Sun-centred Copernican system, however, the problem at first seemed to melt away. The planet's sidereal period represented its equivalent of Earth's year – a full circuit around the Sun – while its synodic period was completed each time that Earth and the planet returned to the same relative positions on their independent orbits of the Sun. Retrograde motion, meanwhile, arose from the fact that Earth orbited the Sun faster than the outer planets. Each time it approached opposition, the relative difference in motions was at its most obvious, as Earth sped past the planet on its inside track, causing the planet to slow and reverse its apparent path across the sky from an earthly stargazer's point of view.

While this sounded simple enough in principle, the messy realities of the solar system meant that observation still obstinately refused to match up with theory – especially where Mars was concerned. In its original form, the Copernican system still insisted upon the idea of perfect circular motion at a uniform rate, merely transferring the centre of that motion from the Earth to the Sun for every body other than the Moon. As a result, it could not hope to account for the complexities induced by the fact that, in reality, the orbits of each of the planets are distinctly elliptical, with the planets moving at different speeds at different points, depending on their proximity to the Sun. This situation affects the planets Mercury and Mars most of all.

In fact, the Red Planet's distance from the Sun can vary by some 20 per cent between its closest point, or perihelion, and its distant aphelion. Moreover, since an opposition can occur when Mars is at any point along its orbit, the size, brightness and relative speed of Mars in the sky can be dramatically different from one opposition to the next. Oppositions with Mars near perihelion occur roughly every sixteen years and can bring Mars within 60 million km (37 million miles) of Earth, while aphelion oppositions barely approach within 100 million km (62 million miles). It was Johannes Kepler's (1571–1630) attempts to explain these motions, which had been meticulously recorded over decades by his late mentor Tycho Brahe (1546–1601), that led him to first formulate his laws of elliptical planetary motion in the early 17th century, providing the missing piece of the puzzle that enabled Copernican astronomy to reach its full potential.

By Doppelmayr's time, the elliptical nature of planetary orbits was well established and he was able to provide accurate shapes for the orbits of all three superior planets in this plate. Combined with a knowledge of Earth's own (relatively modest) changes in distance from the Sun throughout the year, and its consequent changes in orbital speed, these allowed him to provide a detailed explanation for the tracks of the planets through Earth's skies between early 1708 and late 1709.

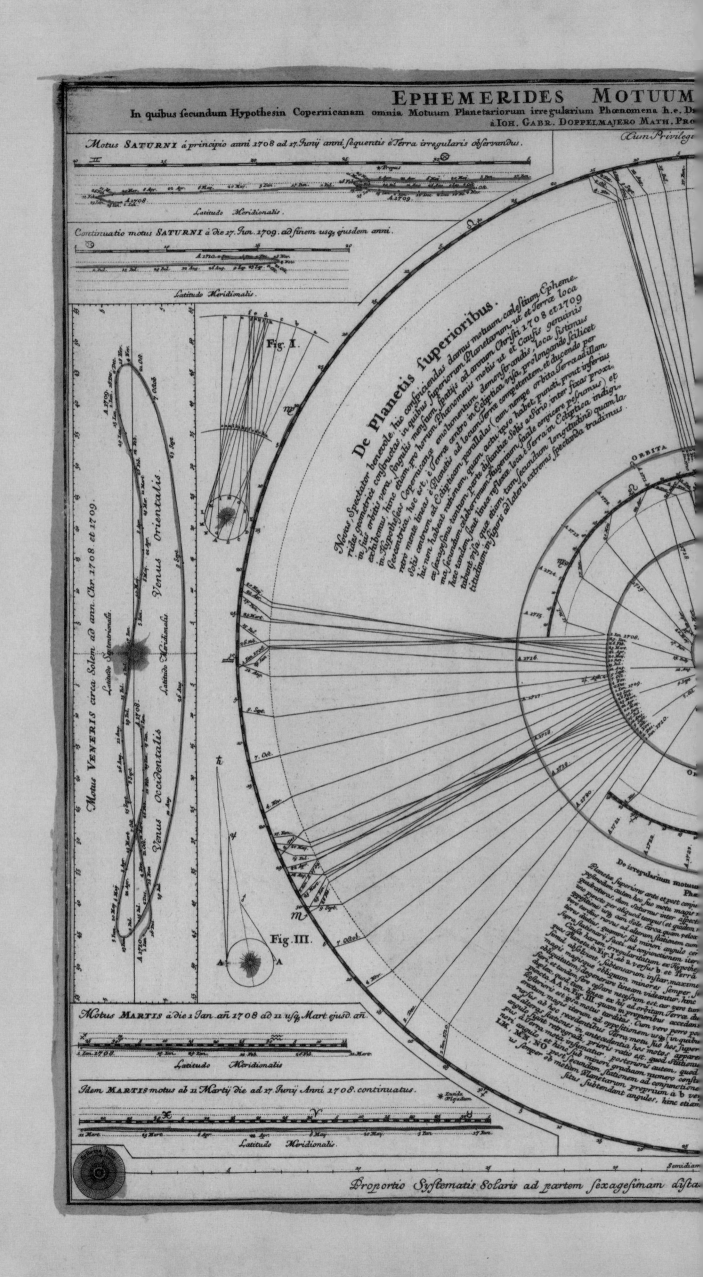

EPHEMERIDES MOTUUM

In quibus secundum Hypothesin Copernicanam omnia Motuum Planetariorum irregularium Phœnomena h.e. D
à IOH. GABR. DOPPELMAJERO MATH. PRO

Motus SATURNI á principio anni 1708 ad 17 Iunij anni sequentis è Terra irregularis observandus.

Latitudo Meridionalis.

Continuatio motus SATURNI à die 17 Iun. 1709 ad finem usq; ejusdem anni.

Latitudo Meridionalis.

Fig. I.

De Planetis superioribus.

Fig. III.

ORBITA

De irregularium motu

Motus MARTIS á die 1 Jan. añ 1708 ad 11 usq; Mart. ejusd. añ.

Latitudo Meridionalis.

Idem MARTIS motus ab 11 Martij die ad 17 Iunij Anni 1708. continuatus.

Latitudo Meridionalis.

Proportio Systematis Solaris ad partem sexagesimam dista

Motus VENERIS circa Solem ad ann. Chr. 1708. et 1709.

Venus Orientalis.

Venus Occidentalis.

Latitudo Septentrionalis.

Latitudo Meridionalis.

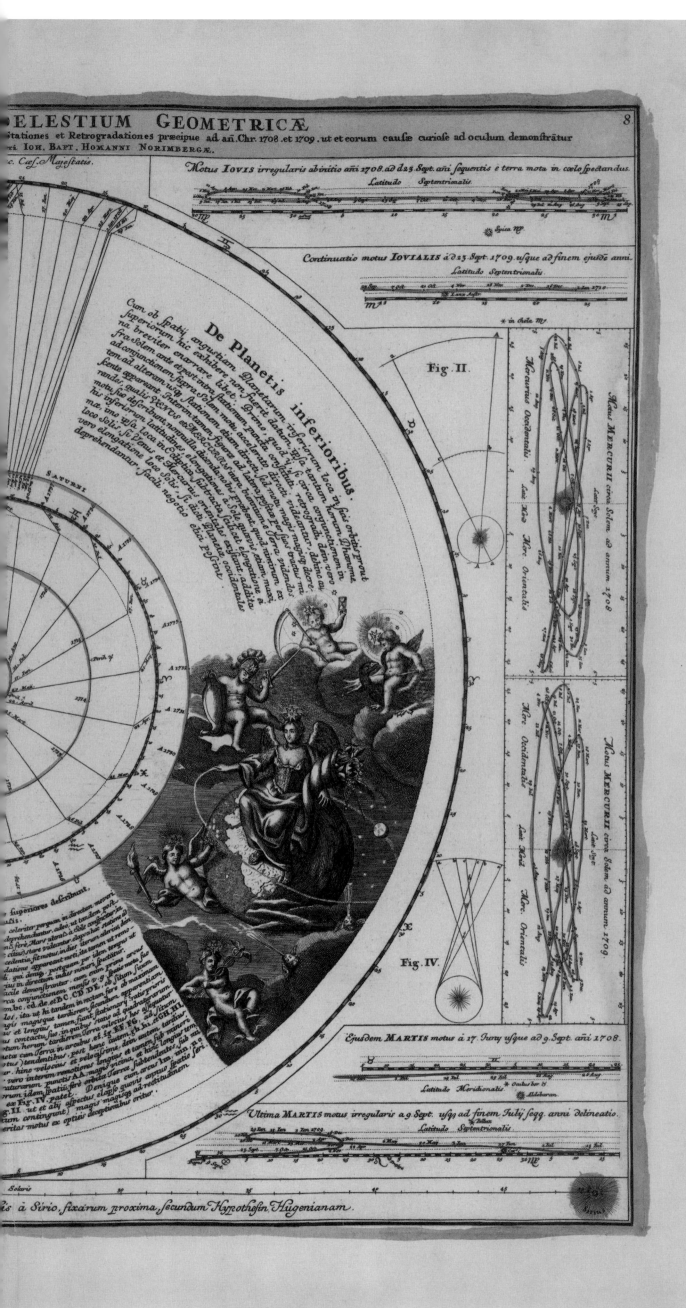

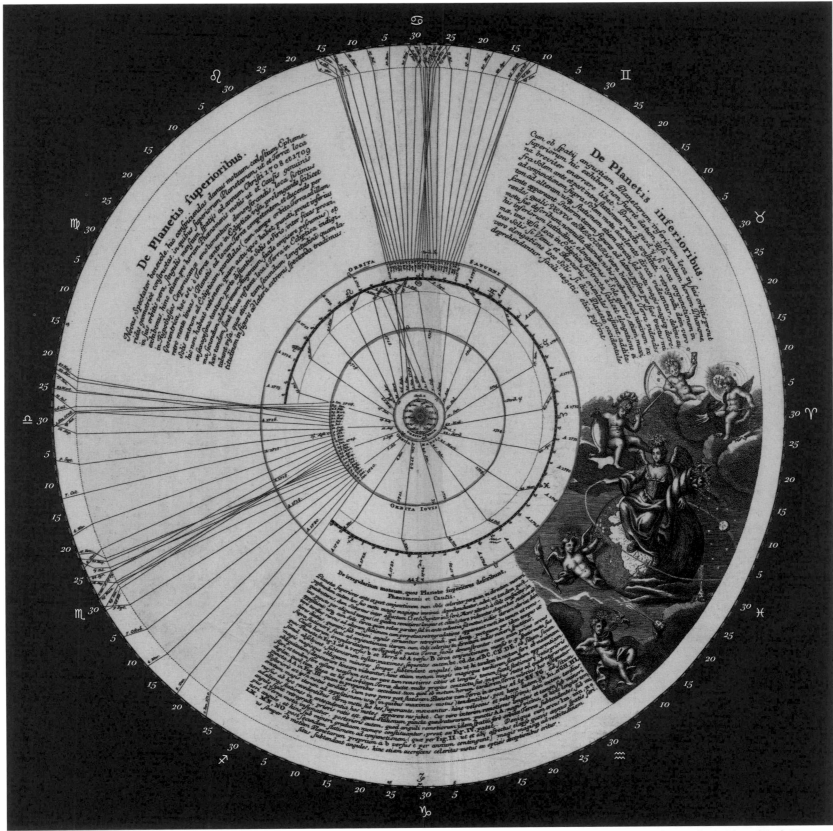

MOTIONS OF MARS, JUPITER AND SATURN.

At the centre of Plate 8, this diagram follows similar principles to that on Plate 7, showing the positions of Earth, Mars, Jupiter and Saturn on their orbits through the years 1708 to 1710 and projecting them onto the surrounding ring of the ecliptic to show how their shifting relative locations give rise to their wandering paths through Earth's sky. To avoid overlapping Mars's complete circuit of the sky with the slower moving tracks of Jupiter and Saturn, Doppelmayr plots its ecliptic positions on a separate ring between the orbits of the two giant planets.

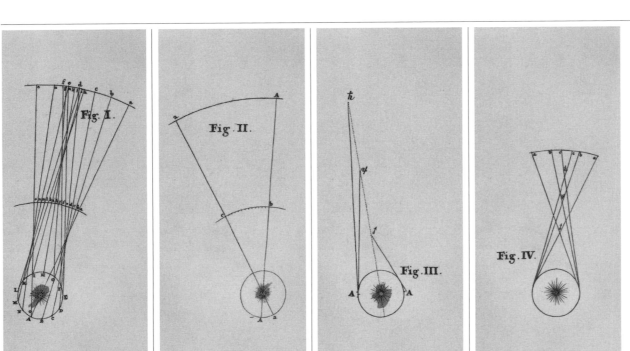

RETROGRADE MOTION. (FIG. 1.)

Doppelmayr demonstrates how the relative speeds of Earth and an outer planet create the illusion of retrograde motion when the two planets align on the same side of the Sun at opposition.

DISTANCE EFFECTS. (FIGS. 2–4.)

These diagrams illustrate various effects to be taken into account in measuring the true positions of the planets. From left to right, they are the greater dimension of an arc subtending a certain angle depending on distance, the effect of a planet's proximity on 'quadrature' (when it appears to be 90 degrees from the Sun in the sky) and the differing effects of parallax (the shift in viewing angle from either side of Earth's orbit).

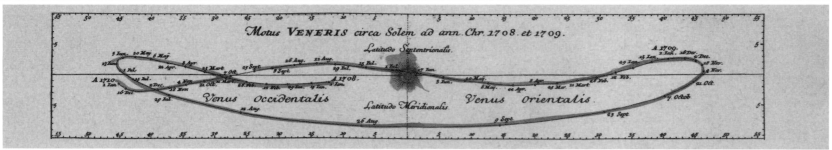

MOTION OF VENUS.

Here, Doppelmayr plots the motion of Venus relative to the Sun from 1 January 1708 to 1 January 1710. He records a complete orbital cycle, including eastern and western elongations, and inferior (close) and superior (distant) conjunctions with the Sun.

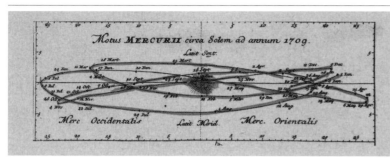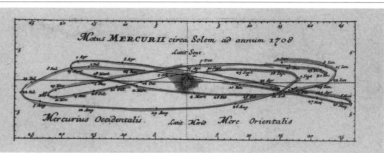

MOTION OF MERCURY.

The short orbital period of Mercury means it appears to circle the Sun in Earth's skies several times a year, although always with significant differences from one cycle to the next. For clarity, Doppelmayr shows one chart for 1708 and one for 1709.

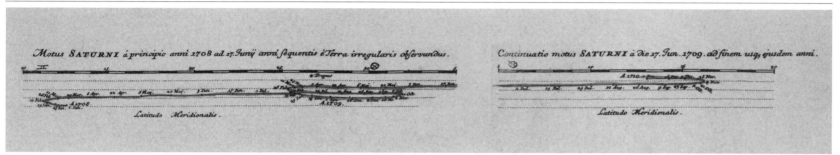

MOTION OF SATURN.

Doppelmayr tracks the slow-moving path of Saturn against the ecliptic in 1708 to 1709 across a pair of charts. They encompass the end of one retrograde loop, a second in its entirety and the beginnings of a third.

MOTION OF JUPITER.

This pair of charts plots the motion of Jupiter along the ecliptic from 1708 to 1709, as it moved from the zodiac constellation of Virgo through Libra to Scorpio, encompassing two near-complete retrograde loops.

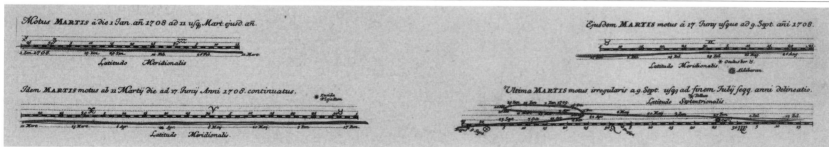

MOTION OF MARS.

The slower paths of Saturn and Jupiter remain resolutely below and above the ecliptic, respectively, but the faster motion of Mars sees it move southwards and then northwards. All planets 'wobble' about the ecliptic due to a difference between their orbital inclinations and that of Earth.

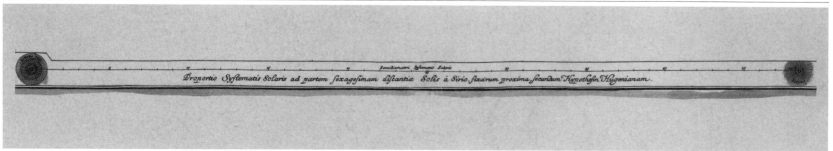

DISTANCE TO SIRIUS ACCORDING TO HUYGENS.

Following an ingenious estimate of the distance to Sirius in Christiaan Huygens's *Cosmotheoros* (1698), Doppelmayr attempts to illustrate the star's distance relative to the scale of the solar system. Even with the entire solar system rendered at the size of a fingertip, the distance must be reduced by a factor of sixty to fit on the plate.

PLATE 9.

SPIRAL MOTIONS
IN THE HEAVENS.

(MOTUS IN COELO SPIRALES)

*Doppelmayr explores the motions of the Sun and the inferior planets
in relation to Earth, according to the theories of Tycho Brahe.*

Given the fact that Doppelmayr and his Nuremberg contemporaries were avowed followers of Nicolaus Copernicus (1473–1543) and Johannes Kepler (1571–1630), it is perhaps surprising to find that – in addition to his initial presentation of the Tychonic system on Plate 3 – two further plates are also seemingly devoted to the motions of the planets in Tycho Brahe's (1546–1601) Earth-centred cosmology. The orbits of the inferior planets Mercury and Venus are shown on Plate 9, but we will save the explanation of

these spiralling paths for discussion alongside those of the superior planets, Mars, Jupiter and Saturn, on Plate 10. First, it is worth exploring the often overlooked successes of the Tychonic model, and how debate around its merits delayed the widespread acceptance of the Copernican cosmology.

In Tycho's theory, the Moon and the Sun orbit Earth while the planets circle the Sun. On page 45, we saw how his ideas were inspired partly by a desire to adhere to biblical descriptions of a moving Sun, but equally by what seemed at the time to be common-sense doubts about the practicality of a massive body like Earth moving through space. However, it was a third factor, generally known as the 'star size argument', that kept the theory alive even after Kepler's elliptical orbits provided a solution to the problems of the Copernican model.

Even given his status as the observer extraordinaire of his day, one of Tycho's most impressive feats must surely be his measurement of the relative sizes of the planets. A generation before the first telescopes, the instruments on his island observatory at Hven, Sweden, were able to measure the angular sizes down to as little as half a minute of arc ($\frac{1}{120}$ of a degree): enough to measure the varied angular diameters of the planets as they moved in Earth's skies. Coupled with a model for the relative distances of the different bodies, he was able to show that Jupiter and Saturn were far larger than Earth, Venus about the same size, Mars somewhat smaller and Mercury smaller still.

After this extraordinary result, it is unsurprising that Tycho then attempted the same method with the stars. His measurements convinced him that Procyon, one of the brightest stars in the sky, had approximately the same angular diameter as Saturn, and that fainter stars had progressively smaller diameters until they went beyond the limits of his instruments.

Meanwhile, the accuracy of Tycho's instruments allowed him to calculate the closest possible distance at which the star could lie if

FIG. 1.
The frontispiece of Riccioli's *New Almagest* of 1651 depicts a muse of astronomy combining archetypal features of Astraea and Urania. Using the scales of the justice goddess Iustitia, she is weighing the Copernican model of the universe against Riccioli's own modified Tychonic system. Perhaps unsurprisingly, the scales are tipped in Riccioli's favour. The older pure Ptolemaic system sits, already discarded, on the ground.

FIG. 1.

Copernicus's theory were true. A moving Earth would inevitably give rise to a 'parallax shift' – a back-and-forth shift in the apparent positions of the stars as seen from either side of Earth's orbit, six months apart. No one was better placed than Tycho to measure such a shift, and yet he found no sign of it. The absence of stellar parallax had been used as an argument against a heliocentric universe ever since it was first proposed, and the standard retort (put forward by Copernicus himself in *On the Revolutions of the Heavenly Spheres*, 1543) was that the stars must simply be immensely far away (as, indeed, they are). With no evidence available on either side, there seemed to be no way of resolving this debate. But Tycho's measurements of star size seemed to tip the balance.

In a letter to the German Copernican Christoph Rothmann (*c.* 1560–*c.* 1600), Tycho pointed out that if a star like Procyon appeared as large as Saturn, and yet lay so far away that he could not detect its parallax, it would have to be hundreds of times the size of the Sun. The same argument went for every star in the sky – even the faintest. If the stars were truly other Suns as the Copernicans argued, why should the Sun itself be uniquely tiny?

Rothmann's only counter-argument took the form of an appeal to religion; nothing lay beyond God's power, so why should He not create such vast Suns? Who was to say that God's creation had to conform to human understanding? Given the later tendency to see the rise of Copernicanism as a triumph of rationality over religious dogma, it is somewhat ironic that the theory's early supporters frequently fell back on such arguments against the reasoned objections of doubters. In contrast, by placing the stars in a spherical halo at a relatively small distance beyond Saturn, Tycho's measurements could make their sizes comparable to that of the Sun.

The star size argument remained a major problem for the Copernican theory for several decades – and the invention of the telescope did little to resolve it. Telescopic measurements reinforced the apparent impossibility of detecting parallax, while at the same time apparently 'proving' that the stars had measurable diameters. Under the most perfect conditions, skilled observers reported that the brightest stars definitely showed significant discs. The argument was one of many used by Giovanni Battista Riccioli (1598–1671) in his *New Almagest* of 1651, where he devoted a significant section to debating the merits of the Copernican and Tychonic systems, putting forward arguments for and against both and ultimately settling on an essentially Tychonic view of the cosmos.

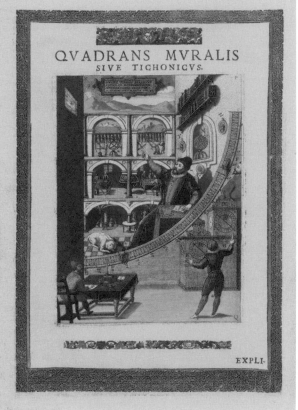

FIG. 2.

A few years after Riccioli's work was published, however, two different sources emerged to cast doubt on the star size argument. In a 1659 work on Saturn, Dutch astronomer Christiaan Huygens (1629–95) noted that observing the stars through differently tinted glasses could dramatically reduce their apparent size while leaving the planets unaffected. In 1662, notes by Jeremiah Horrocks (1618–41) of observations made decades earlier were finally published. In these, Horrocks reported watching the Moon occulting (blocking out) the bright members of the Pleiades star cluster, and seeing that each star disappeared instantaneously, rather than gradually as one might expect with a disc of light. Intrigued by this, many other astronomers watched with interest for occultations of bright zodiac stars such as Aldebaran in Taurus and Spica in Virgo, and found the same result.

By Doppelmayr's time, astronomers such as Edmond Halley (1656–1742) were happy to dismiss apparent star sizes as an 'optic fallacy', but it was not until the 19th century that their true cause – a result of the way that small telescopes in particular diffract and bend light – was properly understood. The reality is that the Sun is an average-sized star, and even though a few true giants do exist, even these are so far away that they appear as nothing more than points of light from Earth.

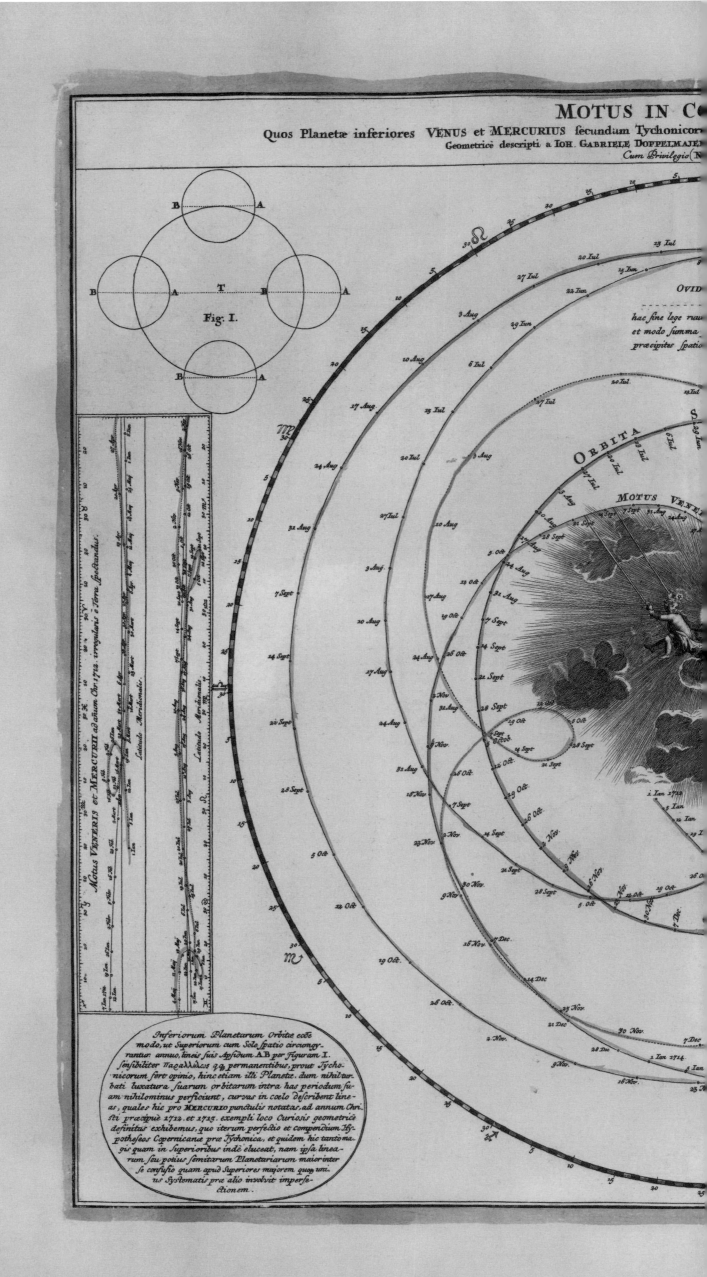

MOTUS IN C

Quos Planetæ inferiores VENUS et MERCURIUS secundum Tychonicor
Geometricè descripti a IOH. GABRIELE DOPPELMAJE
Cum Privilegio N

Fig. I.

OVID

hac fine lege ruu
et modo summa
præcipites spatio

ORBITA

MOTUS VENE

Inferiorum Planetarum Orbitæ eodè
modo, ut Superiorum cum Sole spatio circumgy-
rantur annuo, lineis suis Apsidum A B per Figuram I.
sensibiliter Παραλλάκος 9.4 permanentibus, prout Tycho-
nicorum fert opinio, hinc etiam illi Planetæ, dum nihil tur-
bati luxatura suarum orbitarum intra has periodum suam nihilominus perficiunt, curvas in cœlo describent lineas, quales hic pro MERCURIO punctulis notatas, ad annum Christi præcipuè 1712. et 1713. exempli loco Curiosis geometricè
definitas exhibemus, quo iterum perfectio et compendium Hy-
potheseos Copernicanæ præ Tychonica, et quidem hic tantò magis quam in Superioribus indè eluceat, nam ipsa linea-
rum seu potius semitarum Planetariarum maior inter
se confusio quam apud Superiores maiorem quoq uni-
us Systematis præ alio involvit imperfe-
ctionem.

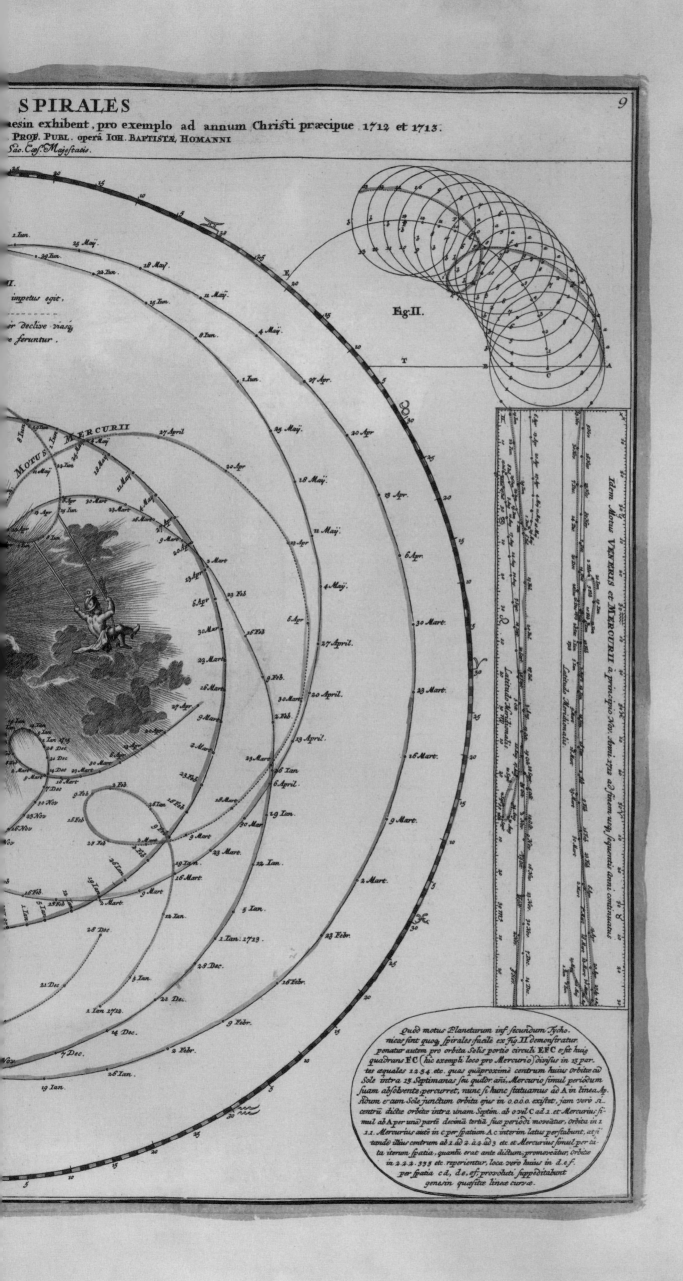

SPIRALES

aesin exhibent, pro exemplo ad annum Christi præcipue 1712 et 1713.
PROF. PUBL. operâ IOH. BAPTISTÆ. HOMANNI
Sac. Cæf. Majestatis.

Fig. II

MOTUS MERCURII

Quod motus Planetarum inf secundum Tychonicos fint quoq spirales facile ex Fig. II demonstratur.
ponatur autem pro orbita Solis portio circuli EFC efse huic
quadrans FC (hic exempli loco pro Mercurio) divisus in 13 partes aequales 1 2 3 4. etc. quas quâproximè centrum huius orbitae cū
Sole intra 13 Septimanas seu quâdr añi, Mercurio simul periodum
suam absolvente percurret, nunc si hunc statuamus ad A in linea Ap
ſidum et cum Sole junctum orbita ejus in o. o. o. o. exiſtet, jam vero ſi
centru dictae orbitae intra unam Septim. ab o vel C ad 1. et Mercurius simul ab A per unâ partê decimâ tertiâ suae periodi moveatur, orbita in 1
11. Mercurius autê in c per spatium A.c interim latus perscabunt, atſi
tandê illius centrum ab 1 ad 2. ū 2. ū 3. etc. et Mercurius semul per câ
ta iterum spatia, quantū erat ante dictum promovebitur, Orbita
in 2. 2. 2. 3. 3. 3. etc. reperientur, loca verò huius in d. e. f.
per spatia c d, d. e. e. f. provoluti suppeditabunt
genesin quaesitae lineae curvae.

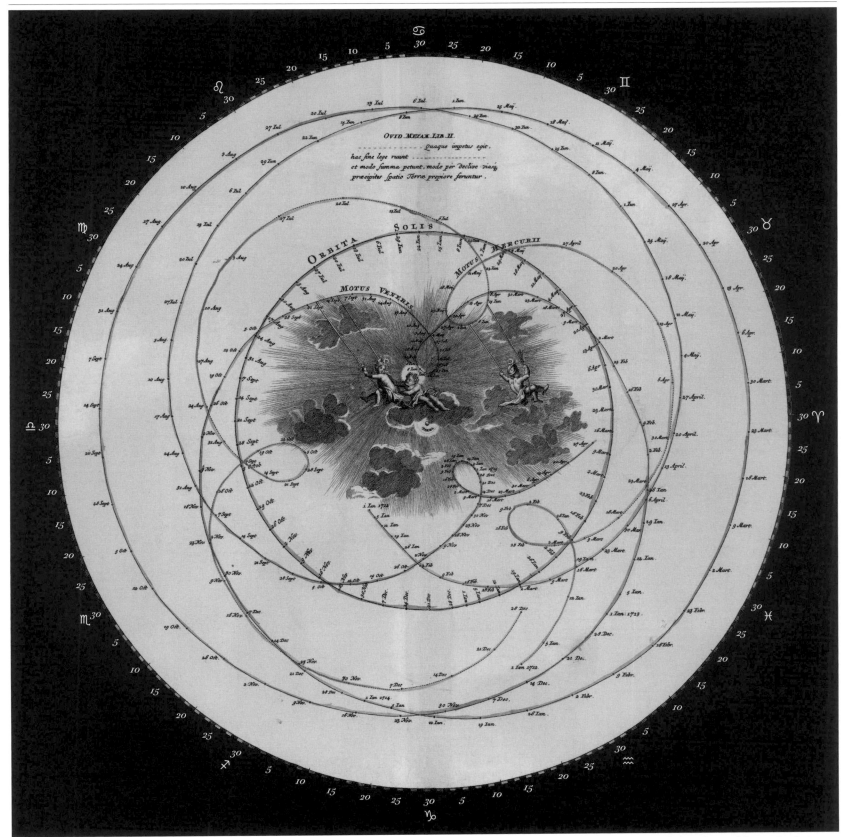

SPIRAL MOTIONS OF THE INNER PLANETS
IN THE TYCHONIC SYSTEM.

The centrepiece of Plate 9 elegantly constructs the spiralling motions and changes in relative distance of the planets Venus and Mercury from a static Earth. The relative positions of the Sun throughout each year, and the inner planets from 1712 to 1713, are depicted, while a cherubic Helios pushes a playful Venus and Mercury on swings at the centre. The quote at the top is from Ovid's *Metamorphoses* (8 CE), recounting Phaethon losing control of the chariot of the Sun.

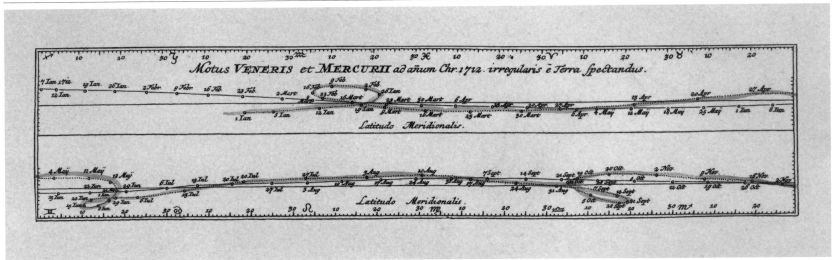

MOTIONS OF VENUS AND MERCURY.

Twin charts show the positions of Venus and Mercury in the heavens through most of 1712. In order to correlate with the central chart, these data are not measured east or west of the Sun, but using coordinates that track the planets' positions relative to the ecliptic, and against the background of zodiac constellations.

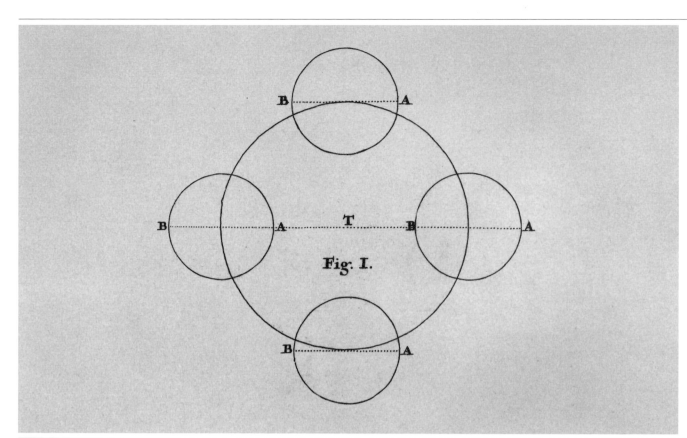

Doppelmayr uses a simple diagram
to explain the effects of a planet
following a circular orbit around the
Sun on its own circular orbit around
Earth. It shows how a parallel line,
AB, across its orbit will indicate
both apogee and perigee (closest
approach and furthest distance
from Earth) at certain points.

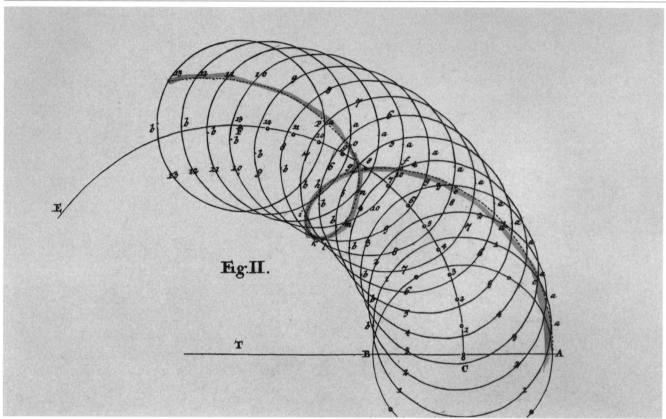

CONSTRUCTING THE SPIRAL
PATTERNS. (FIG. 2.)

Doppelmayr demonstrates how the
looping patterns in the centrepiece
illustration arise by showing the
combined effect of a planet moving
along a circular orbit, whose centre
is itself following a circular path.
The resulting pattern, known as
a 'trochoid' (from the Greek for
'wheel'), does not repeat perfectly
because the circumference of the
planetary orbit does not fit into the
'guide path' a whole number of times.

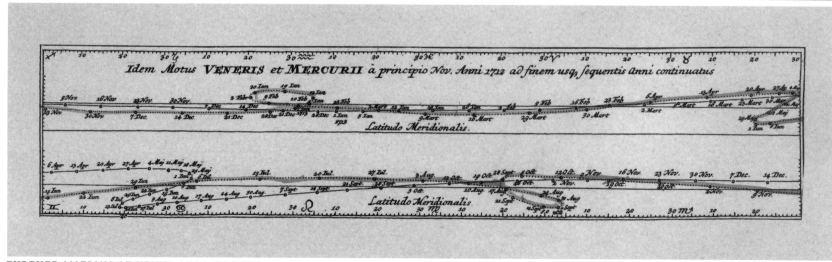

FURTHER MOTIONS OF VENUS AND MERCURY.

A second set of charts continues to plot the locations of the inferior planets from late 1712 through
1713. The uneven break in time is due to the way that Venus completed slightly more than a whole
circuit of the ecliptic in 1712, and slightly less the following year.

PLATE 10.

MOTION OF THE SUPERIOR PLANETS.

(MOTUS PLANETARUM SUPERIORUM)

Doppelmayr explores the spiral paths of superior planets, according to the theories of Tycho Brahe.

Plates 9 and 10 of the *Atlas* form a matching pair for the inferior and superior planets, both exploring the implications of an apparently Tychonic system by considering the positions of the planets relative to Earth as they move on their orbits around the Sun, while the Sun itself is simultaneously carried on its circular track around the Earth.

Doppelmayr's motivation in taking this approach is not in fact to bolster the model of Tycho Brahe (1546–1601) himself, but rather to shift his reader's perspective – from imagining an all-seeing Copernican overview where Earth is the third planet orbiting the Sun on a circular (or slightly elliptical) orbit, to considering what this really means in relation to Earth. For an Earthbound stargazer, the direction, distance and brightness of other objects in the solar system are indeed constantly changing. While the spirals illustrated in these somewhat esoteric-looking plates may seem alien to our modern view of the cosmos, they are not, in fact, some unique quirk arising from the application of Tycho's two-centred system; in reality they are simply a means of understanding the relative motion of the planets.

The standard vision of Tycho's cosmology, shown earlier on Plate 3, depicts Earth as the static centre of the universe, circled by the Moon and the Sun, with the planets then following circular paths around the Sun. The orbits of Venus and Mercury are arranged so that their radii are smaller than the Earth–Sun distance, allowing them to pass between the two centres of motion and (as seen from Earth) remain always in the general direction of the Sun. Meanwhile, the paths of the other planets are large enough to encompass Earth as they circle the Sun, enabling them to appear at opposition on the other side of the sky.

In order to explore how this model relates to the planetary movements seen in the night sky, Doppelmayr traces out what the Tychonic model means from an Earth-centred point of view. The graceful overlapping circles shown in overviews are all very well, but how do the distance and direction of the planets relative to Earth – the two factors that affect their brightness and position in the sky – actually change over time?

The solution is the beautiful pattern known in mathematical terms as an epitrochoid – the result of a point moving on the outside of a rotating circle that rolls along the circumference of another rotating circle. The elegance of the pattern is clearest for the outer planets Saturn and Jupiter, which circle the Sun in such long periods that they form a series of tight spiralling

FIG. 1.
This summary of the Tychonic system is from French cartographer Alain Manesson Mallet's (1630–1706) five-volume atlas *Description de l'Univers* (1683). Several of Mallet's illustrations – including this one – were reproduced and sold as individual prints in 1719 by Frankfurt publisher Johann Adam Jung.

FIG. 1.

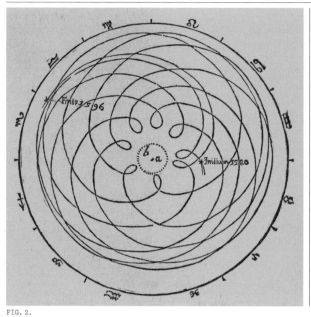

FIG. 2.

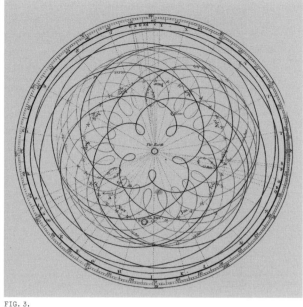

FIG. 3.

FIGS. 2–3.
Two examples of the epitrochoid patterns that describe the relative distances of the planets from Earth. At far left, Johannes Kepler's well-known illustration of the path of Mars between 1580 and 1596, published in his *New Astronomy* (1609), marks the first appearance of these complex diagrams in print. At near left is an illustration from Scottish writer James Ferguson's (1710–76) *Astronomy Explained upon Sir Isaac Newton's Principles* (1756).

loops around Earth. While circling Earth in a generally anti-clockwise direction (as seen from above the plane of the solar system), each planet enters a retrograde loop around opposition, when it is lined up on the same side of the Sun as Earth. This is also when it appears at its brightest, simply due to its proximity to Earth.

Doppelmayr shows the construction of his spiral paths in smaller figures at top left and top right of Plate 10 (and in a similar position on Plate 9). In the centre of the plate, he maps out each planet's position on the path throughout an entire sidereal period (the time it takes a planet to complete its orbit, relative to the position of the stars) in the early 18th century. It is worth noting that the ends of each spiral do not connect because the sidereal periods of the other planets are not simple multiples of Earth's year – each time a planet completes a sidereal period, Earth will be in a different location and so the planet will be at a different distance. For the sake of clarity, Doppelmayr limits his spirals to a single circuit of the Sun rather than continuing and overlapping them. However, in the case of Mars, this results in a single truncated loop: at around 780 days, the interval between Martian oppositions is longer than its 687-day year.

For the inferior planets shown on Plate 9, the Tychonic system results in similar patterns of approach and retreat, with each loop marking the inferior conjunction, at which point the planet lies between Earth and the Sun. To illustrate the path of Mercury, Doppelmayr confines himself to the calendar year of 1712 – enough time for the fast-moving planet to execute three retrograde loops and close in on a fourth. Venus, however, requires a considerably

longer period to tell its story – more than three years, from January 1712 to April 1715. During this period, it executes two long lazy loops in relation to Earth.

Had Doppelmayr extended his plot of Venus further, he would have revealed a curious aspect of the brightest planet's motion, known as the 'pentagram of Venus'. Successive inferior conjunctions of Venus (or, indeed, any successive identical events) are separated from each other by an angle of 144 degrees in the heavens. Therefore, after five such events, Venus returns to its original location among the stars (because 5 × 144 = 720 degrees, or two complete circles). The phenomenon arises because Venus takes almost exactly eight Earth years to complete thirteen orbits of the Sun. This pentagram of Venus has been recognized since ancient times – a pattern in which the planet's conjunctions (alignments to the Sun) and greatest elongations skip around the sky between five different points on the zodiac before returning (almost precisely) to their original settings. Civilizations as diverse as the ancient Sumerians and the Maya frequently linked Venus and its associated deities to the pentagram pattern.

At first glance, then, the Tychonic and Copernican systems seem to arrange the orbits of the planets in radically different ways. However, the fact is that – when motions are considered from the point of view of observers on Earth – they actually produce near-identical results. This startling demonstration serves as a salutary reminder of just how well Tycho's model addressed the challenge of the moving planets in its time, and just why it was able to remain a plausible alternative to Copernicanism for so long.

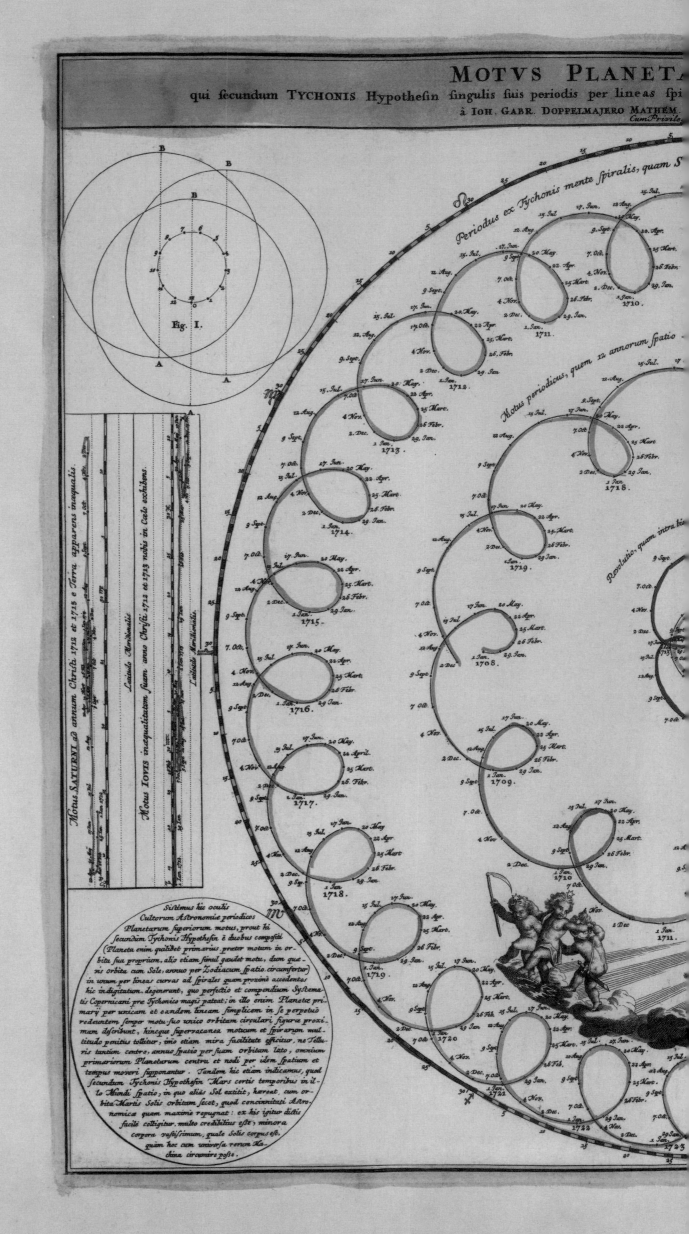

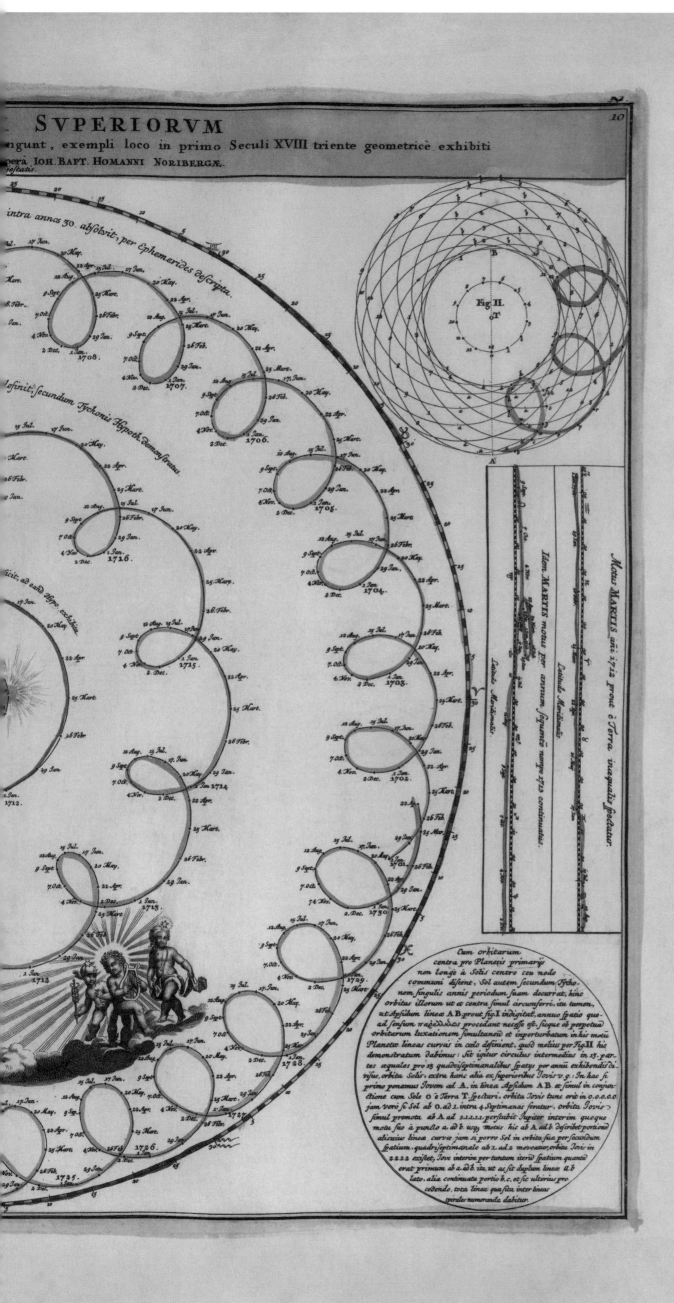

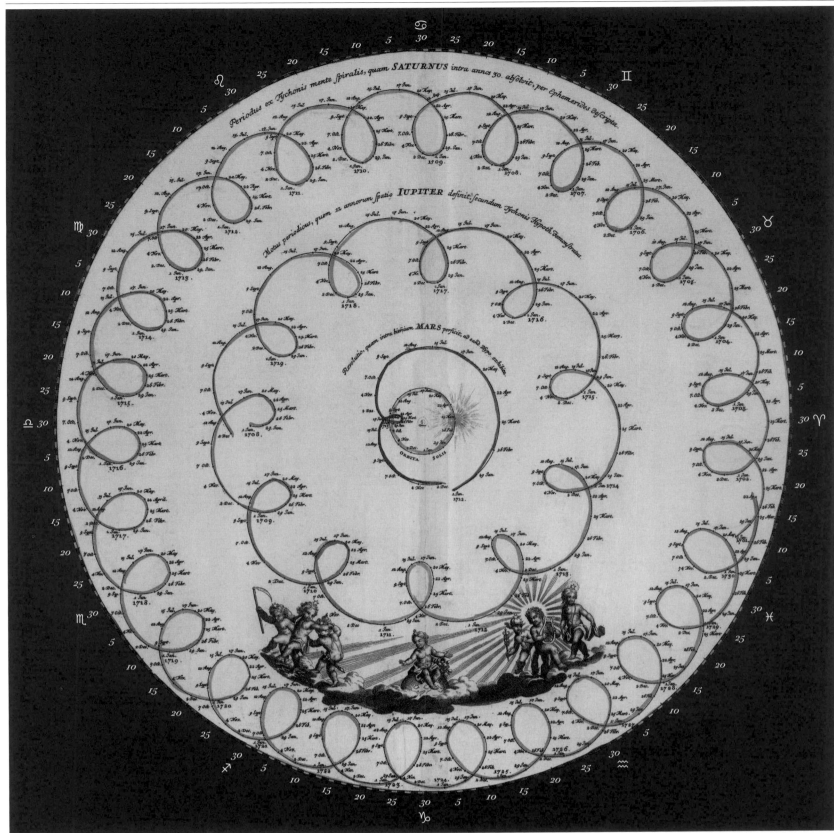

SPIRAL MOTIONS OF THE OUTER PLANETS IN THE TYCHONIC SYSTEM.

Plate 10's centrepiece demonstrates the spiral motions throughout a single circuit of the heavens relative to Earth, for each of the outer planets: Mars, Jupiter and Saturn. Saturn's 29.5-year circuit begins in January 1701, Jupiter's covers twelve years beginning in 1708, and Mars's circuit covers less than two years. As previously noted, the varying distances of the planets from Earth in the Tychonic system did almost as good a job as the fully Copernican model when explaining their changing brightness.

MOTION OF SATURN AND JUPITER.

Doppelmayr plots the locations of Saturn (in Leo and Virgo) and Jupiter (in Capricornus, Aquarius and Pisces) relative to the ecliptic in the years 1712 and 1713. Note that each chart encompasses two retrograde cycles around successive oppositions – the outer planets move slowly enough that Earth's own motion becomes the dominant factor, and little more than a year separates successive oppositions.

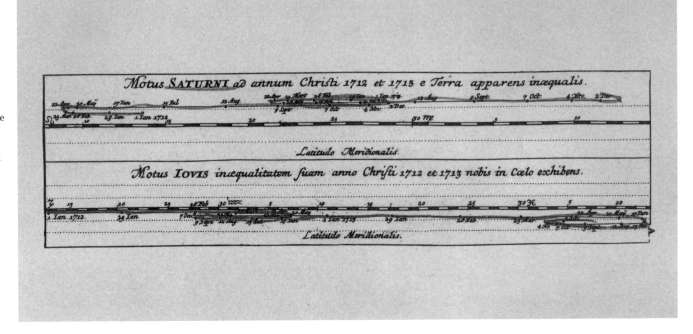

SHIFTING ORBITS.
(FIG. 1.)

This diagram demonstrates the principles behind the construction of the spirals, by first showing how the centre of a planetary orbit in the Tychonic system shifts with the supposed orbit of the Sun around the fixed Earth. Three examples of the planet's orbit at a particular moment are constructed around the locations of the Sun on the central circle, corresponding to points 3, 9 and 13.

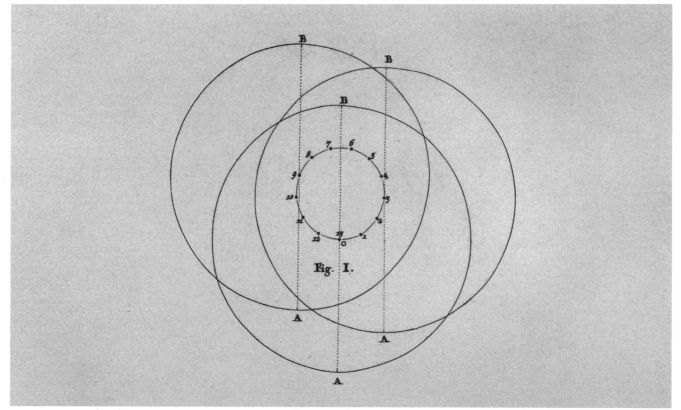

CONSTRUCTING THE SPIRALS. (FIG. 2.)

Here, Doppelmayr demonstrates the effects that produce his spiral paths, by drawing a circular orbit centred on the Sun's own orbital location at monthly intervals. He then shows how the planet's progress around its orbit interacts with the 'motion' of the orbit itself. Numbers on the outer circles link them to their particular centre point, while the early steps of the resulting trochoid pathway are marked with small circles and an alphabetic progression beginning at *B* near the bottom of the diagram. The static Earth (*Terra*) at the centre of everything is marked *T*.

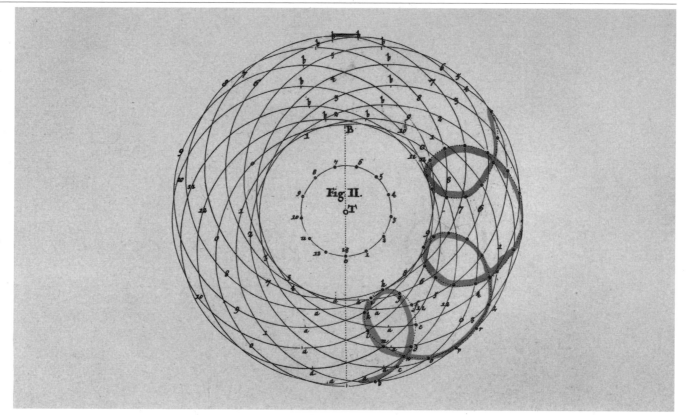

MOTION OF MARS.

A pair of charts show the track of Mars along and around the ecliptic in 1712 to 1713, beginning in Capricornus. In contrast to the slow-moving outer planets, the relatively short Martian orbital period of 687 days combines with Earth's own orbit to produce Martian oppositions and retrograde periods every 780 days. The elliptical shape of the Martian orbit creates further complications.

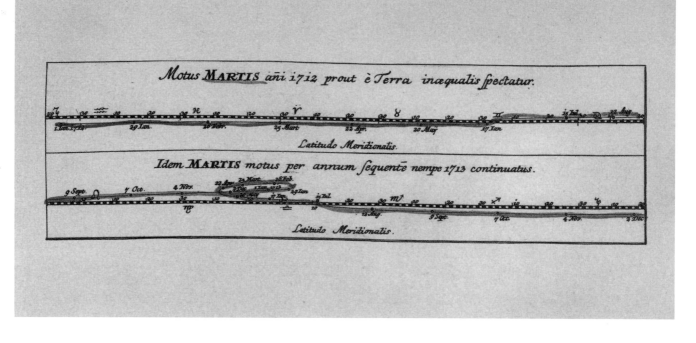

PLATE 10.

SPHAERAE COELESTIS ET PLANETARUM DESCRIPTIO
(c. 1470.)

These stunning illuminations are from the treatise commonly known as *De Sphaera*, attributed to the deaf-mute Lombardy miniaturist Cristoforo de Predis. Commissioned for the court of the Sforza family, who had recently become rulers of Milan, this slender textbook describes both the astrological significance of the celestial bodies and constellations, and the state of contemporary astronomical knowledge regarding how best to predict their motions. Alongside introductory diagrams and verse commentary attributed to the humanist poet

Francesco Filelfo, the major illuminations depict allegorical figures
associated with the Sun, Moon and planets, alongside the houses of
the zodiac over which they are said to 'rule' and scenes of activities
associated with them – both integrated beneath the major figures
and on the facing pages. The five miniatures reproduced here depict
Mercury ruling over Gemini and Virgo, Venus with Libra and Taurus,
Mars with Scorpio and Aries, Jupiter with Pisces and Sagittarius, and
Saturn with Aquarius and Capricornus. The Sun, meanwhile, is said
to rule Leo, while the Moon rules Cancer.

PLATE 11.

SELENOGRAPHIC TABLE.

(TABULA SELENOGRAPHICA)

Doppelmayr depicts the principle markings on the surface of the Moon, following the naming schemes of his time.

The near side of the Moon is a familiar sight to everyone on Earth. With a diameter of roughly half a degree in the sky, the Moon is large enough for both bright and dark markings on its surface to be visible without a telescope. Many cultures have told stories and invented pictures of these patterns. The most popular are the 'Man in the Moon' (seen as either a face or a full figure) and the eastern 'Rabbit in the Moon'.

Despite these differences in interpretation, stargazers from classical Greece to India, China and beyond recognized early on that the Moon's changing phases are governed by how much of the visible surface is illuminated by the Sun, but the nature of the surface markings was long disputed. As early as the mid-5th century BCE, Greek philosopher Democritus (*c.* 460–370 BCE) attributed the markings to mountains and valleys on the lunar surface. This early insight

was largely forgotten in later centuries, however, as the Aristotelean model of the universe became widespread. While the Moon occupied the innermost of the heavenly spheres and was thus subject to more change than the other celestial bodies, Aristotle (384–322 BCE) nevertheless viewed it as an unchanging and perfect sphere, created by the mixing of fire from the uppermost 'sublunary' sphere and aether from the realm of the heavens.

The changing lunar phases and the relationship between the Sun, Earth and the Moon provided an ingenious method of estimating their distance and scale. Since at its first or last quarter (when exactly half of its disc is illuminated) the Moon must sit at the right-angled corner of a triangle linking it with Earth and the Sun, the observed angle between the Sun and the Moon will indicate their relative distances and the scale of the entire system. If the Sun was infinitely distant, then this angular separation would be precisely 90 degrees, but in the mid-3rd century, Greek

FIG. 1.
The first map of the Moon to include a system of nomenclature was published by Dutch cartographer Michael van Langren in 1645. Few of the roughly 300 names introduced by van Langren have survived and those that persist are now mostly applied to different features.

FIG. 2.
Athanasius Kircher's (1601–80) *Typus Corpus Lunaris* (1669) incorporates Kircher's own observations with those of Christoph Scheiner.

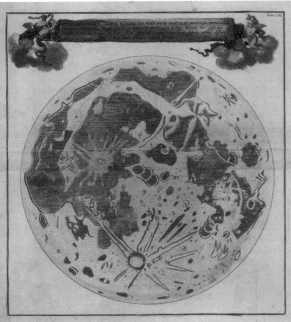

FIG. 1.

FIG. 2.

astronomer Aristarchus of Samos (*c.* 310–230 BCE) estimated it to be just 87 degrees. From this, he was able to calculate that the Moon was about twenty Earth radii away, and that the Sun was twenty times further away and twenty times the Moon's size, since they appear roughly equivalent in the sky. Because Earth's dimensions were already known, it was simple to prove that the Moon was, therefore, a substantial body in its own right, and that the Sun must be larger than Earth itself.

Aristarchus's theory may have been right, but he was working long before the telescope, hence his estimates of precisely when the Moon was half-illuminated and his measurements of the angle separating it from the Sun were significantly off. Today, we know that the Moon's distance is closer to sixty Earth radii, and the Sun is about 400 times further still. Regardless of its precise value, the fact that the Sun was clearly much larger than Earth raised significant questions for classical and medieval thinkers attempting to model an Earth-centred solar system. Indeed, it was enough to convince Aristarchus that the Sun, rather than Earth, must be the centre of everything and led to one of the first attempts at a heliocentric cosmology.

As to the Moon's physical nature, Aristotle's idealized sphere theory benefited by association from the widespread adoption of his entire paradigm of physics, but it took some time to see off its rivals. As late as the 2nd century CE, the Greek philosopher Plutarch (46–119 CE) wrote a remarkable essay in which he argued that the Moon was a world not dissimilar to Earth, with markings created by its landscape features.

With the spread of Christianity, the incorporation of Aristotle's physics and cosmology into the teachings of the Catholic Church ensured that a 'perfect sphere' Moon became the accepted view among European scholars for more than a millennium. It was only at the start of the 17th century that advances in technology provided startling evidence that Aristotle had been wrong.

Beginning in 1609, both Galileo Galilei (1564–1642) and the English observer Thomas Harriot (*c.* 1560–1621) studied the Moon and made sketches of its appearance through the newly invented telescope. Harriot's drawings went unpublished, while Galileo incorporated them, along with other groundbreaking discoveries, in his *Starry Messenger* (1610). Galileo not only recorded details on the surface of the Moon, but also showed how their appearance varied with the lunar phases, according to the angle of sunlight striking them and the length of the shadows they cast. Such shadows could only be explained by differences in elevation; the Moon

FIG. 3.

must have hills, valleys and circular pits on its surface. Galileo suggested that the dark and largely smooth areas might be seas, with land forming the brighter areas that separated them. The discovery of Earth-like lunar relief features, coupled with the observation of shifting spots on the Sun (another supposedly unchanging body) helped shake faith in the old Aristotelean ideas almost as much as the theories of Nicolaus Copernicus (1473–1543) and Johannes Kepler (1571–1630).

As the century progressed and telescopes improved, stargazers of varying talents attempted to map the lunar surface. The earliest to be published was that of Michael van Langren (1598–1675) in 1645, but Doppelmayr chose to reproduce two slightly later maps that became standard authorities for more than a century. The first is from Polish astronomer Johannes Hevelius (1611–87) and it was published in his *Selenographia* (1647), the first work dedicated to lunar theory; the second is the work of Jesuit priests Giovanni Battista Riccioli (1598–1671) and Francesco Maria Grimaldi (1618–63), who published it as part of Riccioli's *New Almagest* of 1651. Comparison of the two maps will immediately show that two different naming schemes are in play, but it is Riccioli's map from which many of our modern names for the lunar markings (in particular those denoting the 'seas' or *maria*) derive.

FIG. 3.
A plate from Kircher's *Great Art of Light and Shadow* (1646) depicts the twenty-eight distinct phases that describe the lunar month. They go from New Moon, through crescent, first quarter and waxing gibbous states to reach Full Moon, and then back through waning gibbous, last quarter and decrescent to the next New Moon.

TABULA SE[I...]

Lunarium Macularum exacta I[...]
Praestantissimo[...]

HEVELII [...]

Curiosis Rei Sid[...]

IOH. GABR. DO[...]

IOH. BAP[...]

CumPri[...]

...e autem, cum Sol illas à latere illuminat. quam maximè conspicuae red-
...ur, cum è contrario à quadraturis ad oppositionem superficies Lunae, duem...
...isce inaequalitatis magis magisque verticaliter immanere pergit, et omne...
...quid umbrosum ante fuit, pedetentim illuminat, aliam semper exhibeat fu-
...ut tandem luminosa et albicans appareat.

...hoc fundamento bina nostra Schemata in delineatione macularum no-
...um etiam differentiam involvunt, eò quod primum, HEVELIANUM puta.Lu-
...oppositione cum Sole existente, hoc est, in plenilunio designatum. alterum
RICCIOLINUM scilicet, è pluribus Lunae phasibus in unum corpus fuerit...
...tum. In denominationibus macularum, utpote signis et significationibus arbi-
...s, dictos Auctores inter se differre hìc in aperto videmus, cum Hevelius no-
...marium, regionum, fluminum et montium nostrorum imitatus. Ricciolus au-
...illustrium et de re siderea optimè meritorum Astronomorum, complurium
...tim suae Societatis Mathematicorum nomina pro usu Astronomico sibi e-
...xit.

...ini circa Lunam limbi, se vicissim secantes nihil aliud, quam motus alicuius in
...à libratorii terminos, intra quos perpetua deprehenditur librationis variatio,
...ndicant; qui hodie demum per Tubos è diversa macularum nonnullarum mu-
...me observatus, nec Veteribus olim notus fuit: eandem quippe nobis faciem
...antissimè semper Lunam obvertere existimantibus: per agit autem haec mo-
...fium libratorium per quatuordecim circiter dies trigesima sexta tantum.

diametri suae parte in plagam superiorem ab Austro. Corum versus, dum Luna versatur in descendentibus signis, in ascendentibus autem per idem tempus et
spatium secundum Hevelii et aliorum observationes retrorsum iterum. et sic
porro vacillare videtur

Eodem tempore, menstruo nempe spatio Lunam quoque orbitam suam, dum
porro et retro librationem absolvit. peragrare, et pro vario situ. diversas phases, hoc
est, luminis figurationes varias, prout figura media inferior B. subindicat: simul ex-
hibere deprehendimus, cum pars Lunae illuminata mox crescere, mox decrescere.
pro maiori vel minori Lunae a Sole distantia debeat. quae sane luminis non propri-
sed à Sole mutuati signa sunt indubia; interim non obstante, quod lumen quoddam de-
bile haud multo ante et post novilunia Lunae quasi innatum, de quo olim multa inter Astronomos movebantur lites. maculas Lunares nonnihil reddat conspicuas. cum
extra omne dubium sit positum. hoc suam originem à Terrae nostrae superficie duo-
decies, et quod excedit, maiori quam illa Lunae, radiis Solis tunc temporis omnium
copiosissimos in illam reflectente. habere, eò quod hac reflexione cessante, ipsum etiam
putativum lumen nonnunquam planè cum ipsa Luna in Eclipsibus disparuerit.

Ultimo denique loco duplices pro Lunae Mensurae longitudinaria notande quoque ve-
niunt. quarum unam pro distantiis et magnitudine macularum ut et diametro Luna-
ri. quae secundum Hevelium 494 mensuratur milliaribus. per Germanica milliaria
definiendis, alteram pro quantitate Eclipsium Lunarium tam secundum digitos E-
clipticos quam eorum partes exactè describenda. huic tabulae apposuimus.

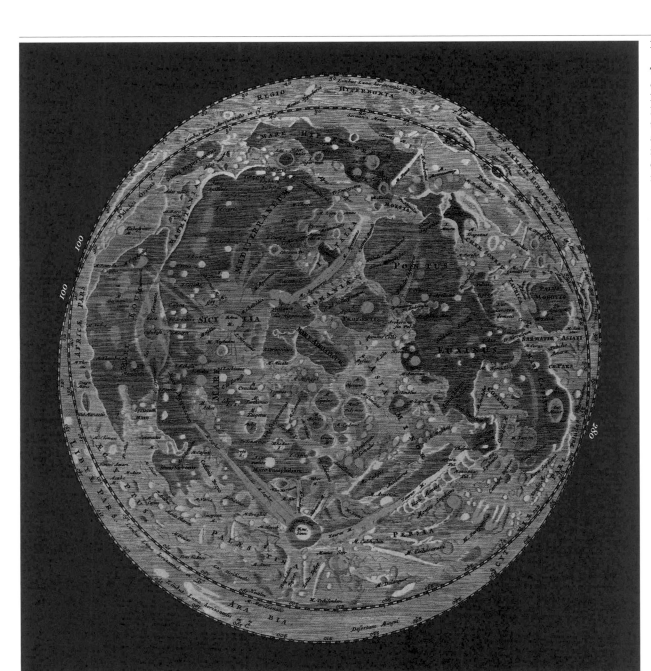

LUNAR MAP AFTER HEVELIUS.

The left-hand side of Plate 11 reproduces one of the three large lunar charts from Hevelius's *Selenographia* of 1647. The book – the first dedicated entirely to the astronomy of the Moon – introduced a system of nomenclature in which features were named largely after classical attributes and geographical regions on Earth.

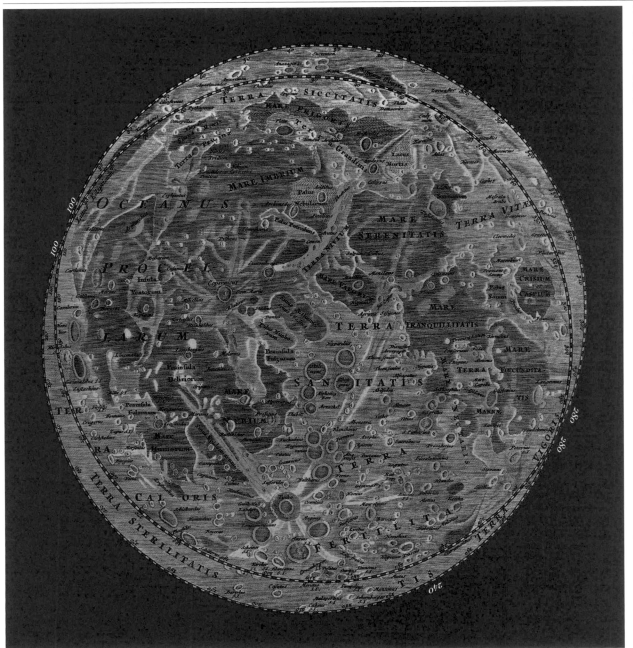

LUNAR MAP AFTER RICCIOLI.

The other side of Plate 11 is dominated by a map drawn from Riccioli's *New Almagest* of 1651. This map was, in fact, compiled by another Jesuit astronomer, Francesco Maria Grimaldi. In contrast to Hevelius's first-hand work, it draws on several different sources. The resulting map bears a far stronger resemblance to the Moon as most people saw it. This no doubt encouraged the wide adoption of Riccioli's naming system, in which seas bore the name of abstract concepts, and other features were named after scientists and philosophers both ancient and modern.

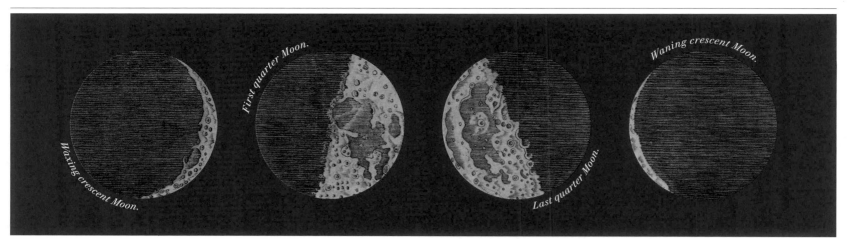

PHASES OF THE MOON.

The four corners of Plate II are adorned with miniature maps representing the Moon in its waxing crescent (*crescentis*), first quarter (*prima quadratura*), last quarter (*ultima quadratura*) and waning crescent phases (described by Doppelmayr as *luna senex*, the Old Moon).

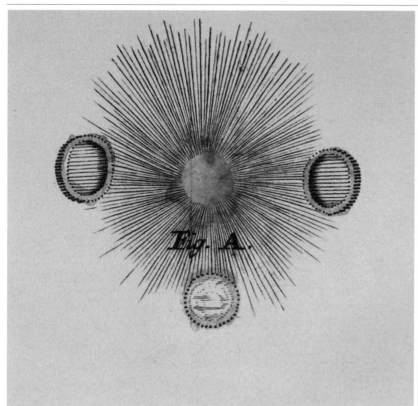

SHADOWS ON THE LUNAR SURFACE. (FIG. A.)

This small diagram illustrates the way in which shadows are cast across craters on the lunar surface in different directions, depending on the orientation of the Sun.

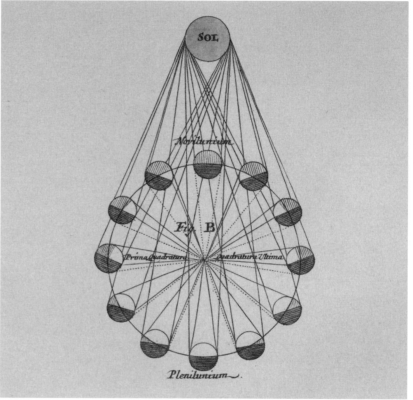

LUNAR LIBRATION. (FIG. B.)

This diagram explains the principle of libration. Due to our changing perspective and the Moon's relative proximity to Earth, different regions come into view along the limb of the Moon at different points in its orbit.

SCALE, GERMAN MILES.

A map scale shows distances in German *Landmeile* (roughly equivalent to 7.5 km/ 5 miles).

ASTRONOMICAL DIGITS.

This now-obsolete measuring system divided the face of the Moon (or Sun) into twelve 'digits', each of which was further divided into sixty minutes.

ALLEGORY OF THE MOON.

At upper left, a group of cherubs play with an ungainly telescope in order to spy on the face of the Moon.

THE MOON'S PITTED FACE.

A cherub illuminates the pockmarked face of the Moon goddess, while accompanying text comments: 'Although Diana is tabby, yet she is beautiful.'

PLATE 11.

*MICROGRAPHIA STELLARUM PHASES LUNÆ ULTRA 300
(1693–98).*

Maria Clara Eimmart (1676–1707), the daughter of Nuremberg
observatory founder Georg Christoph Eimmart, was barely a year older
than Doppelmayr and perhaps the finest observational astronomer
working in the city. Over six years in the late 17th century, she produced
a series of stunning illustrations of the phases of the Moon, usually
rendered in pastels on dark blue card. Eimmart's illustrations capture

PLENILVNIVM,
pinxit ad Archetypum M.C.Eimmarta Norimb:

the changing fall of light and shadow as the near side of the Moon is
illuminated to differing degrees and at varying angles. Eimmart also
illustrated a range of other astronomical phenomena, including the
solar eclipse seen at Nuremberg in 1706. That same year, she married
Johann Heinrich Müller, professor of physics at the *Aegidianum*
and her father's successor as director of the Nuremberg observatory.

PLATE 12.

LUNAR THEORIES.

(THEORIA LUNÆ)

Doppelmayr describes the interpretation of anomalies in the Moon's orbit, the causes of various lunar phenomena, and the nature of the Moon's surface.

Doppelmayr's second plate on the subject of the Moon is one of the latest in the *Atlas*, probably compiled at some point after 1735. It presents a medley of information about the Moon, with two main focuses – the appearance of a range of phenomena on the lunar surface, and theories put forward to explain various aspects of its motion. The illustrations of the lunar surface immediately draw the eye, but it is the group of diagrams that are arguably more significant, because they highlight a crucial debate of Enlightenment astronomy.

Although the Moon's monthly cycle of phases is usually said to repeat after 29.53 days, this is an average calculation, and astronomers from ancient times became aware that the period could vary slightly from month to month. The Moon's speed against the background stars also varied depending on its position. Around New and Full Moon, when Earth, the Moon and the Sun are aligned in space (an alignment called syzygy), the Moon's speed is fastest, while at first and last quarters, when the Moon and Sun are in quadrature (at right angles to one another), it moves slowest.

Astronomers of the early Enlightenment identified recurring patterns in the Moon's misbehaviour and were able to separate them into distinct 'inequalities' with different periods. However, they struggled to understand these patterns in terms of uniform circular motion. Even the great observer Tycho Brahe (1546–1601) could only refine and add yet further inequalities to a growing pile when his lunar theory was published posthumously in 1602.

A crucial breakthrough, however, came a generation later through the insights of the young English astronomer Jeremiah Horrocks (1618–41). His lunar theory, developed in letters to friends in the late 1630s but not published until 1673 (long after his untimely death at the age of twenty-two), concluded for the first time that the Moon's orbit is an ellipse rather than a perfect circle. This idea had been considered and then abandoned by Johannes Kepler (1571–1630) in his own attempts to solve the lunar problem. Horrocks revived the idea and suggested that the main perturbations in the Moon's motion are caused by the orbit's apsidal line (connecting its perigee and apogee, or nearest and furthest points from Earth) wobbling back and forth in a twice-yearly cycle, while the actual eccentricity of the lunar orbit flexes slightly in a similar period.

Horrocks attempted to fit his explanation to Kepler's laws of planetary motion (in which the speed of an orbiting world is governed by its distance from the object that it orbits)

FIG. 1.
Galileo made this series of watercolours of the Moon in November and December 1609. When these sketches were reproduced as engravings in his book *Starry Messenger* (1610), the obvious shadows and other relief features they revealed instantly disproved the Aristotelean ideal of the Moon as a perfect sphere.

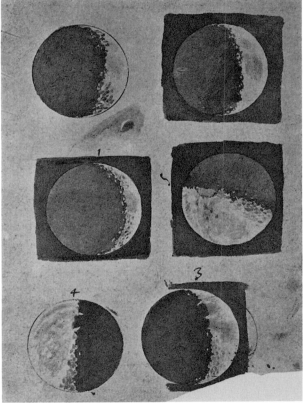

FIG. 1.

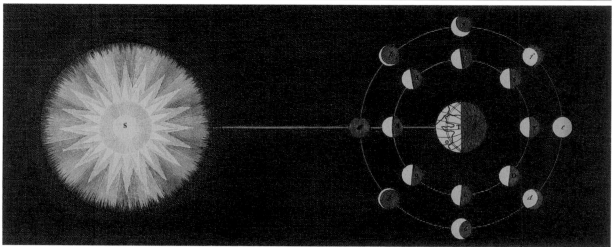

FIG. 2.

but, without a complete picture of the various inequalities involved, he found only limited success. Towards the end of the 17th century, however, his ideas helped inspire Isaac Newton (1643–1727) to develop a lunar theory of his own. Newton explained and refined the principal lunar inequalities as consequences of the Sun accelerating the motions of the Moon and Earth by varying amounts at different points along their orbits. Ironically, however, his dense mathematical treatment of the Moon's motion skipped over the actual cause of this acceleration – the shifting balance in the forces of gravity between the Sun, the Moon and Earth. As the true explanatory power of Newton's own laws of motion and universal gravitation became apparent, the physicists and mathematicians of Doppelmayr's time and beyond devoted a great deal of time and energy to showing how lunar inequalities could, indeed, be explained by gravitational effects.

Elsewhere on Plate 12, Doppelmayr explores somewhat less obscure topics. There are charts showing the causes of lunar libration (the 'wobble' that brings parts of the hidden far side of the Moon into view). Libration in longitude arises as the steady pace of the Moon's monthly rotation slips out of step with its varying speed along its elliptical orbit, while libration in latitude is due to the equivalent of lunar seasons, linked to a very slight tilt in the Moon's axis of rotation. Other diagrams include a two-part 'cycloid', showing the Moon's path through space relative to Earth's in the course of a month, and an explanation of 'Earthshine' – the secondary illumination received by dark areas of the Moon thanks to sunlight reflecting off Earth.

At lower right, two diagrams of lunar occultations, in which the Moon passes in front of a more distant astronomical body, offer some historical curiosities. One shows the typical occultation of one of the brightest stars in the sky, Aldebaran in the constellation of Taurus the Bull – Doppelmayr calls the star by its obsolete ancient name Palilicium. The other shows a 1720 occultation of Venus, in which some observers reported seeing evidence for the refraction (bending) of light by a lunar atmosphere.

Finally, we can turn to this plate's most attractive features: a pair of early sketches exploring lunar geology. At the top sits a reproduction of a renowned drawing made by the English scientist Robert Hooke (1635–1703) through a 9-m (30-ft) telescope in October 1664. The drawing focuses on a small area around the crater known as Hipparchus, and it is the earliest detailed illustration of a single lunar feature. It was published alongside Hooke's more famous microscope studies in *Micrographia* (1665), and it demonstrates great draftsmanship in its depiction of a realistic landscape. Hooke thought carefully about the nature of bowl-shaped depressions observed on the Moon, and even conducted experiments to show that they could be formed by either impacts from space or volcanic eruptions. Ultimately, he came down on the side of volcanism and the idea of the Moon as a living and evolving world in its own right – an idea that his contemporaries eventually followed.

At the bottom of the page is a 1725 sketch by Italian philosopher Francesco Bianchini (1662–1729), focused on the area between the craters Plato, Aristotle and Eudoxus, where he discovered a great valley cutting across the mountain range known as the Lunar Alps. The inset image is a striking impression of the great crater Plato, depicting a reddish glow Bianchini claimed to have seen on the crater floor. Today, such elusive lights are thought to be linked to the escape of dust-laden gas from beneath the lunar surface.

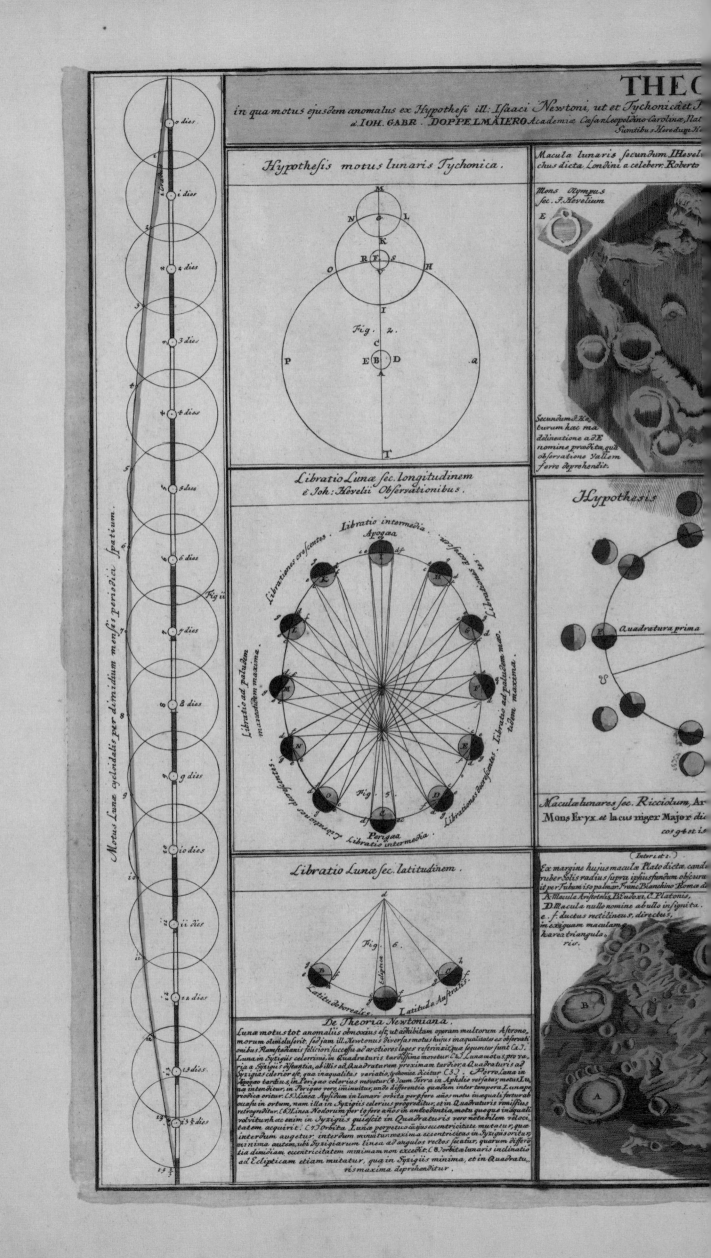

THE C

in qua motus ejusdem anomalus ex Hypothesi ill: Isaaci Newtoni, ut et Tychonicâ et T
a IOH. GABR. DOPPELMAIERO Academiæ Cæsar Leopoldino Carolinæ, Nat
Sumtibus Heredum Ho

Hypothesis motus lunaris Tychonica.

Libratio Lunæ sec. longitudinem
è Ioh: Hevelii Observationibus.

Libratio Lunæ sec. latitudinem.

De Theoria Newtoniana.

Macula lunaris secundum IHevel
chus dicta, Londini a celeberr. Roberto

Mons Olympus
sec. J. Hevelium.

Hypothesis

Quadratura prima

Maculæ lunares sec. Ricciolum, Ar
Mons Eryx et lacus niger Major die
cos q4 et is

Ex margine hujus maculæ Plato dictæ, cand
ruber Solis radius supra ipsius fundum obscura
it per Tubum iso palmar Franc Blanchino Romæ de
A. Macula Aristotelis B. Eudoxi, C. Platonis,
D. Maculæ nullo nomine a nullo insignita
e f ductus rectilineus, directus,
in exiguam maculam
h area triangula
ris.

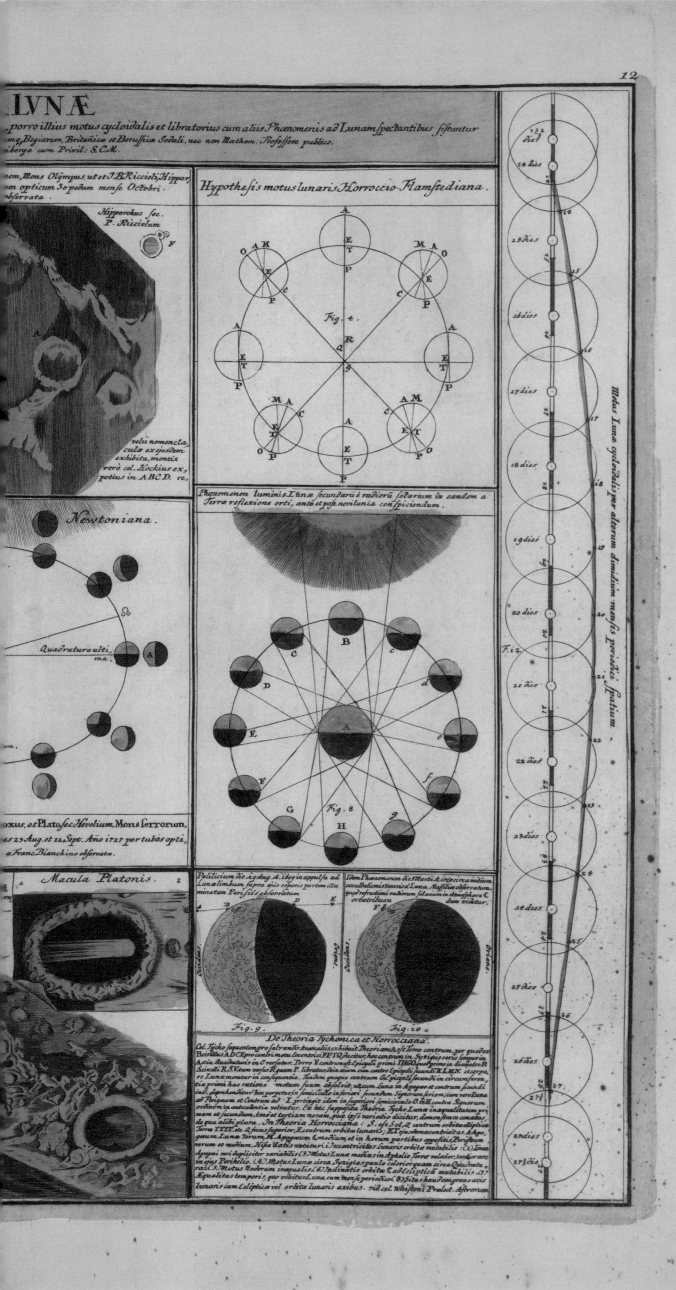

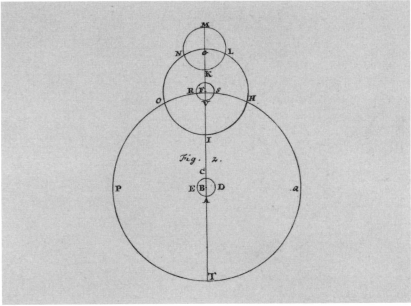

TYCHO'S LUNAR THEORY.
(FIG. 2.)

Tycho's theory required both a primary orbit for the Moon and a pair of epicycles, with each element centred on the last.

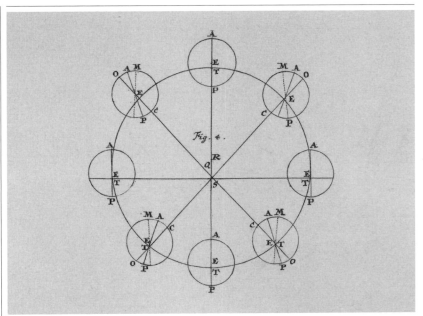

FLAMSTEED/HORROCKS THEORY.
(FIG. 4.)

Doppelmayr illustrates Horrocks's proposals of a twice-yearly oscillation in both the eccentricity of the Moon's elliptical orbit and the orientation of its perigee point to Earth.

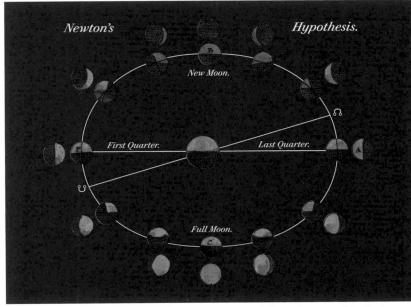

NEWTONIAN MODEL. (FIG. 1.)

Newton's lunar theory used the Sun's varying gravitational attraction on Earth and the Moon at different points in their orbits to explain why the Moon accelerates at certain points in the month and year, and slows down at others.

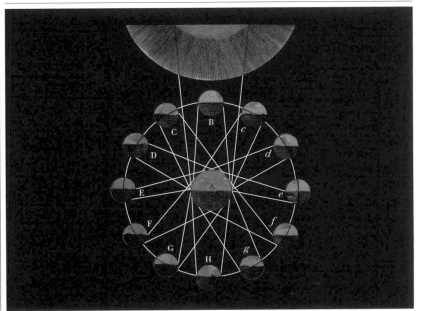

EARTHSHINE. (FIG. 8.)

Doppelmayr shows how sunlight reflecting from Earth can illuminate all or a portion of the Moon's own night side. This effect is known as Earthshine.

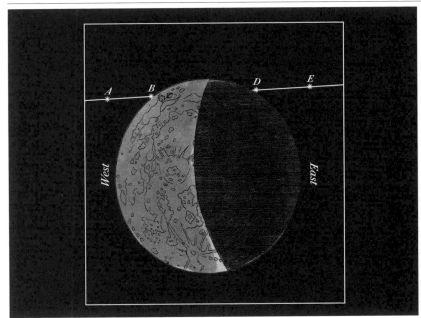

OCCULTATION OF ALDEBARAN/
PALILICIUM. (FIG. 9.)

This diagram shows the path of a lunar occultation of Palilicium in Taurus (modern Aldebaran). Timing the length of onset, path and duration of such events can confirm details about both the Moon and the occulted objects.

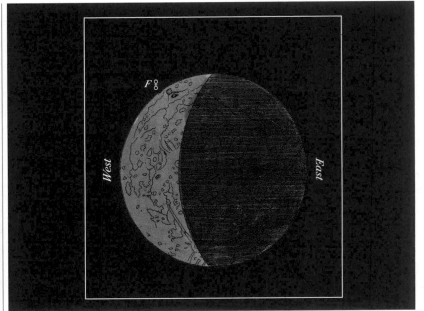

OCCULTATION OF VENUS.
(FIG. 10.)

This diagram shows the contact point in a 1720 occultation of Venus. The time it took planets to be fully hidden in such events offered an early indication of their true angular diameters.

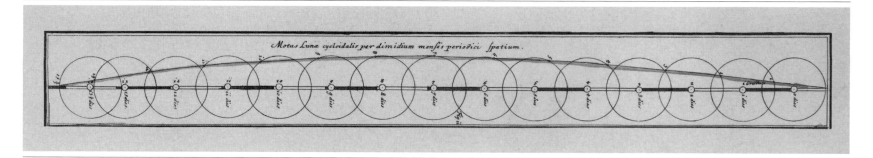

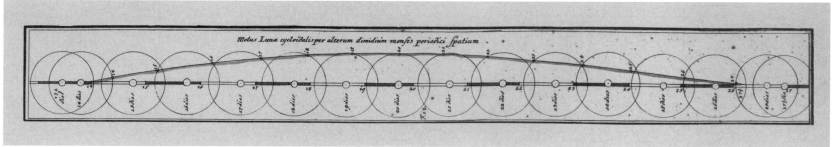

CYCLOIDAL LUNAR MOTION.
(FIGS. 11 AND 12.)

Two charts construct the Moon's position in relation to Earth's orbit as a cycloid: the path described by a point on a circle as it rolls along a straight line. The Moon takes 27.32 days to complete a full orbit, but (thanks to the Sun's slower motion) 29.53 days to return to the same position relative to the Sun in Earth's skies.

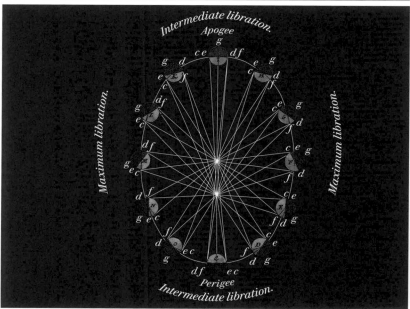

LIBRATION IN LONGITUDE.
(FIG. 5.)

Changes in the Moon's speed between apogee and perigee couple with its steady rotation to bring the eastern and western extremes of the lunar far side into view.

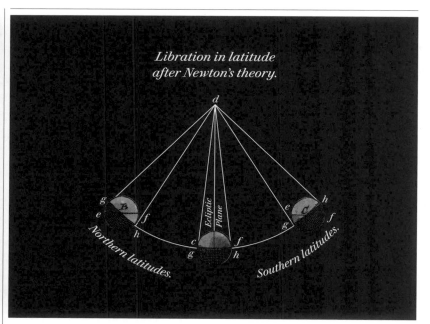

LIBRATION IN LATITUDE.
(FIG. 6.)

Doppelmayr illustrates how the slight inclination of the Moon's orbit relative to the ecliptic brings the far side into view over its north and south poles.

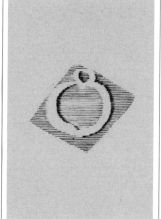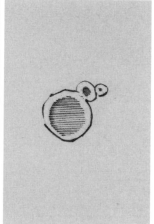

HOOKE'S VIEW. (FIG. 3.)

Doppelmayr copies Hooke's illustration of the crater Hipparchus from his *Micrographia* (1665), one of the first detailed views of lunar surface relief.

SIMPLE INTERPRETATIONS.

These depictions of the same region are from the maps of Hevelius and Riccioli. Hevelius interpreted the region as a mountainous feature he called Mons Olympus, while Riccioli saw it as a valley or crater and was the first to name it Hipparchus.

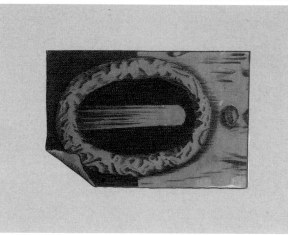

THE LUNAR ALPS. (FIG. 7.)

This illustration reproduces a print by Bianchini. It shows the ridge of highland mountains now called the Lunar Alps, surrounded by craters – including Aristotle (*A*), Eudoxus (*B*) and Plato (*C*) – and cut by a straight slash called the Alpine valley.

A RARE SIGHTING.

Doppelmayr reproduces Bianchini's view of the floor of Plato, showing its dark surface and the red streak he saw there in 1725 – an early example of the mysterious 'transient lunar phenomena'.

PLATE 12.

Phasium Lunæ Icones, quos Anno Salutis 1634 et 1635 pingebat, ac Sculp Aquis Sextiis Claud Mellan Gall? præsentibus at Stagirantib? Illustrib Viris Gassendo et Peyreschio.

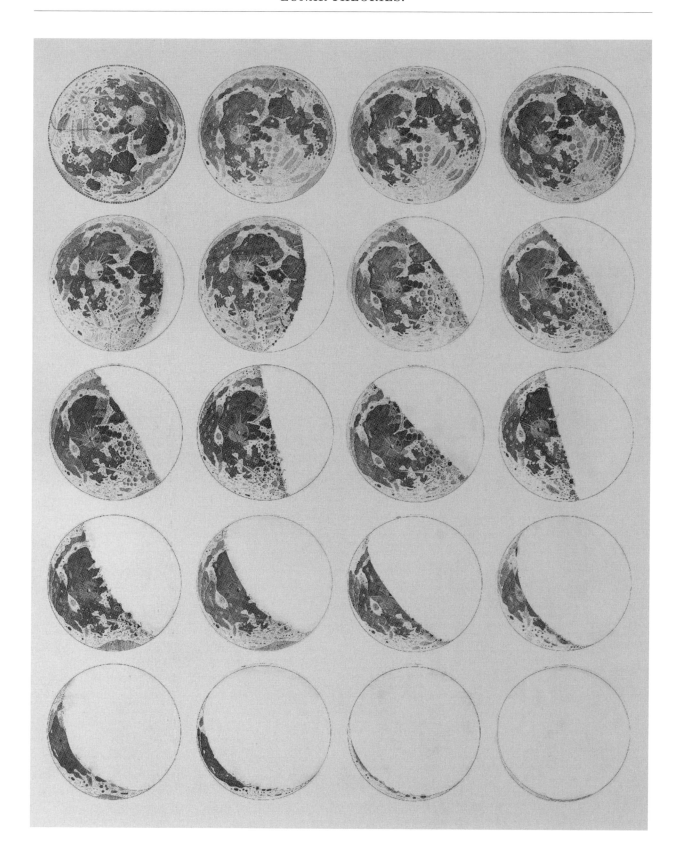

THREE VIEWS OF THE MOON (1635).

Shown on the page opposite, these three depictions of the Moon
at Full Moon (top), last quarter (bottom left) and first quarter
(bottom right) were engraved by Claude Mellan (1598–1688)
from telescopic observations made at Aix-en-Provence in
1635. Mellan's skilled eye allowed him to capture the Moon in
unprecedented naturalistic detail for his sponsors, the humanist
scholar Nicolas-Claude Fabri de Peiresc and Catholic priest
and astronomer Pierre Gassendi.

SELENOGRAPHIA, SIVE LUNAE DESCRIPTIO (1647).

A sequence of illustrations (shown above) from Hevelius's
groundbreaking work *Selenographia, sive Lunae descriptio*
(*Selenography, or A Description of the Moon*) charts the
disappearance of features from visibility on the Moon as its
phase dwindles from Full back to New Moon in (on average)
a little under fifteen days. While Hevelius followed the ancient
tradition of referring to the Moon's darker patches as 'seas'
and other bodies of water, he generally borrowed his names
from geographical features on Earth. He believed this idea
would save effort, but ultimately it ensured the success of
Riccioli's rival naming system.

PLATE 13.

THEORY OF ECLIPSES.

Doppelmayr explores the details of both solar and lunar eclipses.

Plate 13 is thought to be one of the latest in Doppelmayr's *Atlas*, perhaps compiled specifically for this project. The centrepiece revisits an earlier map he produced in 1707, depicting the path of the total solar eclipse as it swept across Europe on 12 May 1706. Surrounding the central chart are a host of other illustrations explaining the causes of eclipses and providing a visual record of key observations, including occultations of bright stars by the Moon, transits of Mercury across the Sun's face, and even the motions of sunspots.

Astronomers had been able to predict solar and lunar eclipses with at least moderate accuracy for centuries. According to the Greek historian Herodotus (*c.* 484–*c.* 425 BCE), the earliest alleged prediction was that of the philosopher Thales of Miletus (*c.* 624/23–*c.* 548/45 BCE). Herodotus claims that Thales predicted an eclipse in 585 BCE whose abrupt onset brought an end to fighting between the Medes and the Lydians in his native Turkey. Many historians doubt such a prediction would have been achievable at the time, but Babylonian astronomers (and others in independent civilizations such as China and Mesoamerica) had certainly already begun to recognize repeating cycles in the occurrence of eclipses.

Both solar and lunar eclipses can only occur when the three bodies – the Sun, Moon and Earth – align precisely in space. Naturally, this only happens at Full Moon (when the Moon and Sun are on opposite sides of Earth) or at New Moon (when the Moon and Sun lie in the same direction). However, the tilt of the lunar orbit, at some 5.1 degrees to the plane of the ecliptic (the apparent path of the Sun over the course of a year), means that, on most of these occasions, a direct alignment is missed. Eclipses are only possible when a Full or New Moon happens to occur just as the Moon is crossing the ecliptic, at points known as the 'nodes' of its orbit.

Just as Earth moves around the Sun each year while its axis of rotation remains pointing in the same direction in space (giving rise to the seasons, see Plate 6), so the Moon's orbit maintains roughly the same orientation in space as it circles Earth. This means that the Sun can align with the line between the nodes twice each year, producing 'eclipse seasons' where full alignments become possible for approximately the span of a lunar month. If the Full Moon crosses the node on the opposite side from the Sun, there is a good chance that it will pass through the large tunnel of shadow cast by Earth across space, resulting in a lunar eclipse. If the New Moon crosses the node on the same side of Earth as the Sun, its own shadow may sweep across Earth. However,

FIG. 1.
This idiosyncratic illustration from Peter Apian's *Cosmographia* (first published in 1524) demonstrates a concept that dates back to Aristotle: the idea of using the shape of the shadow cast across the Moon during a lunar eclipse as evidence that Earth is spherical.

FIG. 1.

this requires a far more precise alignment because, thanks to its smaller size, the Moon's shadow is much narrower.

The Sun's gravity adds a further complication: by tugging at the Moon when it is around its furthest point from Earth, it causes the entire lunar orbit to rotate, pulling the orientation of the nodes westward in a phenomenon known as 'nodal precession'. This causes the eclipse seasons to slowly shift from year to year, so that events occur at different times and locations. However, after a little more than eighteen years (a period known as a 'saros'), precession will return Earth, the Sun and the Moon to the same relative positions and a near-identical cycle of eclipses can begin.

A final wrinkle is added by Earth's own rotation – the saros period is 6,585 days and eight hours, so three saroi must pass before eclipses will recur at the same time of day.

The saros is just one of several cycles required to accurately predict an eclipse, and since its starting point is arbitrary there are effectively multiple saroi overlapping each other at any one time. However, while a practical understanding of its cause had to wait for the acceptance of Copernican astronomy (and Newtonian gravity), that did not prevent the stargazers of late antiquity from putting it to use. Hipparchus of Nicaea (*c.* 190–*c.* 120 BCE), best known today for his estimates of the sizes and distances of the Moon and Sun, laid much of the groundwork in the 2nd century BCE. It is possible that the remarkable Antikythera mechanism – a complex astronomical calculator rescued from an Aegean shipwreck in 1901 – is a practical tool for applying his ideas.

Even though predicting when eclipses were likely became routine, an accurate means of modelling the arrangement of the Sun, Moon and Earth in order to track the relatively narrow path of a total solar eclipse remained elusive. While the region of Earth from which the Moon's disc intrudes onto that of the Sun and causes a partial eclipse may be up to 6,400 km (3,977 miles) wide, the region where the Moon precisely blocks the entire Sun ranges from 160 km (99 miles) across to less than half of that. This 'region of totality' sweeps across Earth at hundreds of miles per hour, partly due to the Moon's inexorable drift in front of the Sun, but mostly due to Earth's own relentless spin beneath the relatively static shadow. A slight inaccuracy in the modelled location of the Sun or Moon can, therefore, push the path a long way to the north or south.

The eclipse of May 1706 was particularly significant as it became the subject of the first predictive eclipse maps. In 1705, Amsterdam

FIG. 2.

mathematician Symon van de Moolen (1658–1741) and Rotterdam navigator Andreas van Lugtenberg (dates unknown) both published accounts of the forthcoming eclipse, showing that totality would sweep diagonally across Europe from the southwest to the northeast. The eclipse, when it duly occurred, excited much astrological comment because it happened to coincide with the defeat of the French 'Sun King' Louis XIV's (1638–1715) forces at Barcelona during the War of the Spanish Succession.

Lying close to the central line of the eclipse path, Nuremberg, Germany, enjoyed some four minutes of totality, and the following year Doppelmayr and Johann Baptist Homann (1664–1724) commemorated the event by producing a detailed map of Europe upon which the path of the eclipse was projected. Constructed using observations gathered from across Europe, it showed not only the limits of totality, but also the varying degrees of partial eclipse seen in other regions. On Plate 13, Doppelmayr revisits the topic from a different perspective, showing the eclipse's track as seen from above the North Pole in order to trace its path across an entire hemisphere.

FIG. 2.
In May 1715, long before the fulfilment of his prediction that a comet seen in 1682 would return in 1758, Edmond Halley pulled off a remarkable demonstration of the power of Newtonian physics by predicting the path of a total solar eclipse across England and Wales, and its time to within four minutes. Halley collaborated with cartographer John Senex to publish a broadside print advertising the event in advance, and followed up with this corrected second edition, which added the path of a forthcoming eclipse in 1724.

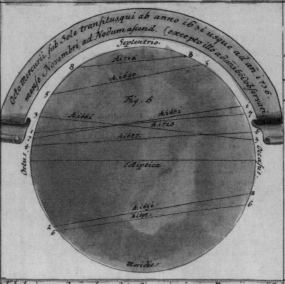

Acta Mercurii sub Sole transitusqui ab anno 1651 usque ad an 1736. (excepto illo ad occidentis Alexandr. mense Novembri ad Nodum ascend.

Septentrio.

Fig. 6

Ecliptica

Meridies.

THEOR

in qua variæ Solis occultationes, obscurationes Terræ et Lunæ ve—
a IOH: GABR: DOPPELMAIERO, Acad. Cæsar. Leopoldino Carol. Nat. Curios

Sumtibus Hered

Typus eclipsis Lunæ partialis.

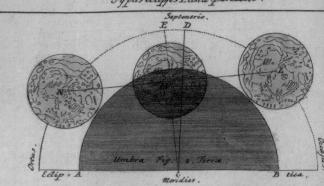

Tabula, in qua ad. anos supra-datos dies conjunctionum Mercurii cum Sole, harum Observatores et Observationum loca exhibentur.

Num. ord.	Anno	Tempus conj. Sen.	Observatores.	Loca Observat.	Num. ord.	Anno	Tempus Conj. Sen.	Observatores.	Loc.Observationum.
1	1651	27. Nov.	Pet: Gassendus	Parisii	5	1690	2. Nov.	Wurselbaum G. Kirch.	Noriberga Erfordia
2	1651	23 Nov.	Jer. Shakerlaus.	Surata in India	6	1697	2. Nov.	Wurselbaur et Cassinus	Noriberga Parisii
3	1661	23 May	Joh. Hevelius Hugenius Severus.	Dantiscum Londinum.	7	1723	9. Nov.	Edm. Halleus Jac. Cassinus	Londinum Parisii
4	1677	d. 7 Nov.	Halletius Edm. Halleus	Axenio Ins. Helena	8	1736	d. Nov.	Manfredius Marinonius Chr. Kirchius	Bononia Viena Berolinum

Graphica designatio orbis retris fere totius, per cujus maximam partem, et quidem per uni st: corr. tam totalis quam partialis spectat

DIAGRAMMA HIPPARCHICUM
pro Eclipsibus Solis et Lunæ.

Fig. 7.

Fig. 3.

Centra, Semidiametri vera, distantia, aliaque, in hac Theoria sequentia notanda.
5. Centrum Solis; T Terræ; L Lunæ novæ; p centrum in confinio Lunæ plenæ perigææ; a Lunæ apogææ;
S B. Semidiameter vera Solis; TE Terræ; L N Lunæ; p Quæ bra Terræ perigææ; a Quæ bra apogææ.
S T. Distantia. Solis à Terra; S L à Lunæ; S C. à mucrone umbra terrestris; S F L ife. Solis à mucrone penumbra.
L T Distantia Lunæ à Terra, quæ varia usq. Distantia à Terra Lunæ perigææ; a T. diste. à Terra Lunæ apogææ.
T C E. Conus Terræ umbrosus; M T N. Conus Lunæ umbrosus, quietam variant; fu. Sisontius; D E F. A F B. conus Terræ
penumbrosus H G I. A G F. conus Lunæ penumbrosus.

Maculæ Solis insignes à die 9. Novembris usque ad 15 A. 1700 Parisiis observata.

Fig. 8.

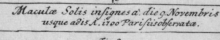
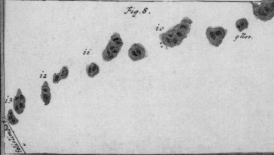

De eclipsibus Lunæ.
Eclipses lunares ex interpositione Terræ inter Solem et Lunam ori—
untur. (vid. Fig. 1. ad B Icum Luna plena circa quasnodos in um—
bræ terrestris conum motu suo interdum volvitur; et vel ex parte, vel
totaliter. (prout illius distantia à nodo proxima variat) lumine
suo mutuato privatur. Secundum hoc duplicis generis eclip—
ses obtutui oculorum se præbent, partiales (fig. 1. Jet totales, et
hæ, vel cum mora magna, vel sine mora. (fig. 3. Nulla prorsus
plenilunii tempore, hujus generis conspiciuntur eclipses, cum
nempe axis umbræ terræ magis is gradibus à nodo distat; tunc
enim Lunæ latitudo majorquam suma semidia metrorum Lunæ
et umbræ terrestris, et sic illa invisibilis, deprehenditur.

De eclipsibus Solis.
Eclipses Solis, seu potius ejus occultationes à Luna, circa
nodos, dum hæc novilunii tempore, inter Solem et Terram
media est. (vid. Fig. 1. sud A D diversas species exhibent; quædam
enim sunt partiales (fig. 1. et figu. ad b quædam totales cum mo—
ra. (fig 2. ad b. Icum terra circa Aphelium, et Luna circa Peri—

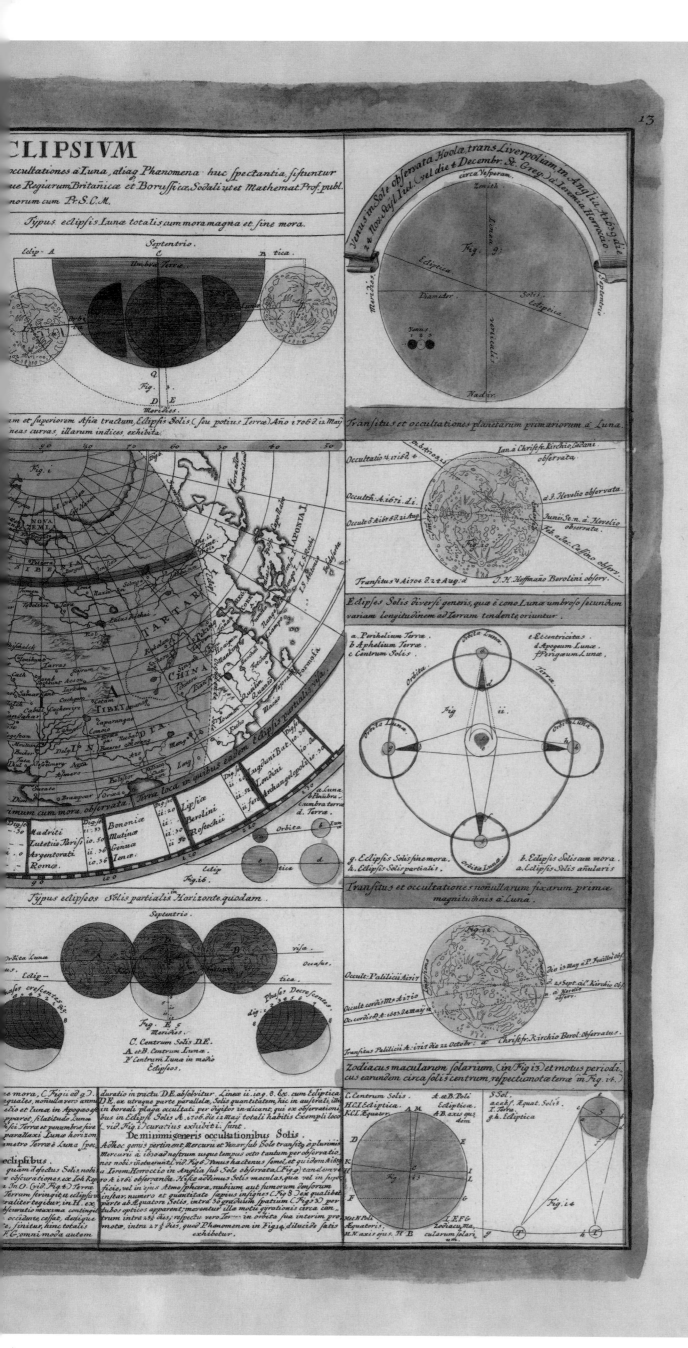

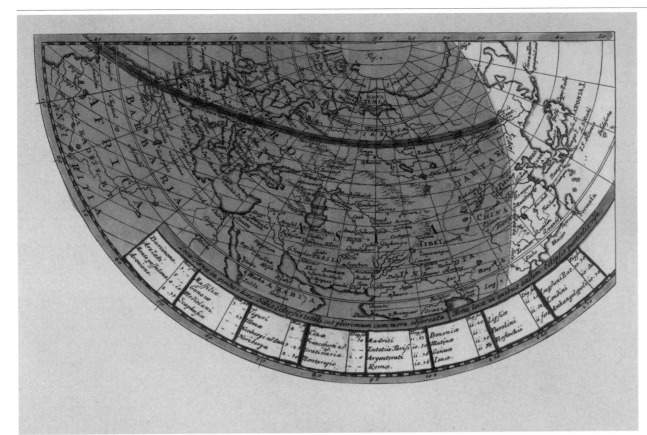

ECLIPSE PATH OF 1706.
(FIG. 1.)

Rather than revisit his previous best-selling chart of the solar eclipse in May 1706, Doppelmayr shows its path on a polar projection of Earth. The tinted central line indicates the track of totality across northwest Africa and Europe into Asia, while parallel lines to either side show the reach of the partial eclipse. Around the bottom of the chart, the length of totality observed from various sites is listed, as well as the degree of partiality at certain locations, measured in digits. Note that Doppelmayr takes the opportunity to correct the track of the eclipse compared to the earlier map (reproduced on page 19), particularly at its southwestern and northeastern extremes.

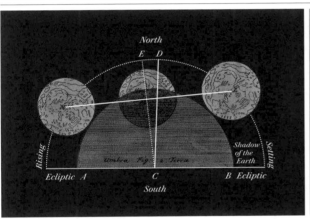
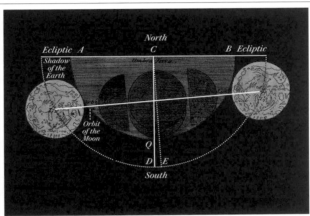

TOTAL AND PARTIAL LUNAR
ECLIPSES. (FIGS. 2-3.)

A pair of diagrams shows key points in the passage of the Moon through Earth's shadow, during both a partial (far left) and total (near left) lunar eclipse.

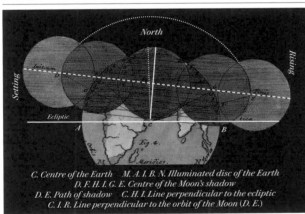
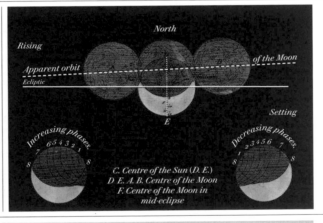

TOTAL AND PARTIAL SOLAR
ECLIPSES. (FIGS. 4-5.)

These illustrations show the track of the Moon's shadow across Earth during a solar eclipse, and the progression of partial solar eclipses as seen from Earth.

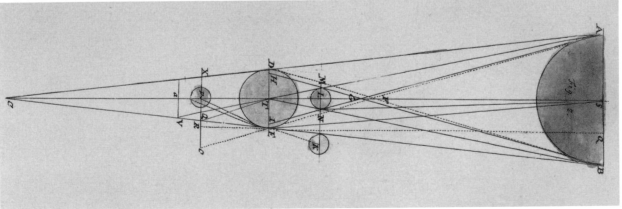

ECLIPSE GEOMETRY.
(FIG. 7.)

This complex diagram describes the precise geometrical alignments that govern whether eclipses will occur at a certain New or Full Moon, and how differences in the Moon's apparent direction from different points on Earth give rise to different forms of eclipse.

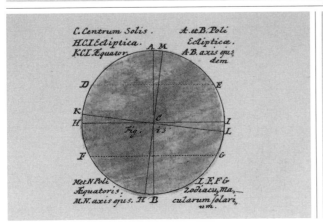
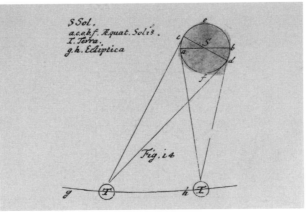

THE 'ZODIAC OF SUNSPOTS'.
(FIGS. 13-14.)

In this pair of diagrams, Doppelmayr describes the zones in which sunspots occur to either side of the Sun's equator. He also shows how Earth's motion with respect to the Sun must be taken into account when using them to measure the Sun's rotation.

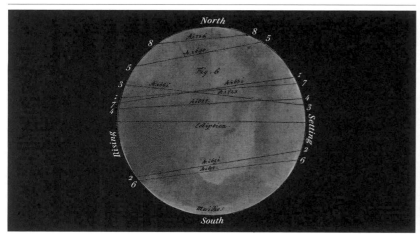

TRANSITS OF MERCURY.
(FIG. 6.)

The track of Mercury across the disc of the Sun is plotted for eight 'transit' events between 1631 and 1736.

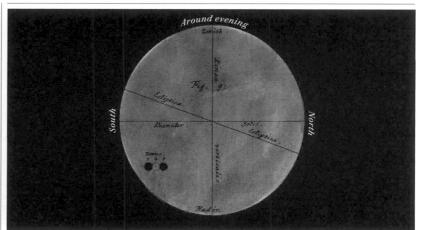

THE 1639 TRANSIT OF VENUS.
(FIG. 9.)

Doppelmayr reproduces Hevelius's somewhat inaccurate 1662 rendering of the transit of Venus predicted and recorded by Jeremiah Horrocks a generation earlier.

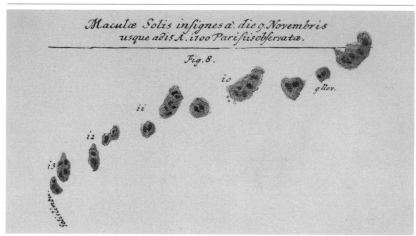

SPOTS ON THE SUN.
(FIG. 8.)

Illustrating a large group of sunspots seen in November 1700, Doppelmayr speculates that they are either surface markings or dense clouds or fumes in the solar atmosphere.

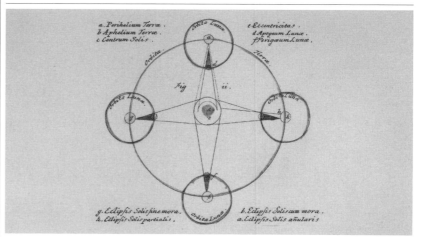

ANNULAR ECLIPSES.
(FIG. 11.)

This diagram demonstrates how, if a solar eclipse occurs with the Moon near apogee, the lunar disc may not entirely block the Sun, thus producing a ring-shaped or annular eclipse.

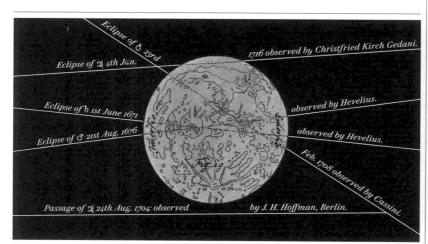

PLANETARY OCCULTATIONS.
(FIG. 10.)

This chart illustrates a number of occultations in which the Moon passes in front of a planet.

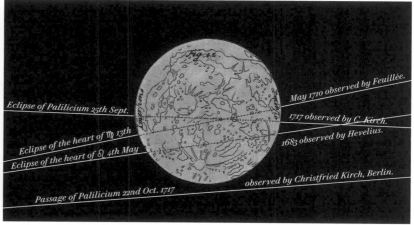

OCCULTATIONS OF FIRST-MAGNITUDE STARS.
(FIG. 12.)

This chart records lunar occultations of the brightest stars. Precise measurements of such events (and even near misses) allowed refinement of the Moon's orbit.

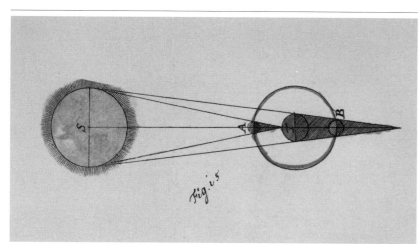

GEOMETRY OF ECLIPSES.
(FIG. 15.)

A simple diagram introduces the basic concept of the two types of eclipse. These are elaborated on in more complex form elsewhere.

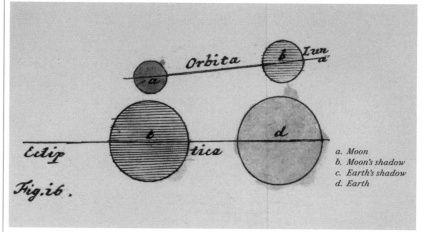

ORIENTATION OF MOON'S ORBIT TO ECLIPTIC. (FIG. 16.)

Doppelmayr illustrates the tilt of the Moon's orbit at 5.1 degrees to the ecliptic, which means the New and Full Moons usually 'miss' a perfect alignment with Earth and the Sun.

PLATE 13.

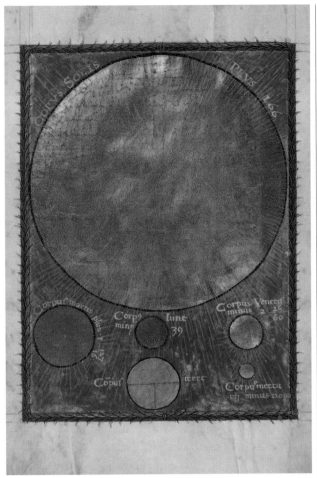

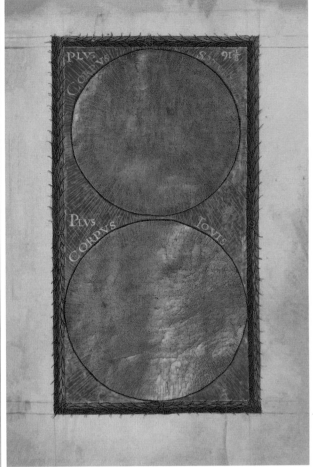

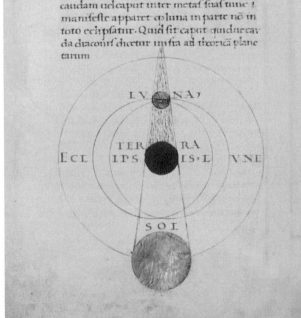

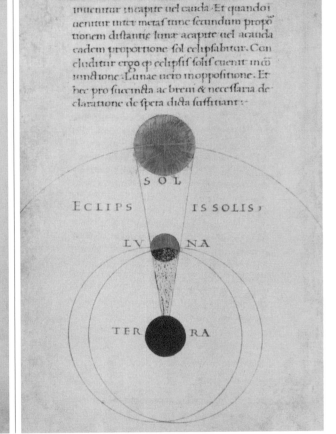

ASTRONOMIA (1478)

Tuscan scholar Christianus Prolianus's treatise of 1478 on geography and astronomy was produced in several editions, with the original manuscript ornamented with stunning illuminations using gold leaf. Among its many illustrations is this series of images documenting eclipses and other astronomical phenomena. The two illustrations at upper left depict the relative sizes of the Sun and other celestial bodies, calculated on the basis of their estimated distances in the Copernican system, their respective brightnesses, and evidence from solar and

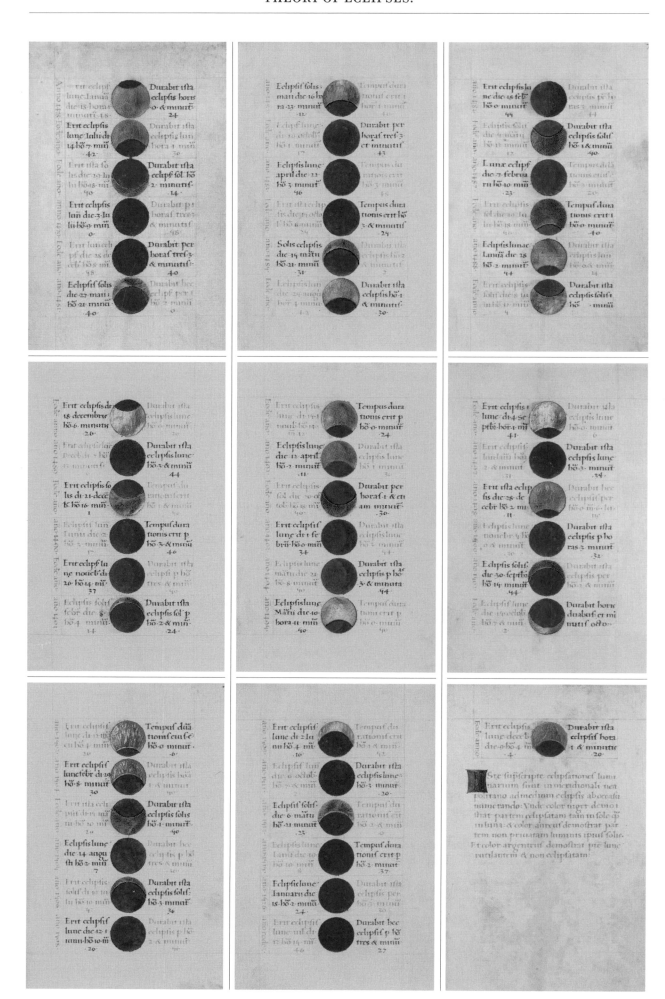

lunar eclipses. This scheme, therefore, recognizes the considerable challenge presented to the Ptolemaic view by the fact that the Sun, Saturn and Jupiter were all considerably larger than Earth and yet were supposed to orbit around it. The lower pair of diagrams describe the two types of eclipse and the further evidence they present for the relative sizes of the Moon and the Sun. On the right-hand side is a remarkable calendar, chronicling eclipses predicted to occur at the meridian of Naples (where Prolianus worked at the court of Ferdinand II) from 1478 to 1515, including their extent, start time and duration.

PLATE 14.

THEORY OF THE SATELLITES
OF JUPITER AND SATURN.

(THEORIA SATELLITUM IOVIS ET SATURNI)

Doppelmayr summarizes the known information
about the moons of the giant planets.

Galileo Galilei's (1564–1642) discovery that Jupiter had four satellites of its own is widely regarded as a key moment in the history of astronomy, proving for the first time that there was more than one centre of motion in the universe and, therefore, fatally undermining one of the key claims of Aristotle's (384–322 BCE) physics.

The discovery was one of the first made by Galileo with a new telescope that delivered an unprecedented twenty times magnification. On, or shortly before, 7 January 1610, he turned this device towards Jupiter and saw that the planet was flanked by three faint stars – two to the east and one to the west. On 8 January, he looked again and found that the three stars were still there, but now they all lay to the west of Jupiter. With no means of gauging the exact positions of these objects, he naturally assumed that Jupiter alone was moving past these more distant fixed stars. However, when he returned on following nights and found the arrangement different each time, he became intrigued. By 12 January, he was convinced that these 'stars' were, in fact, somehow related to Jupiter itself, and the following night he spotted a fourth point of light that would prove to have similar motions. Within a little over a week, he had reached the conclusion that he was seeing the motion of four objects orbiting Jupiter, moving back and forth because their orbits lay almost edge-on to Earth.

Seeking to curry political favour, Galileo named his discoveries the 'Medicean stars' after the powerful Medici family, but the satellites are today widely known as the 'Galilean moons'. Their names – Io, Europa, Ganymede and Callisto, from the closest to the furthest – were proposed in 1614 by the German astronomer Simon Marius (1573–1625), with whom Galileo soon became entangled in a bitter dispute over plagiarism and priority. Galileo himself refused to use these names and instead simply referred to the moons by number. This scheme remained in widespread use until the discovery of countless smaller moons orbiting both inside and outside the Galilean orbits rendered it impractical and Marius's names were revived. Doppelmayr, therefore, somewhat clumsily refers to the first, second, third and fourth satellites of Jupiter, rather than use today's familiar proper names.

One consequence of the fact that the moons orbit on a plane aligned with Earth is that we can see them pass behind and in front of both each other and Jupiter, creating a series of mutual eclipses. As instruments improved, it was not only possible to spot the moons as they disappeared behind Jupiter or came so close

FIG. 1.
Galileo recorded these observations of the moons of Jupiter in March and April 1612. He produced this chart (along with an ephemeris describing the orbital periods of the satellites and predicting their future motion) as part of an increasingly irate exchange of public letters with Jesuit priest Christoph Scheiner about the newly discovered sunspots. Scheiner insisted that the moons were unpredictable; hence, similarly unpredictable bodies passing in front of the Sun could be the cause of the apparent markings on its surface.

FIG. 1.

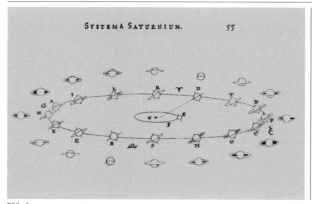

FIG. 2.

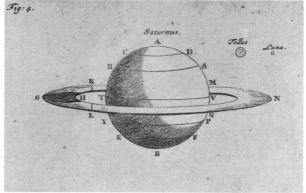

FIG. 3.

FIG. 2.
Huygens's illustration of 1659 shows how the visibility of Saturn's rings varies as it orbits the Sun.

FIG. 3.
This more detailed view of Saturn (with Earth and the Moon for scale) is from Huygens's *Cosmotheoros* (1698).

together that they briefly merged into a single star, but also to track the shadows they cast as they passed across the disc of the giant planet.

Galileo (who had a keen eye for commercial applications to his discoveries and inventions) soon realized that once the orbital periods of the moons were accurately known, the precise timings of these events could be predicted. Since they could be observed all over the world, they could, therefore, be used as a means of determining longitude on Earth's surface. We will explore this further on the next plate, but the first challenge was simply to refine the orbital periods.

The most precise measurements of the 17th century were made by Giovanni Domenico Cassini (1625-1712) from Rome, and published in 1668. In them, Cassini noted some curious discrepancies depending on the relative locations of Jupiter and Earth as the eclipses took place. He toyed with the revolutionary idea that light might have a finite speed, meaning that the light from mutual eclipse events would arrive 'early' when Jupiter was at opposition and closest to Earth, and 'late' when Jupiter and Earth lay on opposite sides of their orbits. While Cassini soon abandoned the idea as absurd, a few years later, Danish astronomer Ole Rømer (1644-1710) used Cassini's published data to make the first determination of the speed of light.

In the meantime, Saturn had revealed its own set of companion satellites. Christiaan Huygens (1629-95) discovered the largest and brightest in 1655, while Cassini discovered four more between 1671 and 1684 following his move to the Paris Observatory. Huygens simply referred to his satellite as 'the Moon of Saturn', while Cassini followed Galileo's example by cannily naming his discoveries the *Sidera Lodoicea* (Stars of Louis) in honour of his patron, Louis XIV (1638-1715). Huygens's moon turned out to be the fourth of these five in order from Saturn, and so astronomers of Doppelmayr's day and well beyond simply referred to them

by the numbers 1 through 5. The modern names (Tethys, Dione, Rhea and Iapetus for Cassini's quartet, and Titan for Huygens's giant) were suggested by astronomer John Herschel (1792-1871) in 1847. Since the entire Saturn system is usually tilted at an angle with respect to Earth, rather than being permanently edge-on, mutual eclipses and related events are far more rare.

Doppelmayr's plate is mainly concerned with describing the orbits of the satellites around their parent planets, and how these give rise to their motions as seen from Earth. He lists the orbital periods established by Cassini, and also the maximum angular separations of each moon from its parent planet, thereby allowing him to construct proportional plans of each satellite system. Other diagrams describe the positions of the satellites relative to Earth as their planet moves along its orbit (in a similar vein to the Tychonic diagrams of Plates 9 and 10), and the geometry of eclipses involving Jupiter and its moons.

Sadly overlooked here, however, is one of the few clues to the physical properties of these moons to emerge prior to the 20th century. While most of the satellites of the outer solar system show uniform brightness and colour as they circle their parent planets and, therefore, give away little about their composition, Saturn's fifth satellite, now known as Iapetus, is a notable exception. Cassini first identified it to the west of Saturn in 1671, but failed repeatedly to observe it on the eastern side of its orbit. When he finally tracked it down using an improved telescope in 1705, he confirmed that it is significantly fainter while on the eastern side of Saturn. Assuming (correctly) that Iapetus is like our Moon and orbits with one face permanently turned towards Saturn, Cassini realized that its leading hemisphere (the one that faces forward in its orbit) must be much darker than its trailing hemisphere – a difference since confirmed by space probes, and thought to be due to a dusty residue picked up on the leading hemisphere of this naturally bright moon as it orbits Saturn.

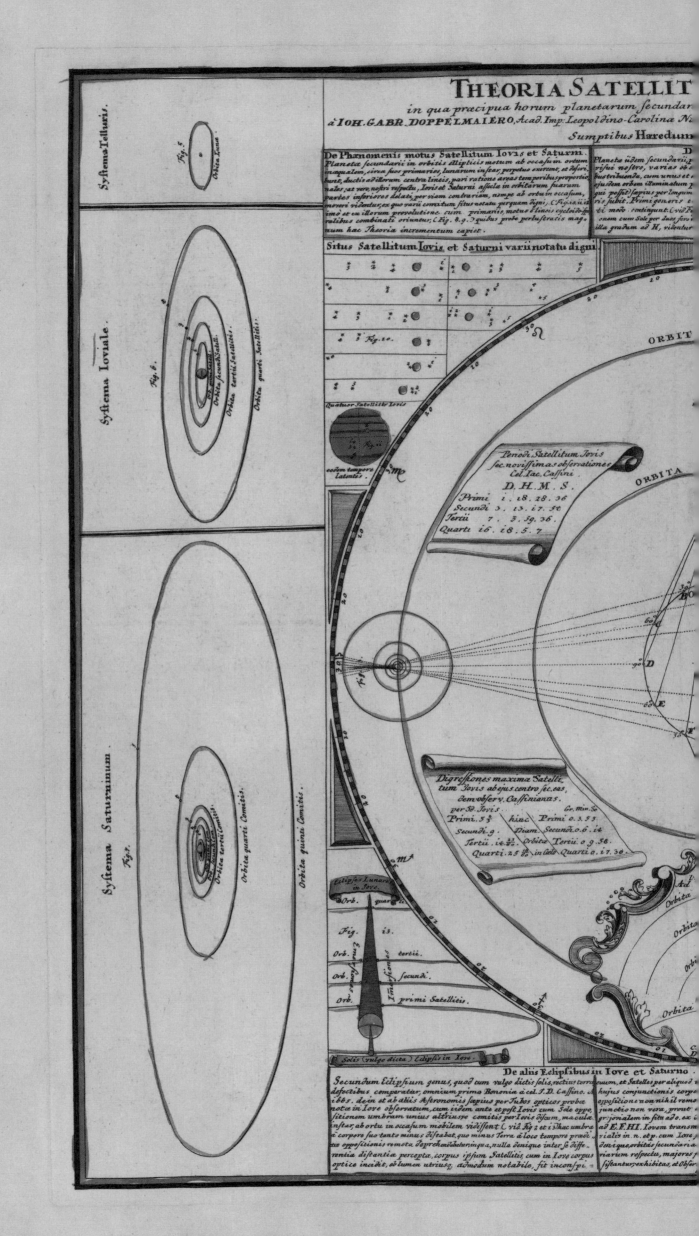

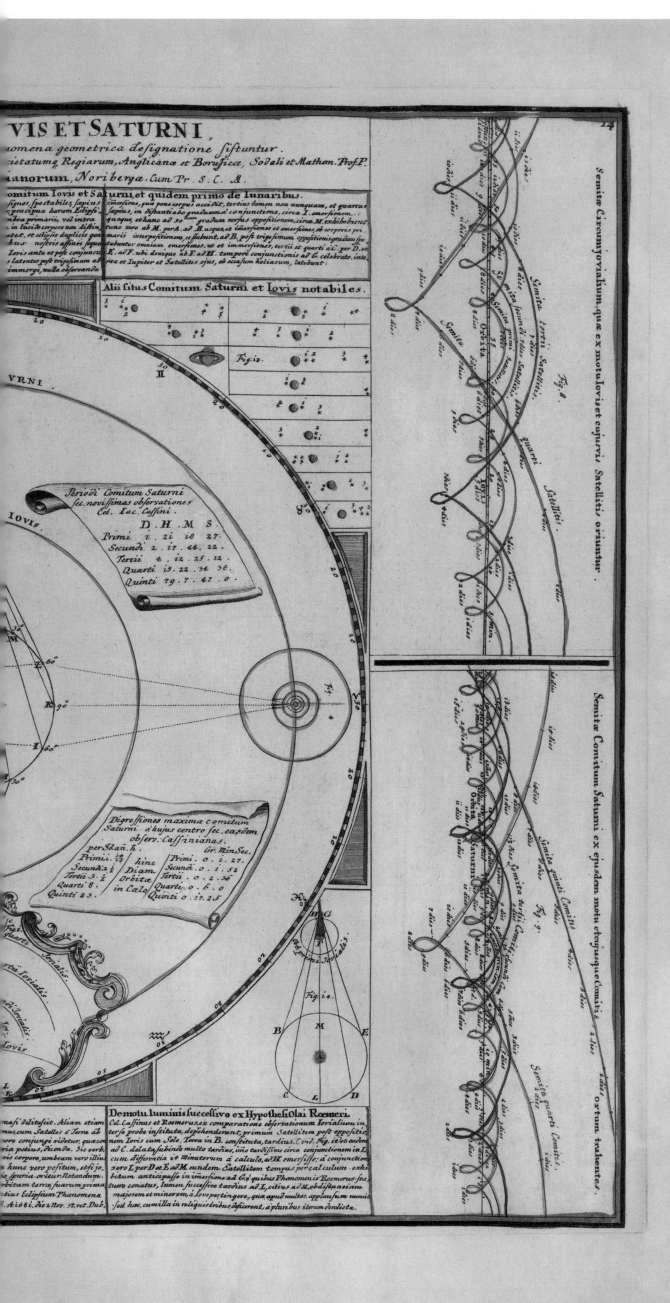

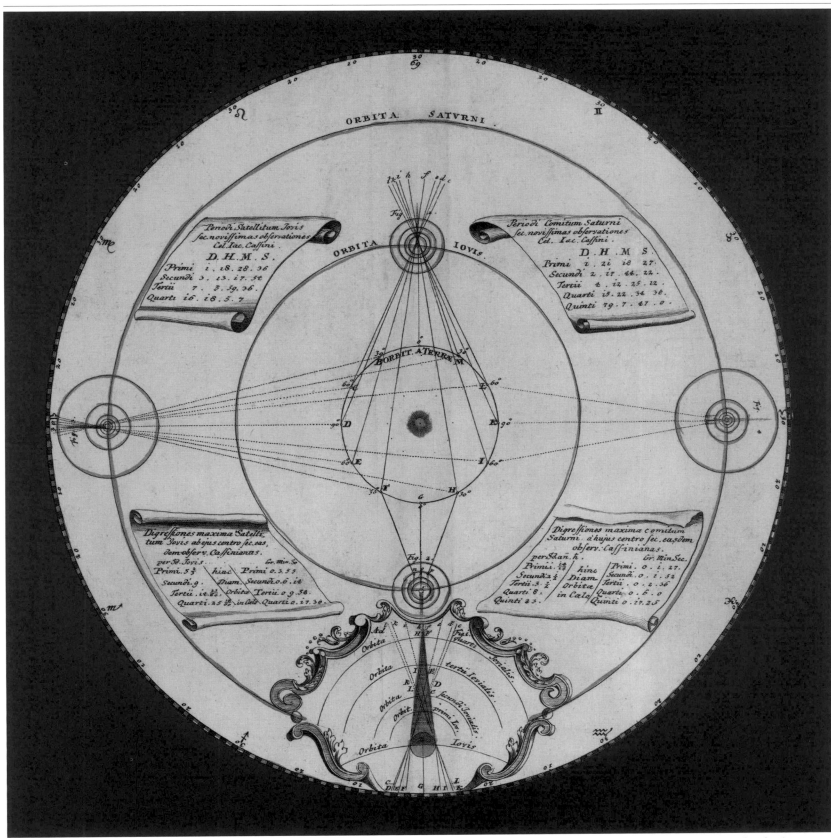

ORBITS OF THE SATELLITES. (FIGS. 1–4.)

The main illustration on Plate 14 demonstrates how Earth's changing position in its orbit must be taken into account when measuring the positions of the Jovian and Saturnian satellites relative to their planets. A magnified vignette at the bottom shows how the Jovian satellites may be hidden from Earth but not necessarily in Jupiter's shadow, and vice versa. The upper pair of scrolls provide Cassini's estimates of the satellites' orbital period, while those below show the greatest separation between each satellite and its planet, measured in both planetary radii and degrees, minutes and seconds.

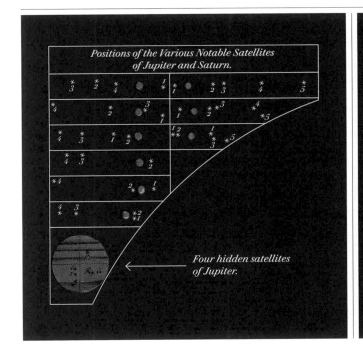

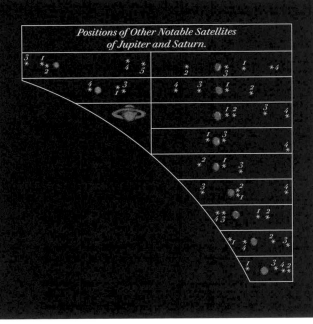

NOTABLE OBSERVATIONS OF THE SATELLITES OF JUPITER AND SATURN. (FIGS. 10 AND 12.)

Doppelmayr depicts the shifting appearance of the satellite systems through copies of a series of individual sketches. Numbers below the satellites show how their observed positions are often out of step with their physical distance from their planets.

HIDDEN SATELLITES. (FIG. 11.)

On rare occasions, the satellites all fall into line with Jupiter and briefly disappear, as in this depiction of an event observed by William Molyneux (1656–98) in 1681 in Dublin.

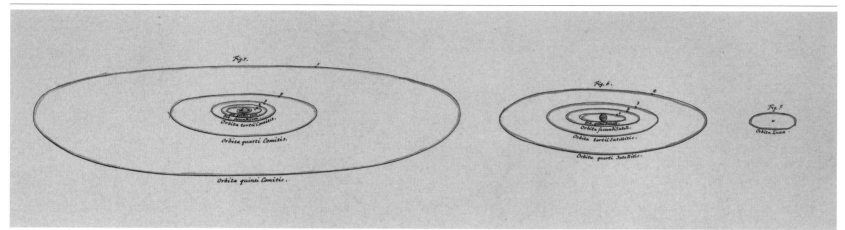

COMPARISON OF SATELLITE SYSTEMS.
(FIGS. 5-7.)

Here, Doppelmayr attempts to render the three known satellite systems to scale with each other. For comparison with modern figures, the Moon's orbital radius is about 384,000 km (239,000 miles), while that of Io (the 'first satellite of Jupiter') is about 422,000 km (262,000 miles) and that of Dione (Saturn's 'second satellite' in Doppelmayr's terminology) is 377,000 km (234,000 miles).

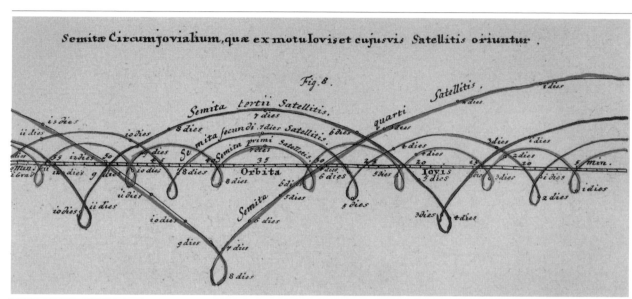

PATHS OF THE JOVIAN SATELLITES. (FIG. 8.)

In this diagram, Doppelmayr plots the changing positions of Jupiter's satellites relative to the planet as it moves along its orbit. Rendering the orbits in this way reveals a mathematical relationship between the periods of the three inner moons, known as a 'resonance'. For every single cycle of the third moon, Ganymede, the second moon, Europa, orbits twice, while the first moon, Io, completes four circuits of Jupiter.

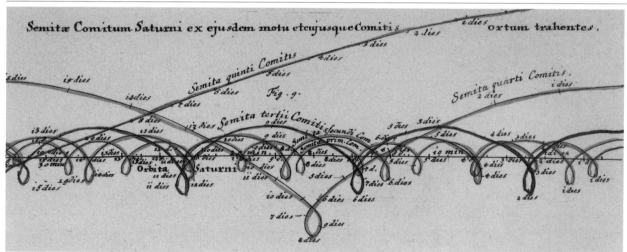

PATHS OF THE SATURNIAN SATELLITES. (FIG. 9.)

Here, the locations of Saturn's satellites are plotted relative to the planet's motion on its orbit, producing patterns that recall both the cycloidal lunar motion on Plate 12 and the trochoid spirals on Plates 9 and 10. The satellites known to Doppelmayr lack the neat orbital relationships of the inner Jovian satellites, although there are some similar patterns among moons discovered later. Such resonances arise through a natural process that minimizes the gravitational influence of one moon on another's orbit, but they are unstable over long time periods.

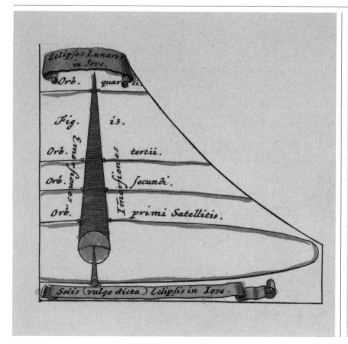

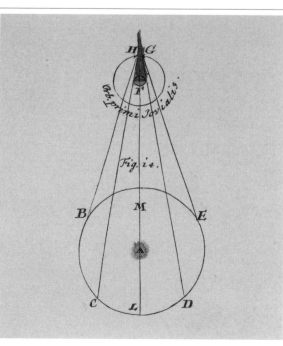

ECLIPSES AND TRANSITS.
(FIG. 13.)

A simple diagram demonstrates the two types of eclipse associated with the Jovian moons. They are true 'lunar' eclipses, where they pass through the giant planet's shadow, and transits, where they cast a shadow onto Jupiter.

RØMER'S HYPOTHESIS.
(FIG. 14.)

Doppelmayr illustrates and describes the first attempt to measure the speed of light. Rømer discovered a variation in the apparent time at which eclipses occurred due to the varying distance between Earth and the Jovian system.

PLATE 15.

ASTRONOMICAL BASIS
OF MODERN GEOGRAPHY.

(BASIS GEOGRAPHIÆ RECENTIORIS ASTRONOMICA)

*Doppelmayr depicts the world in line with
the latest astronomical observations.*

It is startling to realize, looking at the twin hemispheres that comprise Doppelmayr's treatment of Earth's own geography within the *Atlas*, just how vague the knowledge of certain areas remained until relatively recently. The age of European exploration that began in the 15th century still had some way to run before it would transform into the 19th-century age of global trade.

In Doppelmayr's time, some parts of the globe still remained uncertain and ill-defined. Although there is no resort to *Terra incognito*, Antarctica is absent, Australasia and north-western America fade into hopeful outlines, and eastern Asia, while somewhat recognizable, is severely truncated. The areas that are best defined owe their accuracy, in large part, to three astronomical methods that Doppelmayr illustrates in the margins of the plate: the timings of eclipses among the moons of Jupiter and, more widely, the measurement of solar and lunar eclipses.

The major problem facing navigators and map makers of the 18th century and earlier was one of longitude: how locations were positioned relative to each other in an east–west direction. Latitude north or south of the equator, in contrast, was easy to measure because of the way that our view of the celestial sphere changes depending on our location on Earth (see Plate 1). It could be easily and accurately calculated simply by measuring the altitude angle at which the Sun or any known star transited the sky's north–south meridian. However, because all points on the same line of latitude see the same parts of the celestial sphere rising and setting in the course of a day, it is impossible to distinguish easily between them.

The key difference, of course, is time. Observers further to the east see objects rise, cross the meridian and set earlier than their counterparts to the west, and one hour's time difference is equivalent to 15 degrees difference in longitude (since all locations on the globe complete a full 360-degree rotation in twenty-four hours). However, the same time difference also affects the Sun, which was the principal means of measuring the time at any given location. In an age before reliable and portable mechanical timekeepers and rapid communication, how could one make the comparison with time at a baseline location that was necessary to measure longitude?

For centuries, the crude solution for navigation was 'dead reckoning' – paying out a length of line behind a ship in order to estimate its speed at regular intervals, and calculating distance travelled from both the speed and the time at sea. However, more precise methods of navigation had such clear advantages for exploration, trade and basic maritime safety that several large prizes were offered for practical solutions to the 'longitude problem'.

Since ancient times, the most widely used method involved the timing of lunar eclipses. These events were fairly predictable, frequent and visible over a large part of Earth's surface, so the trick was to measure the onset of the eclipse as accurately as possible according to the local (solar) time of observers at two different locations. The eclipse began almost

FIG. 1.
This engraving on the title page of *Introductio Geographica* (1533) by Peter Apian shows astronomers measuring the position of the Moon in relation to the stars, in order to determine their longitude. The astronomers and surveyors all use cross-staffs – a simple instrument for measuring angles.

FIG. 1.

simultaneously for both, which meant that a comparison of the local times of its onset would reveal their longitude separation.

Since lunar eclipses usually occurred at night, a clock set to local time was required, and by the late 17th century the invention of the pendulum clock by Christiaan Huygens (1629–95) had made such devices far more reliable – at least, if set up and properly calibrated on dry land. However, a key limitation lay in the fact that lunar eclipses are somewhat diffuse. The dimming effect of Earth's shadow on the Moon comes on gradually, and while experienced observers were able to time the onset to within ten minutes, this still amounts to more than 2 degrees uncertainty in longitude.

Solar eclipses offered more precision because their onset is unmistakable. They are rarer than lunar ones, however, and required a more flexible approach. The precision of alignment needed for a solar eclipse means that they vary far more in appearance, through varying degrees of partial eclipse to totality in a very limited area. Identifying the precise amount of the Sun eclipsed at a particular location was vital, and led to the system of measuring the solar eclipse in twelve 'digits'. Coupled with more accurate models for eclipse prediction, solar eclipse observations added an occasional second string to the geographer's bow.

German mathematician Johann Werner (1468–1522) proposed an ingenious alternative idea in 1514. Unfortunately, it was somewhat ahead of its time. Werner's 'lunar distance method' measured the time by the Moon's position among the fixed stars, which shifts from hour to hour. However, the challenges of accurate observation, together with the Moon's complex motion through space (see Plate 12) prevented it from being put into operation for a long time.

Along the top and bottom of the plate, Doppelmayr provides a gazetteer with the longitude and latitude of many locations. These are carefully attributed to specific observers, with a crescent moon or the traditional solar symbol (☉) indicating the type of eclipse used in determining the location. The astronomical symbol representing Jupiter (♃) also appears repeatedly, indicating longitudes established by using the eclipses of Jupiter's moons as a universal timekeeper. From the 1670s onwards, French cartographer Jean Picard (1620–82) made great use of this method, relying on Giovanni Domenico Cassini's (1625–1712) accurate tables of predicted eclipses to produce a much-improved map of France, published in 1693. Doppelmayr was, therefore, able to map western Europe with some precision, even if the accuracy of more remote regions is sometimes found wanting.

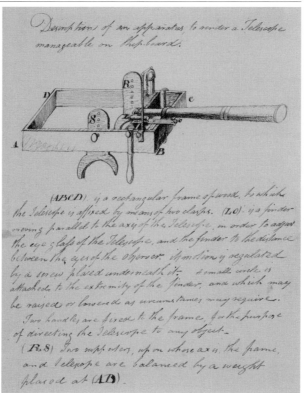

FIG. 2.

FIG. 2.
This ingenious early 19th-century design depicts a later improvement on Galileo's celatone – a device to steady a shipboard telescope in order to observe the moons of Jupiter.

There remained, of course, the problem of navigation at sea, often said to have been troublesome because of the difficulty in taking accurate observations from the pitching deck of a ship. As early as 1616, Galileo Galilei (1564–1642) came up with a possible solution in the form of the celatone, an ingenious telescope attached to a helmet that allowed an observer to keep a stable view on the water. A later and more elaborate version even sat the navigator in a self-levelling chair. However, the ultimate problem in using any of the eclipse methods at sea was simply that the events involved were too infrequent to usefully track a vessel in motion.

In the end, of course, it was technology rather than astronomy that would provide a solution to the longitude problem, in the form of the marine chronometer. These accurate timepieces, which could be set at the home port and run reliably during long sea voyages to allow simple comparisons with local solar time, offered a far simpler means of determining longitude than any of the eclipse methods. Early chronometers, however, remained prohibitively expensive. Somewhat ironically, their acceptance by seafaring authorities in Europe coincided with the final fine-tuning to models of the Moon's motion, which made the (far cheaper) lunar distance method a practical alternative. The two solutions – mechanical and astronomical – would, therefore, be used in parallel until well into the 19th century.

BASIS GEOGRAPHIÆ RE[...]

in qua situs locorum insigniorum geographici ea exactitudine, qua celeberrimi Astronomi[...]

pro certiori Geographia[...]

A. IOHANNE GABRIELE DOPPELMAIER[...]

Cum Privilegio [...]

Nomina Locorum.	Observatores Viri celeberrimi D:D	Longitudo observ.	ex eclips.	Latitudo observ.	Nomina Locorum.	Observatores viri celeberrimi D:D	Longitudo observ.	ex Eclip.	Latitudo observata.	Nomina Locorum	Observ. viri celeb.
		HISPANIÆ.					**GALLIÆ.**				
Madritum	P.P. Cassini e Petro	18. 24. 45	☉☽	40. 26. 0	Lutetia Parisiorum	Picart, Cassini, de la Hire	22. 36. 0	☽☉☽	48. 50. 10.	Ambianum, Amiens	
Hispalis		14. 0. 10	☽	37. 36. 0.	Lugdunum, Lyon	P.P.Bonet e Bosta	24. 54. 45	☉☽	45. 45. 20.	Pupinianum, Perpignan	Cassini[...]
Corduba	Pet. Ant. de Blancas	16. 45. 45	☽	37. 56. 0.	Massilia, Marseille	P. Laval e Chazelles	25. 37. 0	4☽☉	43. 14.33.	Aurelia, Orleans	
Valentia		19. 24. 45	☽	39. 30. 0.	Rupella, Rochelle	Des Hayes	19. 8. 45	☉☽	46. 10. 15.	Bajoma, Bayonne	Picart e[...]
Majorca		32. 40. 10	☽	39. 35. 0.	Arelatum, Arles	Dampard	24. 51. 0	☽	43. 34. 12	Narbo, Narbonne	
Cadix		14. 3. 0.	☽	36. 33. 34	Aqua Sextia, Aix	Gaultier	25. 42. 0	4☉	43. 31. 20.	Pictavum, Poictiers	P.P.Rich[...]
		PORTUGALLIÆ.			Mons Pessul. Mongellier	Plantade e Clapies	24. 2. 30	☉4☽	43. 36. 30	Biturix, Bourges	
Ulyssipo, Lisbona	Couplet	11. 25. 0	4☽	38. 45. 25.	Tolo Martius, Toulon	De la Hire	26. 5. 30	4	43. 6. 40	Diepa, Dieppe	Varin e[...]
					Divionum, Tours	Nonnet	30. 54. 0	☉☽	47. 23. 40	Divio, Dyon	
					Antipolis, Antibe	De la Hire	27. 17. 45	4	43. 34. 12.	Flexia, La Fleche	Picart[...]
					Brestia, Brest	Picart e De la Hire	15. 56. 0	4	48. 23. 0.	Lingones, Langres	Tancard[...]
					Avenio, Avignon	P. Bonfa P.Feuillée	25. 2. 0	☉☽	43. 57. 0.	Rothomagus, Rouen	Des Hay[...]
					Caletum, Calais	De la Hire, Picart	22. 52. 30	4	50. 58. 50.	Namete, Nantes	Picart e[...]

Polus Arcticus

Circulus Arcticus

Tropicus Cancri

Ecliptica

Linea Æquinoctialis sive Æquator

Tropicus Capricorni

Circulus Antarcticus

Polus Antarcticus

Nomina Locorum.	Observatores Viri celeberrimi D:D	Longitudo observ.	ex Eclip.	Latitudo observa.	Nomina Locorum.	Observatores Viri celeberrimi D:D	Longitudo observ.	ex Eclip.	Latitudo observa.	Nomina Locorum.	Observ. Viri celeb.
		GERMANIÆ.					**ANGLIÆ.**				**ITALIÆ.**
Gripswaldia	Pylius	28. 40. 30	☉	54. 14. 0	Londinum	Wright Flamsteed	20. 4. 45	☽4☉	51. 31. 0	Roma	Bianchin[...]
Mons Regius	Linemannus	41. 47. 30	☽	54. 43. 0	Grenovicum	Flamsteed	20. 12. 30	☽☉4	51. 28. 30	Mutina, Modena	P.P.Ricc[...]
Erfordia	Kirchius	51. 17. 15	in ☽	51. 8. 0	Oxonium	Halley	18. 45. 0	4	51. 44. 30	Florentia	
Lincium	Keplerus	55. 10. 0	☽	48. 16. 0			**SCOTIÆ.**			Bononia Bologna	Manfredi e[...]
Tubinga	Maestlinus et Schickard	29. 25. 0	☽	48. 34. 0	Edenburgum		19. 4. 45		55. 58. 0	Ferrara	
Wittberga		33. 10. 0	☽	51. 48. 30			**HIBERNIÆ.**			Genua	Mary de Sa[...]
Rostochum	Brucæus	32. 54. 30	☽	54. 10. 0	Dublinum	Molineux	13. 9. 45	☽	53. 11. 0	Malta Ins.	Chazelles [...]
		BELGII					**DANIÆ.**				**TURCIÆ.**
Amstelodamum	Hortensius	25. 9. 0	☽	52. 23. 0	Hafnia	Picardus Roemer	30. 55. 15	☽4	55. 40. 45	Constantinopolis	Chazelles[...]
Rotterodamum	Castinus junior	25. 7. 0	☽4 io	51. 59. 45	Uranieburgum	Tycho Picardus	28. 2. 30	☽4	55. 54. 15	Smirna	P.Feuillée
Lugdunum Batavorum	D. van Bach	24. 47. 15	4☽	52. 12. 0	Landia		33. 32. 15		55. 42. 10	Aleppo	
Antverpia	Vladislaus	24. 42. 0	☽	51. 2. 0			**POLONIÆ.**			Thessalonica	P.Feuillée
Pinbroca	de la Hire Chazelles	24. 52. 0	4☽	51. 2. 30	Warsovia	Oucerius	42. 42. 0	☽4	52. 14. 0	Trapezus	P.Bot e[...]
Bruxella	Langren	24. 33. 0	☽	50. 50. 0	Dantiscum	Hevelius	38. 42. 15		54. 22. 30	Erzerum	Idem.
Leodium		25. 43. 0		50. 36. 0			**MOSCOVIÆ.**			Melo Ins.	P.Feuillée
		GERMANIÆ			Moscua	Timmermannus	60. 30. 0	☽	55. 36. 10	Canea in Candia	Idem.
Lindavia	Sarpyius	29. 54. 0	☉	47. 48. 0	Petropolis	de l'Isle	50. 30. 0		60. 0. 0	Candia	Idem.
Francofurt ad Mœn		28. 45. 15	☉	50. 4. 0							**TARTARIÆ MAG**
										Siringua	P. Gouye
										Rotscham	Idem.

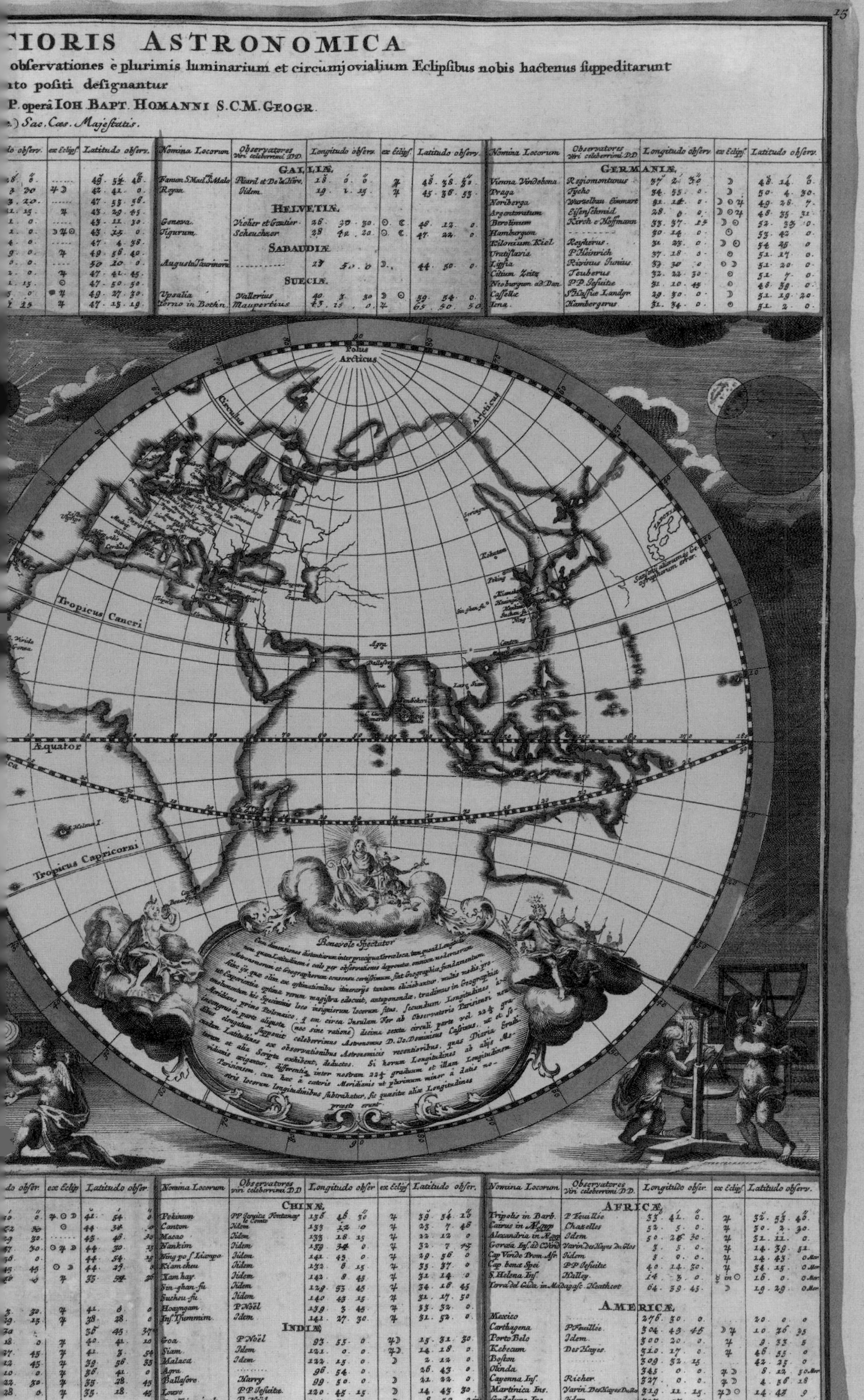
TIORIS ASTRONOMICA
observationes è plurimis luminarium et circumjovialium Eclipsibus nobis hactenus suppeditarunt
nto positi designantur
P. operâ IOH. BAPT. HOMANNI S.C.M. GEOGR.
Sac. Cæs. Majestatis.
Polus Arcticus
Circulus Arcticus
Tropicus Cancri
Æquator
Tropicus Capricorni
Benevole Spectator
GALLIÆ.
HELVETIÆ.
SABAUDIÆ.
SUECIÆ.
GERMANIÆ.
CHINÆ.
INDIÆ.
AFRICÆ.
AMERICÆ.

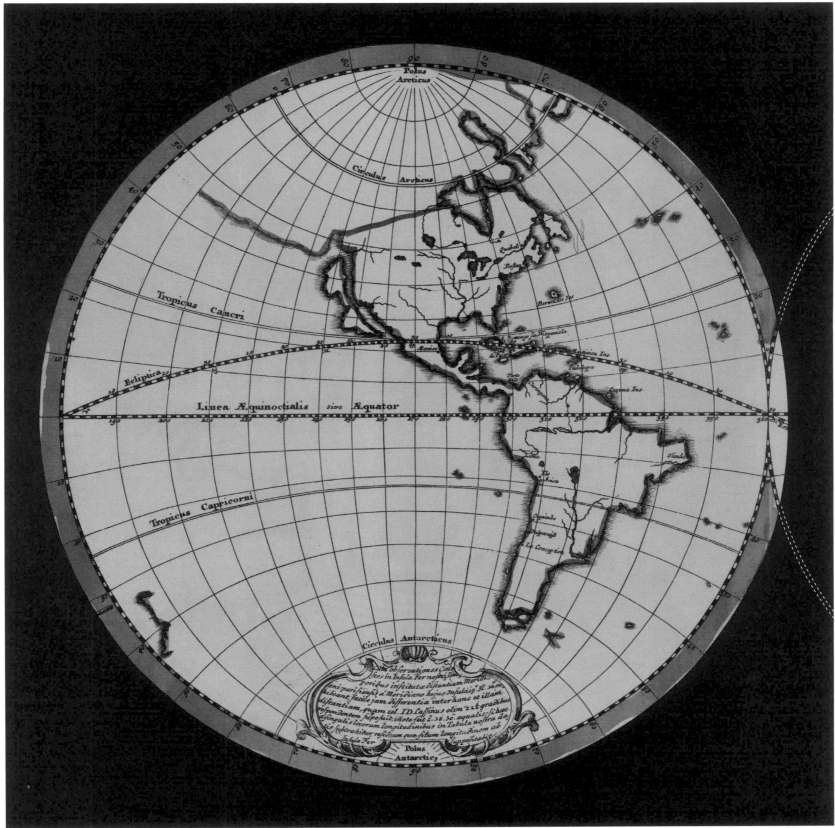

THE NEW WORLD.

The New World hemisphere of Doppelmayr's map includes only a handful of places whose locations had by this time been determined astronomically. Although the shapes of North and South America are recognizable at low latitudes around the equator, elsewhere (particularly in high northern latitudes), the outlines become more speculative. For example, California is depicted as an island.

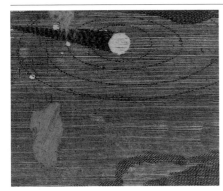

JOVIAN ECLIPSES.

The moons of Jupiter pass in and out of the shadow cast by the giant planet with regular precision.

SOLAR ECLIPSES.

The precise timing and extent of solar eclipses depends on an observer's location on the surface of Earth.

LUNAR ECLIPSES.

While eclipses of the Moon are less dependent on an observer's location, they still offer an agreed standard of time.

THE OLD WORLD.

Doppelmayr's depictions of Europe, Africa and Asia show a larger number of locations pinned down by astronomical observations. However, the outlines of the continents get less accurate with greater distance from the well-charted lands of western Europe. In the far east, the tip of Alaska intrudes from the other hemisphere, and there are two representations of Japan; Australia and New Guinea are linked; and Australia itself is incomplete.

LONGITUDE FROM LUNAR ECLIPSE.

Curious semi-cherubic figures observe the lunar eclipse at top right of Plate 15 through a telescope, using a chart to map its extent and a clock to record local time.

LONGITUDE FROM SOLAR ECLIPSE.

A group of cherubs observe the eclipsed Sun using a projection device, while recording its extent and timing.

LONGITUDE FROM JOVIAN MOONS.

More cherubs observe and record eclipses of the Jovian moons, comparing the local timing of events with ephemeris predictions.

PLATE 16.

THE NORTHERN HEMISPHERE
OF THE HEAVENS.

*Doppelmayr introduces the constellations
and notable stars of the northern sky.*

In the beautifully detailed Plate 16, Doppelmayr depicts the major stars of the northern half of the sky. Just as today, these stars are grouped into constellations, and many of the star patterns shown here are familiar. There is a key difference between today's constellations and those of the past, however: while the astronomers of Doppelmayr's day saw them principally as the stars used to make the 'picture' in the sky, modern astronomers define them as tightly bounded areas of the heavens. This avoids confusion about which constellation newly discovered objects should be attributed to.

People have imagined pictures among the stars since prehistoric times, and the oldest known constellation is probably Taurus, the Bull (depicted at lower right on Doppelmayr's plate). At least one of the famous charging bull paintings at Lascaux in France, dating back around 16,500 years, incorporates dots corresponding to the V-shaped Hyades star cluster that forms the face of the celestial bull and the fish-hook-shaped Pleiades cluster usually interpreted as the hump of its shoulders.

In general, however, the oldest constellations depicted in the *Atlas Coelestis* come from a tradition that can be traced back to ancient Mesopotamia, and from there through neighbouring cultures to the world of the classical Mediterranean. The MUL.APIN – a treatise on Babylonian astronomical knowledge compiled around 1000 BCE but with much earlier origins – lists a series of constellations along the path of the Moon that correspond broadly to the twelve zodiac star patterns and some of their immediate neighbours. Some, such as the Bull of Heaven (Taurus), the Great Twins (Gemini), the Lion (Leo), the Scales (Libra) and the Goat-Fish (Capricorn), have survived unaltered to the present day. Others, including the Crayfish (Cancer), the Loyal Shepherd (Orion) and the Great God Enki are essentially modern star patterns masquerading under different names.

Of the eighty-eight constellations officially recognized today, some forty-eight are considered to be 'classical'. Thanks to their listing in Ptolemy's (*c.* 100–170 CE) *Almagest* (150 CE), they long outlasted the fall of Rome to become the standard constellations of medieval times in both Europe and the Islamic world. Unsurprisingly, given the location of Greek stargazers, the majority lie in the sky's northern hemisphere. Some of them are directly inherited from Mesopotamian patterns, but it is less clear whether others were entirely Greek inventions or imported and repurposed from elsewhere to fit the tales of Greek mythology.

FIG. 1.
Here, zodiac signs are depicted in a 14th-century anthology of Persian poetry. Muhammad ibn Badr al-Din Jajarmi's *Free Man's Companion to the Subtleties of Poems* includes verses covering a wide variety of subjects, such as the astrological influence of the Moon as it passes through the various constellations.

FIG. 1.

Homer's (b. *c.* 750 BCE) *Iliad* and *Odyssey*, thought to have been written down in the 8th century BCE but based on earlier oral traditions, include the earliest references to star patterns such as the Great Bear (Ursa Major) and today's herdsman constellation Boötes, who perpetually chases the bear around the north celestial pole as the sky rotates. However, the earliest systematic list of Greek constellations that has survived comes from around five centuries later. In his *Phenomena* – an epic poem written around 275 BCE that discusses both the weather and celestial objects – the poet Aratus of Soli (*c.* 315/10–*c.* 240 BCE) lists forty-three of the classical constellations later adopted by Ptolemy.

Aratus drew on lost works from the philosopher Eudoxus of Cnidus (*c.* 408–*c.* 355 BCE), written around a century earlier. These described each constellation in terms of where and when it could be seen, what particular significance it had and its mythological origins. The *Phenomena* inspired various other works of 'catasterism' (stories explaining how heroes and other legendary creatures were transformed into constellations). The most notable of these are a lost work by Eratosthenes (*c.* 276–*c.* 194 BCE) from the late 3rd century, a surviving summary of Eratosthenes called the *Epitome Catasterismorum*, and *De Astronomica*, a work of uncertain date originally attributed to Roman historian Gaius Julius Hyginus (*c.* 64 BCE–17 CE).

The culmination of classical astronomy, of course, was the work known today as the *Almagest*, in which Ptolemy not only detailed his own cosmology, but also provided an exhaustive catalogue of stars and constellations visible from the ancient Mediterranean. The name *Almagest* (originally *Almagestum*) is actually a product of the 12th century: the original title was the somewhat more daunting and less poetic *Mathematical Syntaxis*. However, when Ptolemy's book was reintroduced from the Islamic world to western Europe after the better part of a millennium, the new translation came complete with a Latinized version of the work's Arabic name, *al-Majisti* (meaning 'the greatest').

The list of Ptolemy's constellations that survive to the present is usually given as detailed in the table below. However, not all of Ptolemy's constellations have made it through to the present unscathed. Equuleus and Pegasus, for example, were considered a single grouping in the *Almagest*, and a large star pattern in the far southern sky, the ship Argo Navis, is now split into three separate elements. Furthermore, another constellation, Coma Berenices (the hair of Queen Berenice of Egypt, sacrificed to bring her husband safely home from war) is considered classical because it was recognized in the 3rd century BCE, but was omitted from Ptolemy's list.

Doppelmayr captures all of Ptolemy's ancient figures alongside more modern constellations added from the 15th century (some of which have since disappeared once again). The stories behind these more recent patterns are explored in more detail alongside Plate 17, where the *Atlas* turns its focus to the southern sky.

ANDROMEDA *the Chained Woman*	CASSIOPEIA *the Queen*	ERIDANUS *the River*	PERSEUS –
AQUARIUS *the Water Carrier*	CENTAURUS *the Centaur*	GEMINI *the Twins*	PISCES *the Fishes*
AQUILA *the Eagle*	CEPHEUS *the King*	HERCULES –	PISCIS AUSTRINUS *the Southern Fish*
ARA *the Altar*	CETUS *the Whale or Sea Monster*	HYDRA *the Water Snake*	SAGITTA *the Arrow*
ARGO NAVIS *the Ship Argo*	CORONA AUSTRALIS *the Southern Crown*	LEO *the Lion*	SAGITTARIUS *the Archer*
ARIES *the Ram*	CORONA BOREALIS *the Northern Crown*	LEPUS *the Hare*	SCORPIUS *the Scorpion*
AURIGA *the Charioteer*	CORVUS *the Crow*	LIBRA *the Balance*	SERPENS *the Snake*
BOÖTES *the Herdsman*	CRATER *the Cup*	LUPUS *the Wolf*	TAURUS *the Bull*
CANCER *the Crab*	CYGNUS *the Swan*	LYRA *the Lyre*	TRIANGULUM *the Triangle*
CANIS MAJOR *the Great Dog*	DELPHINUS *the Dolphin*	OPHIUCHUS *the Serpent-Bearer*	URSA MAJOR *the Great Bear*
CANIS MINOR *the Lesser Dog*	DRACO *the Dragon*	ORION *the Hunter*	URSA MINOR *the Lesser Bear*
CAPRICORNUS *the Sea-Goat*	EQUULEUS *the Foal*	PEGASUS *the Flying Horse*	VIRGO *the Maiden*

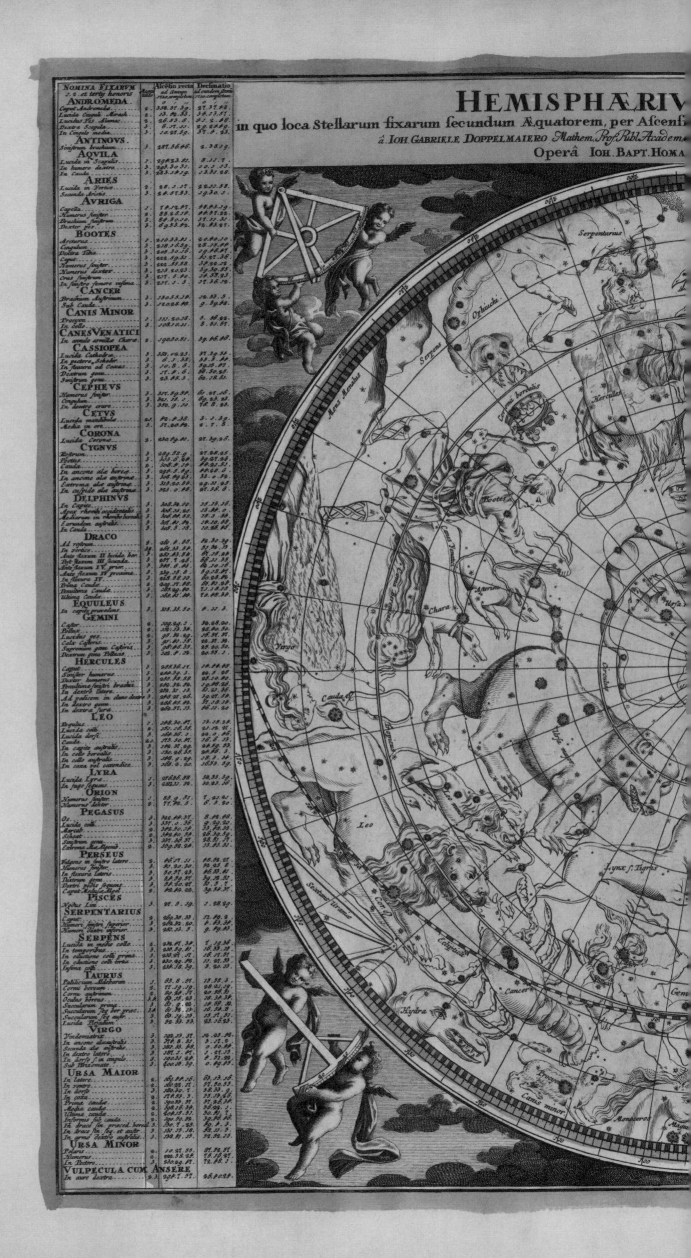

HEMISPHÆRIV

in quo loca Stellarum fixarum secundum Æquatorem, per Ascensi

â IOH. GABRIELE DOPPELMAIERO Mathem. Prof. Publ. Academ

Operâ IOH. BAPT. HOMA

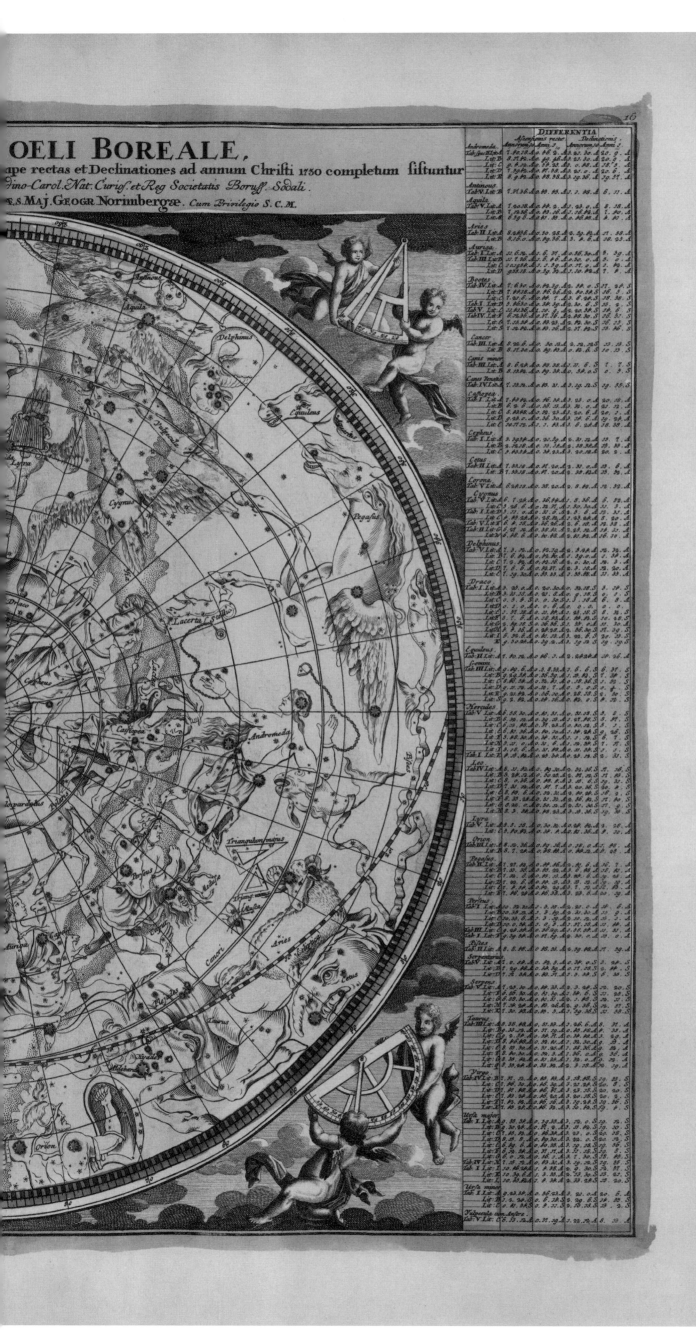

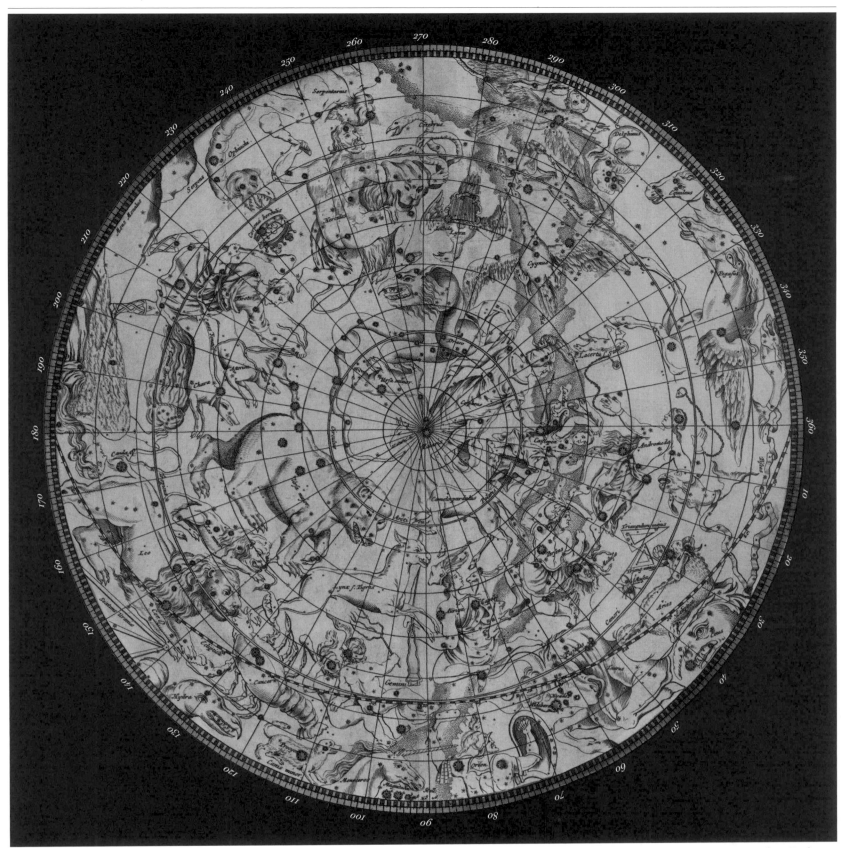

GEOCENTRIC REPRESENTATION OF THE
NORTHERN SKY.

The centrepiece of the plate shows the sky's northern hemisphere as it appears from Earth.
It is centred on the north celestial pole so that the celestial equator runs around the edges of
the map and the northern half of the ecliptic sweeps across the lower half of the map. The chart
uses an azimuthal equidistant projection, in which lines of declination form concentric circles
around the pole, and lines of right ascension radiate from it in straight lines.

QUADRANT.

The corners of the map are decorated
with cherubs holding various instruments
from the Nuremberg observatory: in this
case, a simple quadrant with a sighting bar.
In practice, such instruments included
a plumb line for determining the vertical
direction of the zenith.

SEXTANT.

The astronomical sextant was a more
convenient alternative to the quadrant,
with an arc of 60 degrees. Often mounted
in a frame that allowed its orientation to
be changed, it could be used for determining
angular separations between objects in
any plane.

DIVIDED ARC.

This instrument appears to represent the
'arcus bipartitus'. It has sighting points at
one end and a graduated arc at the other,
along which pins could slide to align
with distant stars and reveal the angular
separation between them.

SEMICIRCULAR ARC.

Although such instruments were less
popular than quadrants and sextants, a
properly graduated arc – encompassing an
entire semicircle – could be used to measure
the separation between objects at angles
of up to 180 degrees.

CONSTELLATION	DOPPELMAYR STARS	MODERN NAMES
ANDROMEDA the Chained Woman	The head of Andromeda	Alpheratz, α Andromedae
	The bright one in the girdle, Mirach	Mirach, β Andromedae
	The bright one in the foot, Alamac	Almach, γ Andromedae
	The right shoulder	δ Andromedae
	In the middle of the girdle	μ Andromedae
ANTINOUS the Altar	The left forearm	δ Aquilae
AQUILA the Eagle	Bright one in the shoulder	Altair, α Aquilae
	In the right wing	Tarazed, γ Aquilae
	In the tail	Okab, ζ Aquilae
ARIES the Ram	The bright one in the corner	Hamal, α Arietis
	Second star of Aries	Sheraton, β Arietis
AURIGA the Charioteer	Capella	Capella, α Aurigae
	Left upper arm	Menkalinan, β Aurigae
	Left forearm	Mahasim, θ Aurigae
	Right foot	Elnath, β Tauri
BOÖTES the Herdsman	Arcturus	Arcturus, α Boötis
	The belt	Izar, ε Boötis
	Right shin	Muphrid, η Boötis
	The head	Nekkar, β Boötis
	Left upper arm	δ Boötis
	Right upper arm	Seginus, γ Boötis
	Left leg	ζ Boötis
	In the left inner thigh	π Boötis
CANCER the Crab	Southern forearm	Acubens, α Cancri
	Below the tail	Tarf, β Cancri
CANIS MINOR the Lesser Dog	Procyon	Procyon, α Canis Minoris
	In the neck	Gomeisa, β Canis Minoris
CANES VENATICI the Hunting Dogs	In Chara's collar	Cor Caroli, α Canum Venaticorum
CASSIOPEIA the Queen	The bright one in the throne	Caph, β Cassiopeiae
	On the chest, Schedir	Schedar, α Cassiopeiae
	At the waist	γ Cassiopeiae
	Right knee	Ruchbah, δ Cassiopeiae
	Left knee	Segin, ε Cassiopeiae
CEPHEUS the King	The left upper arm	Alderamin, α Cephei
	The belt	Alfirk, β Cephei
	In the right leg	Errai, γ Cephei
CETUS the Sea Monster	The bright one in the jaw	Menkar, α Ceti
	In the middle of the mouth	Kaffaljidhma, γ Ceti
CORONA the Northern Crown	The bright one of the crown	Alphecca, α Coronae Borealis
CYGNUS the Swan	The beak	Albireo, β Cygni
	The chest	Sadr, γ Cygni
	The tail	Deneb, α Cygni
	Front of the northern wing	Fawaris, δ Cygni
	Front of the southern wing	Aljanah, ε Cygni
	At the tip of the southern wing	Okab, ζ Cygni
	On the rear of the southern wing	υ Cygni
DELPHINUS the Dolphin	In the head	γ Delphini
	Western corner of the rhombus	Rotanev, β Delphini
	Middle of the north of the rhombus	Sualocin, α Delphini
	Middle of the south of the rhombus	δ Delphini
	In the tail	Aldulfin, ε Delphini
DRACO the Dragon	On the beak	Rastaban, β Draconis
	In the corner	Eltanin, γ Draconis
	Northern bright star at the second coil	Altais, δ Draconis
	After the third coil	Aldhibah, ζ Draconis
	Before the fourth coil, in front	Athebyne, η Draconis
	Before the fourth coil, close	θ Draconis
	In the fourth coil	Edasich, ι Draconis
	First one in the tail	Thuban, α Draconis
	Second-last in the tail	κ Draconis
	Last in the tail	Giausar, λ Draconis
EQUULEUS the Little Horse	The leading one in the head	Kitalpha, α Equulei

CONSTELLATION	DOPPELMAYR STARS	MODERN NAMES
GEMINI the twins	Castor	Castor, α Geminorum
	Pollux	Pollux, β Geminorum
	The bright one in the foot	Alhena, γ Geminorum
	The heel of Castor	Tejat, μ Geminorum
	The brightest in Castor's knee	Mebsuta, ε Geminorum
	The right knee of Pollux	Wasat, δ Geminorums
HERCULES	The head	Rasalgethi, α Herculis
	The left upper arm	Kornephoros, β Herculis
	The right upper arm	Sarin, δ Herculis
	The penultimate star in the left arm	γ Herculis
	On the right side	ε Herculis
	On the right buttock	η Herculis
	On the right knee	θ Herculis
	On the right calf	ι Herculis
LEO the Lion	Regulus	Regulus, α Leonis
	The bright star in the neck	Algieba, γ Leonis
	The bright star on the back	Zosma, δ Leonis
	The tail	Deneobola, β Leonis
	The southern star in the head	ε Leonis
	The northern star in the neck	Adhafera, ζ Leonis
	The southern star in the neck	η Leonis
	On the hipbone	Chertan, θ Leonis
LYRA the Lyre	The bright one in Lyra	Vega, α Lyrae
	The following one in the yoke	Sulafat, γ Lyrae
ORION the Hunter	The left upper arm	Betelgeuse, α Orionis
	The right upper arm	Bellatrix, γ Orionis
PEGASUS the Winged Horse	The mouth	Enif, ε Pegasi
	The bright star in the neck	Homam, ζ Pegasi
	Marcab	Markab, α Pegasi
	Scheat	Scheat, β Pegasi
	The left knee	Matar, η Pegasi
	The tip of the wing, Algenib	Algenib, γ Pegasi
PERSEUS the Hero	The brilliant star on the left side	Mirfak, α Persei
	The left upper arm	γ Persei
	In the bend of the side	δ Persei
	The right knee	ε Persei
	The following one in the right foot	Atik, ζ Persei
	The head of Medusa, Algol	Algol, β Persei
PISCES the Fishes	The knot in the cord	Alrescha, α Piscium
SERPENTARIUS (OPHIUCHUS) the Serpent-Bearer	The head	Rasalhague, α Ophiuchi
	The superior star in the left upper arm	Cebalrai, β Ophiuchi
	The inferior star in the right upper arm	κ Ophiuchi
SERPENS the Snake	The bright star in the middle of the neck	Unukalhai, α Serpentis
	In the temple	γ Serpentis
	In the first extension of the neck	β Serpentis
	In the third extension of the neck	δ Serpentis
	Within the neck	ε Serpentis
TAURUS the Bull	Palilicium, Aldebaran	Aldebaran, α Tauri
	The northern horn	Elnath, β Tauri
	The southern horn	Tianguan, ζ Tauri
	The northern eye	Ain, ε Tauri
	The first in the face	Prima Hyadum, γ Tauri
	The next in the face, to the north, leading	Secunda Hyadum, δ Tauri
	The next in the face, to the south	θ Tauri
	The brightest star of the Pleiades	Alcyone, η Tauri
VIRGO the Virgin	Vindemiatrix	Vindemiatrix, ε Virginis
	Star in the elbow of the southern wing	Zavijava, β Virginis
	Second star in the southern wing	Zaniah, η Virginis
	On the right side	Porrima, γ Virginis
	On the back or girdle	Minelauva, δ Virginis
	Below the aprons	Heze, ζ Virginis
URSA MAJOR the Great Bear	On the side	Dubhe, α Ursae Majoris
	On the belly	Merak, β Ursae Majoris
	On the back	Megrez, δ Ursae Majoris
	On the hip	Phecda, γ Ursae Majoris
	First star in the tail	Alioth, ε Ursae Majoris
	Middle star in the tail	Mizar, ζ Ursae Majoris
	Last star in the tail	Alkaid, η Ursae Majoris
	Formless below the tail	Erroneous duplicate
	In the left paw, leading and northern	Talitha, ι Ursae Majoris
	In the left paw, following and southern	Alkaphrah, κ Ursae Majoris
	Southern star in the right shoulder	θ Ursae Majoris
URSA MINOR the Lesser Bear	Polaris	Polaris, α Ursae Minoris
	On the upper arm	Kochab, β Ursae Minoris
	In the chest	Pherkad, γ Ursae Minoris
VULPECULA CUM ANSERE the Little Fox and Goose	In the right ear	Anser, α Vulpeculi

PLATE 16.

LEIDEN ARATEA (c. 816).

Among the earliest and most beautiful depictions of the constellations
are those from a 9th-century illuminated manuscript now held in the
library of Leiden University, the Netherlands. The *Aratea* is based
on an astronomical treatise by Germanicus Caesar, itself dating to the
early 1st century CE and based on the epic *Phenomena* of Aratus from
the 3rd century BCE. The illuminations are very different from those
of later charts and atlases, concerned more with depicting the various
mythical figures, animals and objects than with matching their

appearance in the sky. Equally, the stars themselves (picked out in gold leaf) bear little relation to the patterns in the sky, and are instead used as aids to decorate and highlight the shapes of the illustrated figures. While the constellations mostly match those of Ptolemy, there are some exceptions, such as the treatment of the Pleiades star cluster as an independent constellation. The manuscript was compiled at the court of Louis the Pious (778–840), king of the Franks and the son of Emperor Charlemagne, but the classical style of the illuminations suggests an original source in late antiquity.

PLATE 17.

THE SOUTHERN HEMISPHERE OF THE HEAVENS.

Doppelmayr describes the constellations and notable stars of the southern sky.

In contrast to the crowded skies of the northern hemisphere, Doppelmayr's introductory map of the southern skies appears quite barren, with large gaps separating some of the star patterns. At the time the map was compiled (at some point between 1716 and 1724), the stars of the southern hemisphere were still quite poorly known, and more than a dozen of the official constellations recognized today had not yet been invented.

To the right of centre, Doppelmayr's chart includes a few members of Ptolemy's canonical list of forty-eight classical constellations that lie surprisingly close to the south celestial pole. These include Argo Navis (the ship sailed by the hero Jason on his quest for the Golden Fleece), Centaurus (the sky's second centaur, alongside Sagittarius, the Archer), Lupus (the Wolf) and Ara (the Altar).

Today, the stars of these constellations are mostly invisible from the Mediterranean, so one might reasonably ask how Ptolemy was aware of them. The explanation is that in classical times these constellations were visible further north, due to an effect known as 'axial precession'. While Earth's axis of rotation points in the same direction in space throughout each year (giving rise to the seasons as described on Plate 6), it slowly 'wobbles' like a spinning top, describing a full circle every 25,800 years. Precession causes the north and south celestial poles to shift slowly around the sky, so that the regions of the celestial sphere visible from a certain location on Earth slowly change over very long periods.

Closer to the pole and to the left of centre, in a region that was hidden from the view of classical astronomers, lie a dozen of the earliest constellations to be added to the sky in post-classical times. In Doppelmayr's interpretation, they are:

APUS, *the Bird of Paradise*
CHAMELEON
DORADO, *the Swordfish*
GRUS, *the Crane*
HYDRUS, *the Little Water Snake*
INDUS, *the Indian*
MUSCA APIS (now MUSCA), *the Fly*
PAVO, *the Peacock*
PHOENIX
PISCIS VOLANS (now VOLANS),
    the Flying Fish
TRIANGULUM AUSTRALE,
    the Southern Triangle
TUCANA, *the Toucan*

These constellations were based on a catalogue of 135 far southern stars compiled by Pieter Dirkszoon Keyser (*c.* 1540–96) and Frederick de Houtman (*c.* 1571–1627) of the Dutch East India Company during an expedition to southeast Asia. They were first depicted on a celestial globe of 1598 by Dutch cartographer Petrus Plancius (1552–1622), alongside another significant addition – the Southern Cross, Crux. This iconic group of stars had been visible in antiquity, but was originally considered part of Centaurus, which lies just to its north.

Plancius was only one of several observers and cartographers to make a habit of adding to the sky. In addition to Crux and the twelve Keyser/de Houtman constellations, he had already inserted Columba (supposedly the Dove sent from Noah's Ark to find land) in the space below Lepus (the Hare) in 1592. He attempted to add eight further constellations in around 1613, but only two of these – Camelopardalis (the Giraffe) in far northern skies and Monoceros (the Unicorn, close to Orion) – have survived to the present day.

The invention of the telescope created an understandable pressure to fill 'gaps' in the sky. The new instrument frequently showed these

FIG. 1.

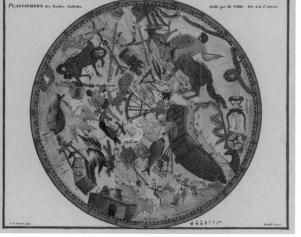

FIG. 2.

FIG. 1.
A 1660 chart of
southern skies by
Andreas Cellarius
incorporates the
constellations of
far southern skies
introduced by
Plancius in the
late 16th century.

FIG. 2.
Jean-Nicolas Fortin's
(1750–1831) *Atlas
Céleste de Flamstéed*
(1776) was among the
first to depict Lacaille's
new additions to the
southern skies.

regions to be anything but empty, and attaching newly catalogued stars to a particular constellation seemed not only scientifically necessary but also useful on a practical level. In the 1680s, the great observer Johannes Hevelius (1611–87) of Danzig (now Gdańsk, Poland) invented ten new constellations in mostly northern skies, seven of which are still used today:

CANES VENATICI, *the Hunting Dogs*
LACERTA, *the Lizard*
LEO MINOR, *the Lesser Lion*
LYNX, *the Lynx*
SCUTUM, *the Shield of his sponsor,*
    King John III Sobieskii
SEXTANS, *the Sextant*
VULPECULA ET ANSER (now VULPECULA),
    the Fox and Goose

While many other astronomers invented constellations that are now forgotten, Hevelius's abandoned trio are worth mentioning because Doppelmayr, drawing heavily on the Polish astronomer's work, includes them here. First, Cerberus represented a serpent being wrestled by the hero Hercules. Perhaps unsurprisingly, it was often depicted with three heads like its namesake, the hound that guarded the entrance to the underworld. Second, Mons Maenalus was a mountain on which the herdsman Boötes stood. Third, Triangulum Minus was simply a smaller triangle beneath the better known Triangulum. Close by in the northern sky, Doppelmayr incorporates a second fly: the now-forgotten Musca Borealis (another Plancius invention). A further lost constellation with older and more obscure origins was Antinous (historically, the lover of the Roman Emperor Hadrian, but here depicted as a sort of Cupid), which lies near Aquila (the Eagle) at the top of Plate 16.

The last major wave of additional constellations was added two decades after the *Atlas Coelestis* was published, courtesy of French astronomer Nicolas Louis de Lacaille (1713–62). Between 1750 and 1754, Lacaille conducted the first detailed survey of southern skies from the Cape of Good Hope, cataloguing almost 10,000 stars. To fill the gaps between existing constellations he added a further fourteen. They are mostly small and somewhat obscure star patterns, with names often inspired by contemporary arts and sciences:

ANTLIA, *the Air Pump*
CAELUM, *the Chisel*
CIRCINUS, *the Drawing Compasses*
FORNAX, *the Furnace*
HOROLOGIUM, *the Clock*
MENSA, *Table Mountain*
MICROSCOPIUM, *the Microscope*
NORMA, *the Set Square*
OCTANS, *the Octant* (encompassing
    the south celestial pole)
PICTOR, *the Painter's Easel*
PYXIS, *the Mariner's Compass*
RETICULUM, *the Crosshairs*
SCULPTOR, *the Sculptor's Workshop*
TELESCOPIUM, *the Telescope*

In addition, Lacaille considered Argo Navis too cumbersome for practical use, and so split it into three separate constellations that survive today – Carina (the Keel), Puppis (the Poop Deck) and Vela (the Sails). The atlas containing these changed and novel constellations was finally published in 1763, a year after Lacaille's death. While further new star groupings were occasionally suggested until well into the 19th century, Lacaille's were the last to be widely accepted, and to be included in the official list of eighty-eight constellations adopted by the International Astronomical Union in 1928.

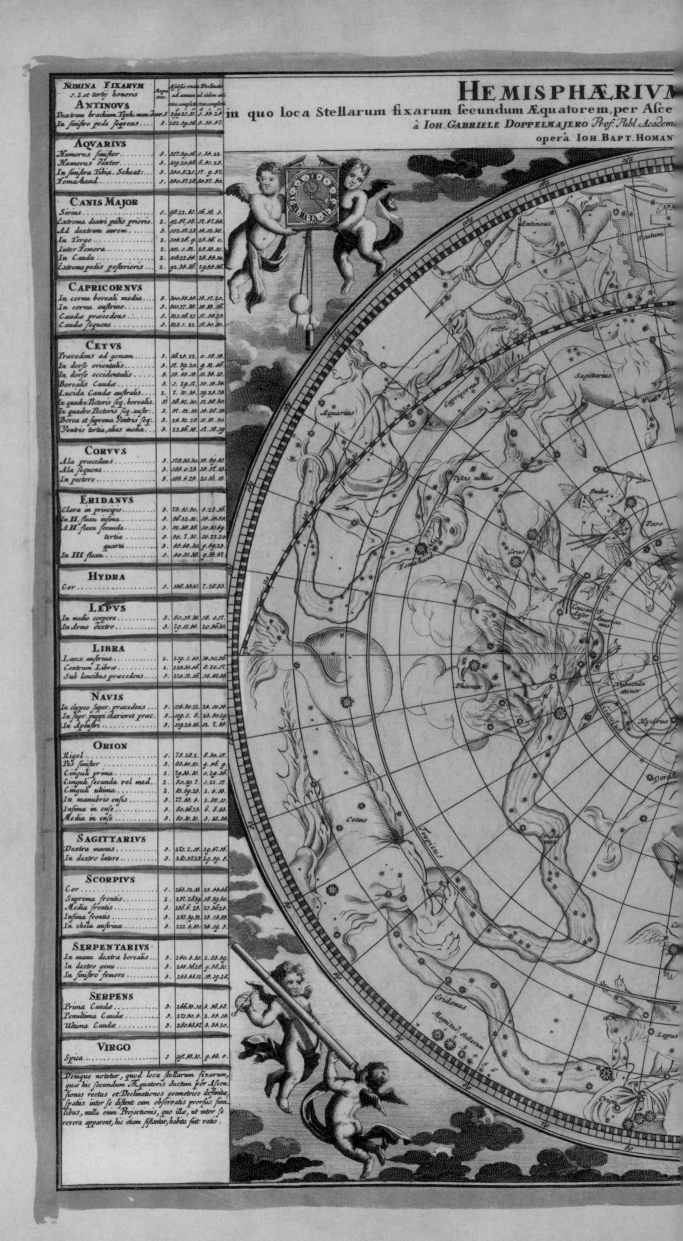

NOMINA FIXARVM 1, 2, et tertij honoris	Magni tudo	Ascensio recta ad annum completum	Declinatio ad idem ann. completum
ANTINOVS			
Dextrum brachium Tych. num Auet.	3	291.11.17	1.11.11
In sinistro pede sequens	3	282.29.16	0.19.57
AQVARIVS			
Humerus sinister	3	327.29.16	1.38.11
Humerus dexter	3	329.20.56	6.11.13
In sinistra Tibia, Scheat	3	340.3.11	17.9.52
Fomahand	1	340.37.18	30.27. &c.
CANIS MAJOR			
Sirius	1	98.11.41	16.18.1
Extrema dextri pedis prioris	2	92.47.18	17.61.56
Ad dextram aurem	3	103.07.13	18.11.30
In Tergo	2	104.16.9	28.46.0
Inter Femora	3	102.1.31	28.33.31
In Cauda	2	108.11.44	18.53.30
Extrema pedis posterioris	2	91.34.36	13.53.36
CAPRICORNVS			
In cornu boreali media	3	300.59.11	13.57.20
In cornu austrino	3	301.11.38	18.43.16
Cauda præcedens	3	312.16.11	17.38.21
Cauda sequens	3	313.7.11	17.40.30
CETVS			
Præcedens ad genam	3	36.14.11	0.18.19
In dorso orientalis	3	13.39.10	9.11.15
In dorso occidentalis	3	13.48.19	11.34.31
Borealis Caudæ	3	1.19.11	10.18.30
Lucida Caudæ australis	2	7.31.39	19.13.38
In quadro Pectoris seq. borealis	3	28.31.30	11.00.30
In quadro Pectoris seq. austr.	3	37.11.38	11.30.31
Borea et suprema Ventris seq.	3	19.31.18	11.31.30
Ventris tertia, alias media	3	11.56.18	11.10.19
CORVVS			
Ala præcedens	3	178.53.30	19.89.41
Ala sequens	3	188.0.18	18.57.33
In pectore	3	189.9.18	11.51.18
ERIDANVS			
Clara in principio	3	73.41.30	8.22.16
In II. flexu infima	3	06.22.41	14.39.34
A.II. flexu secunda	3	53.30.38	10.42.89
tertia	3	50.7.31	10.11.10
quarta	3	45.43.30	9.49.13
In III flexu	3	40.31.38	9.50.91
HYDRA			
Cor	1	138.38.61	7.28.33
LEPVS			
In medio corpore	3	80.25.34	18. 0.17
In Armo dextro	3	79.11.44	20.46.33
LIBRA			
Lanx austrina	2	219.7.41	14.42.46
Centrum Libræ	2	220.30.56	8.10.17
Sub lancibus præcedens	3	220.13.18	18.43.35
NAVIS			
In clypeo super præcedens	3	114.30.13	16.10.34
In super. puppi clarioris præc.	3	119.3. 8.	13.30.29
In Aplustri	3	119.16.49	11. 7. 34
ORION			
Rigel	1	74.18.1	8.30.25
Pes sinister	3	83.40.51	9.46. 9
Cinguli prima	2	79.34.41	0.19.36
Cinguli secunda vel med.	2	80.39.7	1.11.17
Cinguli ultima	2	81.29.18	1. 8.39
In manubrio ensis	3	77.48. 4	3.39.11
Infima in ense	3	80.36.13	6. 5.43
Media in ense	3	80.31.31	3. 33.30
SAGITTARIVS			
Dextra manus	3	271.7.19	19.57.39
In dextro latere	3	181.18.11	19.19. 8
SCORPIVS			
Cor	1	243.11.18	15.46.65
Suprema frontis	2	237.28.39	18.89.40
Media frontis	3	236.6.19	11.36.11
Infima frontis	3	233.49.31	19.14.39
In chela austrina	3	221.9.39	14.19. 3
SERPENTARIVS			
In manu dextra borealis	3	180.0.30	1.11.59
In dextro genu	3	248.36.18	9.38.31
In sinistro femore	3	193.43.11	19.19.18
SERPENS			
Prima Caudæ	3	266.39.10	3.36.13
Penultima Caudæ	3	271.30. 6	1.38.14
Ultima Caudæ	3	280.43.31	1.38.20
VIRGO			
Spica	1	197.40.31	9.43. 0

Denique notetur, quod loca stellarum fixarum, quæ hic secundum Æquatoris ductum per Ascensiones rectas et Declinationes geometrice definita, spatijs inter se distent cum observatis prorsus similibus, nulla enim Projectione, quo illa, ut intuer se revera apparent, hic etiam sistuntur, habita suit ratio.

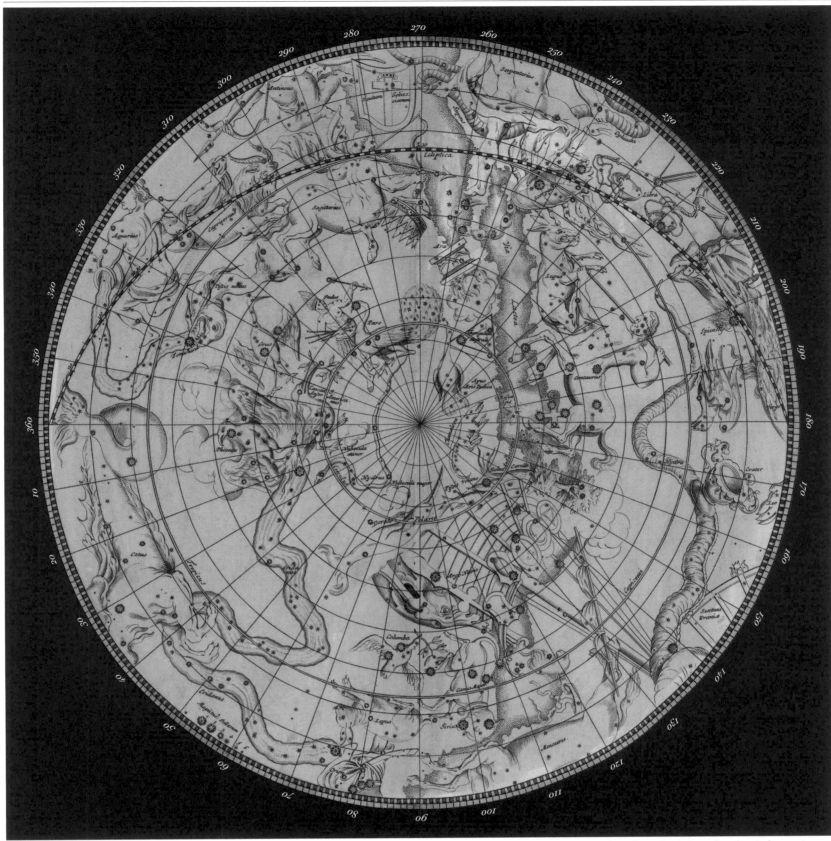

GEOCENTRIC REPRESENTATION OF THE
SOUTHERN SKY.

Matching with its northern counterpart on Plate 16, this dominant chart shows the sky's southern hemisphere, using equatorial coordinates centred around the south celestial pole. The line of the ecliptic through the southern zodiac constellations sweeps across the chart's upper half, but the far southern skies remain remarkably empty, aside from the dozen additions by Plancius in the early 16th century. It is worth noting that Doppelmayr measures right ascension in degrees eastwards of the First Point of Aries, rather than the hours, minutes and seconds commonly used today.

TELESCOPE.

The most important astronomical instrument of the time is represented as a simple tubular refracting telescope, based on one in use at the Nuremberg observatory. Telescopes not only brightened and magnified objects but also permitted far more precise measurements of angle.

TELESCOPIC SEXTANT.

Adding a properly aligned telescope to the alidade (sighting bar) of a sextant allowed angles and positions in the sky to be measured with far greater accuracy, by placing the magnified image in the centre of the field of view.

CONSTELLATION	DOPPELMAYR STARS	MODERN NAMES
ANTINOUS *the Beloved of Hadrian*	*The right lower arm (Tycho's right hand)* *The following star in the left foot*	θ *Aquilae* λ *Aquilae*
AQUARIUS *the Water Carrier*	*The left upper arm* *The right upper arm* *In the left shin, Scheat* *Fomalhault*	*Sadalmelik,* α *Aquarii* *Sadalsuud,* β *Aquarii* *Skat,* δ *Aquarii* *Fomalhaut,* α *Piscis Austrinis*
CANIS MAJOR *the Greater Dog*	*Sirius* *The tip of the right forepaw* *At the right ear* *On the rear* *Between the thighs* *On the tail* *The tip of the hind foot*	*Sirius,* α *Canis Majoris* *Mirzam,* β *Canis Majoris* *Muliphein,* γ *Canis Minoris* *Wezen,* δ *Canis Majoris* *Adhara,* ε *Canis Majoris* *Aludra,* η *Canis Majoris* *Furud,* ζ *Canis Majoris*
CAPRICORNUS *the Sea-Goat*	*In the middle of the northern horn* *In the southern horn* *The leading star in the tail* *The following star in the tail*	*Algedi,* α *Capricorni* *Dabih,* β *Capricorni* *Nashira,* γ *Capricorni* *Deneb Algedi,* δ *Capricorni*
CETUS *the Sea Monster*	*On the cheek* *Eastern star on the back* *Western star on the back* *The northern star in the tail* *The bright southern star in the tail* *The following northern star on the chest* *The following southern star on the chest* *The northern bright lead star in the belly* *The third star in the belly, in the middle*	*Menkar,* α *Ceti* θ *Ceti* η *Ceti* ι *Ceti* *Diphda,* β *Ceti* ε *Ceti* π *Ceti* *Baten Kaitos,* ζ *Ceti* τ *Ceti*
CORVUS *the Crow*	*The leading wing* *The following wing* *On the chest*	*Gienah,* γ *Corvi* *Algorab,* δ *Corvi* *Kraz,* β *Corvis*
ERIDANUS *the River*	*The bright star at the beginning* *Inside the second bend* *The second star in the second bend* *The third* *The fourth* *In the third bend*	*Cursa,* β *Eridani* *Zaurak,* γ *Eridani* δ *Eridani* *Ran,* ε *Eridani* *Zibal,* ζ *Eridani* *Azha,* η *Eridani*
HYDRA *the Water Snake*	*The heart*	*Alphard,* α *Hydrae*

CONSTELLATION	DOPPELMAYR STARS	MODERN NAMES
LEPUS *the Hare*	*In the middle of the body* *On the right shoulder*	*Arneb,* α *Leporum* *Nihal,* β *Leporum*
LIBRA *the Scales*	*The southern dish* *The pivot of the scales* *Beneath the leading dish*	*Zubenelgenubi,* α *Librae* *Zubenelschamali,* β *Librae* ι *Librae*
NAVIS *the Ship Argo*	*In the shield, above and leading* *Brighter leading star above the poop* *In the stern-post*	*Naos,* ζ *Puppis* *Canopus,* α *Carinae* *Miaplacidus,* β *Carinae*
ORION *the Hunter*	*Rigel* *The left foot* *The first star in the belt* *The second star in the belt, in the middle* *The last star in the belt* *The sword handle* *The innermost part of the sword* *In the middle of the sword*	*Rigel,* β *Orionis* *Saiph,* κ *Orionis* *Mintaka,* δ *Orionis* *Alnilam,* ε *Orionis* *Alnitak,* ζ *Orionis* σ *Orionis* *Hatysa,* ι *Orionis* θ *Orionis*
SAGITTARIUS *the Archer*	*The right hand* *On the right side*	*Kaus Media,* δ *Sagittarii* *Ascella,* ζ *Sagittarii*
SCORPIUS *the Scorpion*	*The heart* *Suprema frontis* *Media frontis* *Infima frontis* *In the southern claw*	*Antares,* α *Scorpii* *Acrab,* β *Scorpii* *Dschubba,* δ *Scorpii* *Fang,* π *Scorpii* *Brachium,* σ *Librae*
SERPENTARIUS (OPHIUCHUS) *the Serpent-Bearer*	*The northern star in the right hand* *In the right knee* *In the left thigh*	*Yed Prior,* δ *Ophiuchi* ζ *Ophiuchi* *Sabik,* η *Ophiuchi*
SERPENS *the Snake*	*The first star in the tail* *The second last in the tail* *The last in the tail*	ζ *Serpentis* η *Serpentis* *Alya,* θ *Serpentis*
VIRGO *the Virgin*	*Spica*	α *Virginis*

NOTES ON DOPPELMAYR'S TABLES DISPLAYED ON PLATES 16 AND 17 (pp. 140–41, pp. 148–49).

On Plates 16 and 17, Doppelmayr lists stars of 3rd magnitude or above within each constellation, describing them by position or popular name and giving their right ascension and declination coordinates for 1730. Doppelmayr's listing of bright stars in the southern sky is somewhat idiosyncratic. While ignoring the relatively new and poorly mapped constellations of the far south might seem justifiable, the omission of Ptolemaic constellations, such as Centaurus, Lupus and Crux, is more puzzling. Note, however, that these stars are included in the more exhaustive listing across Plates 20 to 25.

NOTES ON THE EQUIVALENT TABLES DISPLAYED ON PAGE 143 AND ABOVE.

These two tables list English translations of Doppelmayr's star names or descriptions, but do not include Doppelmayr's 1730 ascension and declination coordinates for these stars. The 'Modern Names' column provides the present-day names of Doppelmayr's stars as recognized by the International Astronomical Union, as well as their designations in the 'Bayer' (Greek letter) system that became the de facto standard in more recent times.

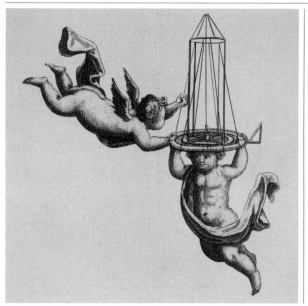

AZIMUTHAL RINGS.

This instrument uses a horizontal sight rotating on the radius of a circle to precisely measure the azimuth (horizontal angle 'around' the sky relative to astronomical north) of celestial objects.

PENDULUM CLOCK.

The corners of this plate are decorated with the key instruments of Enlightenment astronomy. Among the most fundamental was the pendulum clock. The precise timing of transits across the meridian was key to accurately measuring the right ascension of celestial objects.

PLATE 17.

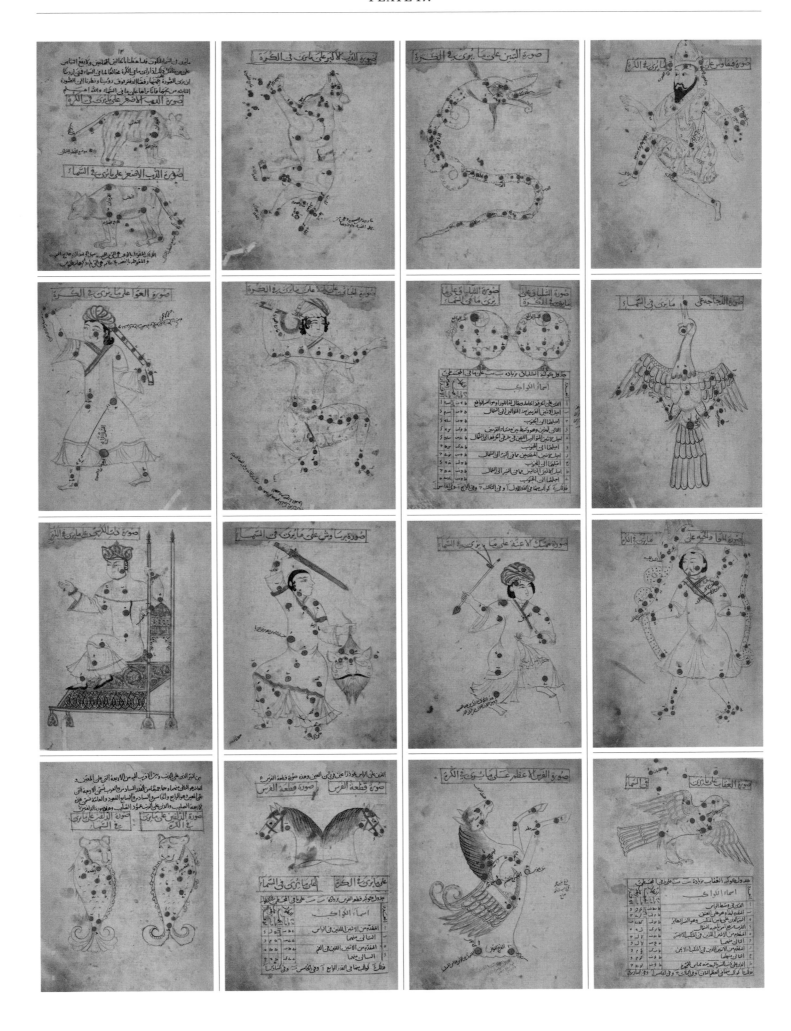

BOOK OF THE FIXED STARS (15TH CENTURY).

Multiple copies of Persian astronomer Abd al-Rahman al-Sufi's *Book of the Fixed Stars* (completed in *c.* 964 CE) exist from the 11th century, although the representation shown here dates from the 15th century. Al-Sufi's book features a distinct innovation: while writers such as Ptolemy frequently referred to unnamed stars by their anatomical positions within the constellation figures, known visual depictions of the sky (such as the 2nd-century celestial sphere on the Roman statue known as the *Farnese Atlas*) mostly disregarded the stars, or (as in

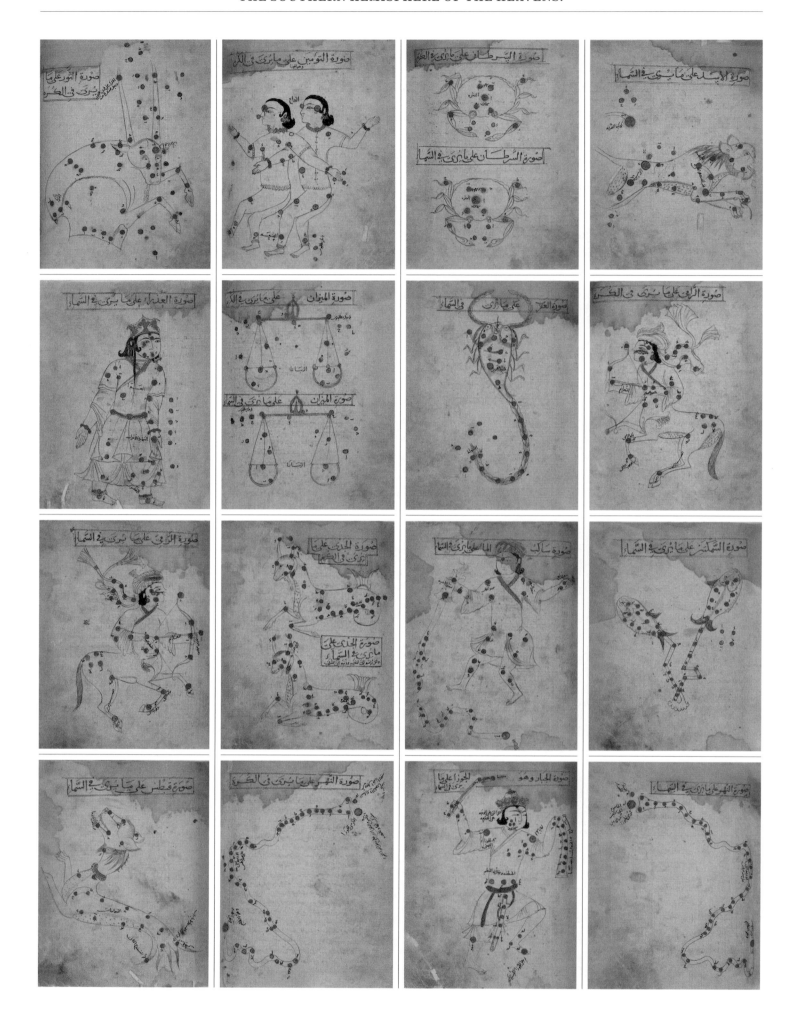

the Leiden *Aratea*) completely disregarded the supposed shape of the
figures. Al-Sufi and his copiers, in contrast, depict constellation figures
overlaid with the more-or-less accurate positions of individual stars.
The constellations are mostly the same as those of Ptolemy (with a few
exceptions), although the figures are, unsurprisingly, depicted in the
Islamic style of the time. Two views of each constellation are provided:
an external one as seen on the outer surface of the celestial sphere,
and a geocentric one as it appears in Earth's skies.

PLATE 18.

THE NORTHERN HEMISPHERE OF THE HEAVENS, DETERMINED IN RELATION TO THE ECLIPTIC.

(HEMISPHÆRIUM COELI BOREALE)

Doppelmayr shows the sky in a projection
based on the Sun's motion through the skies.

At a glance, Plate 18 appears superficially similar to Plate 16's projection of the northern sky – familiar constellations such as the Great Bear, Ursa Major, and the Flying Horse, Pegasus, are all present and correct. A closer look, however, reveals that something different is going on. The twelve constellations of the zodiac form a boundary around the map; the pole star Polaris (on the tail of Ursa Minor) is located some way above the centre. Then we come to the jarring realization that the constellations themselves, and the way they are arranged in relation to each other, are a mirror-image rendering of the way we usually see them.

This chart, and the accompanying Plate 19, are projections of the sky's two ecliptic hemispheres, the stars that lie to the north of the Sun's annual apparent path around the sky, and those that lie to its south. In providing these maps, Doppelmayr was not only following some of the oldest traditions in stellar cartography, but also providing a useful alternative way of looking at the sky, in which the Sun and zodiac took primacy over the celestial poles and equator.

Most modern charts of the sky are projected in the equatorial coordinate system. Objects are located in terms of their declination (how far north or south of the celestial equator they lie) and their right ascension (an equivalent of longitude, measured in terms of how long they take to follow a reference point, known as the First Point of Aries, across an observer's north–south meridian line). This system is usually attributed to the Greek astronomers of the early centuries BCE, but its precise origins are still a matter for debate. Its current dominance, however, is a relatively recent phenomenon, arising in part from its usefulness in conjunction with 'equatorial' telescope mounts that became popular in the 19th century. While previous telescope mounts had to be continuously adjusted in both up–down and left–right directions to keep objects centred in the eyepiece, the equatorial mount tilts the tube's planes of movement so that they correspond to declination and right ascension. Finding an object, therefore, becomes a simple matter of adjusting the settings in these two planes, taking into account the precise local time of observation. What is more, once an object has been found, only the right ascension plane needs to be adjusted to keep up with its motion across the sky (a process that can even be automated using clockwork or electric motors).

One notable disadvantage of equatorial coordinates, however, is that they keep changing. As the phenomenon of precession causes Earth's axis to slowly change direction, the north and south celestial poles, the celestial equator between them, and all the lines of declination and right ascension slowly shift in relation to the stars themselves. To be accurate, then, equatorial coordinates must be calculated for the specific 'epoch' when observations are being made. In practice, this means that modern star atlases are redrawn every fifty years.

In contrast, ecliptic coordinates provide a far more stable alternative. Using the plane of the Sun's apparent motion around the sky as the basis for the system, it is simple to define two 'poles of the ecliptic' at 90-degree angles to it. From these, lines of ecliptic latitude and longitude can be imagined on the sky. Ecliptic latitude is simply an object's angular position north or south of the ecliptic, while ecliptic longitude is measured in degrees eastward from a fixed reference point, the same First Point of Aries used in the equatorial system. The First Point is defined as the location of the northern vernal equinox, where the ecliptic crosses the celestial equator as the Sun enters the sky's northern hemisphere on the first day of northern spring.

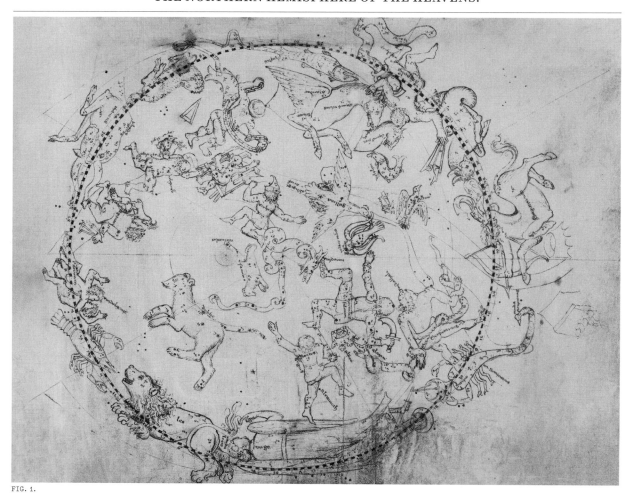

FIG. 1.

FIG. 1.
The earliest surviving
European star charts
are held at the National
Library of Austria in
Vienna. Dating to 1440,
the anonymous maps
show the northern and
southern hemisphere
constellations in
relation to the ecliptic.
The zodiac patterns
form a ring around
the perimeter of the
northern map. Stars
are marked in more
or less their modern
positions within the
constellations, but the
entire sky is projected
from an external
viewpoint, as if looking
at the outer surface
of a celestial sphere.

This system is not immune to the effects of precession, but it does minimize them. An unmoving star's ecliptic latitude remains effectively constant, while its ecliptic longitude increases by about 1.4 degrees each century, carrying it through 360 degrees in the full 25,800-year precession cycle.

Centred as it is on the de facto plane of the solar system, the ecliptic system was more useful than other alternatives for stargazers who were mostly interested in the changing positions of the Sun and planets. Most planetary orbits stay roughly in line with the ecliptic, and plotting their paths in these coordinates produces graphs with obvious repeating oscillations. The system has recently been found in ancient Babylonian tables of planetary conjunctions, and it seems to have been the most widely used across the classical world. Ptolemy (*c.* 100–170 CE) used ecliptic coordinates for the star catalogue in his *Almagest* (150 CE), possibly borrowing measurements from the work of Hipparchus of Nicaea (*c.* 190–*c.* 120 BCE) a few centuries before and attempting to update them to his own epoch. They remained in use throughout the medieval European and Islamic worlds.

Doppelmayr's chart shows a projection of the celestial sphere as viewed from above the north pole of the ecliptic, looking down. Hence, the constellations appear flipped compared to our view on Earth looking up. He traces three lines of equatorial declination corresponding to the celestial equator, the Tropic of Cancer and the Arctic Circle, and divides the entire chart into twelve equal segments of 30 degrees each. These divisions are an approximate representation of the Sun's motion along the ecliptic in the course of a month, corresponding to the twelve astrological signs of the zodiac. The sequence of houses begins at the right-hand edge, where the celestial equator meets the ecliptic, and each 30-degree segment is marked with a zodiac sign, beginning at Aries (♈) and progressing counter-clockwise through Taurus (♉), Gemini (♊), Cancer (♋), Leo (♌), Virgo (♍), Libra (♎), Scorpio (♏), Sagittarius (♐), Capricorn (♑) and Aquarius (♒) to complete the circle at Pisces (♓). However, a comparison with the map figures running along the edge will show the effect of precession over the 2,000 years that separate this map from the system's foundations. The actual constellations (never a perfect match for the zodiac signs due to their varying size) have all drifted notably out of alignment, and the First Point of Aries actually occurs in the neighbouring constellation of Pisces.

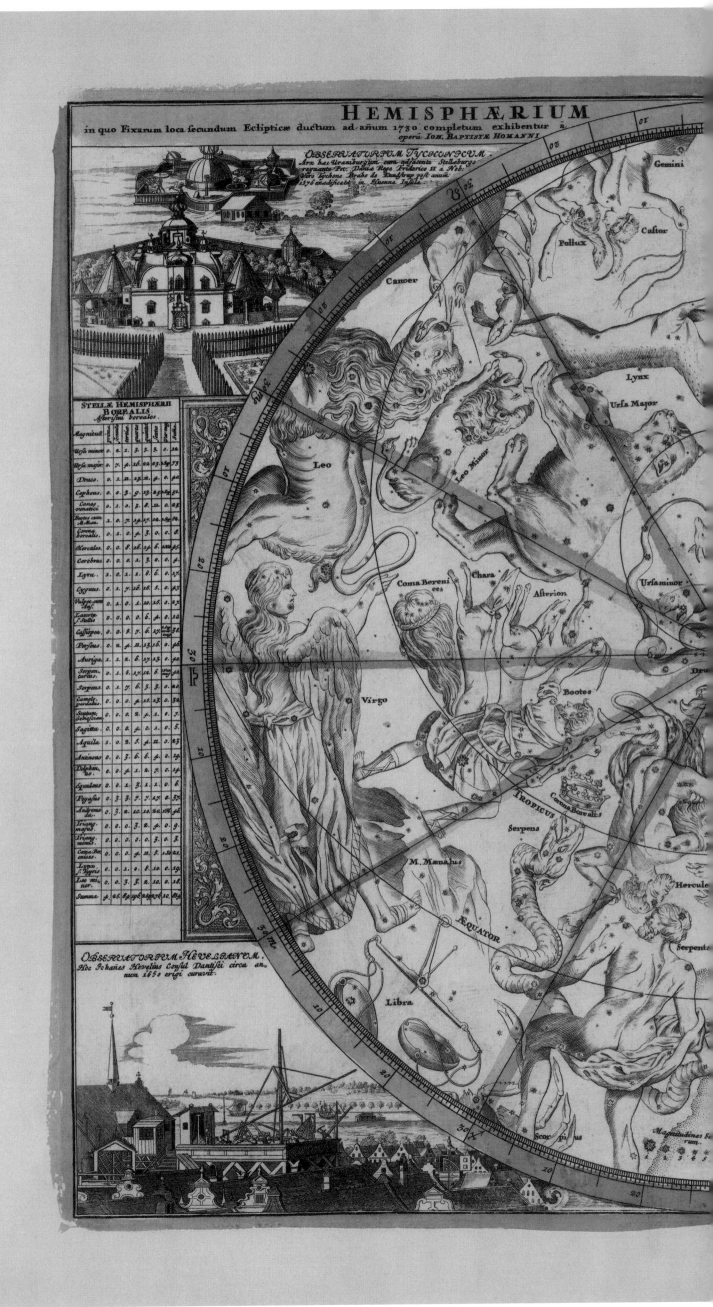

OBSERVATORIUM TYCHONICUM.
Arx hæc Uraniburgium cum ædificiis Stelleburgi-
cognato-Pvc, Dahiæ Rege Friderico II a Nob.
Vira Tychone Brahe de Knudstrup post anū
1576 ædificata in Huenna Insula.

STELLÆ HEMISPHÆRII BOREALIS
Asterismi boreales.

Magnitud.							
Ursa minor	0.	2.	1.	3.	3.	3.	0. 12.
Ursa major	0.	7.	4.	18.	22.	23.	34. 75
Draco	0.	1.	13.	12.	9.	6.	40.
Cepheus	0.	3.	9.	13.	8.	3.	35.
Canes venatici	0.	1.	1.	3.	12.	11.	0. 45.
Bootes cum & Mon.	1.	1.	7.	19.	17.	22.	28. 54.
Corona borealis	0.	1.	1.	6.	3.	0.	0. 8.
Hercules	0.	0.	8.	18.	36.	5.	24. 69
Cerberus	0.	0.	2.	2.	3.	0.	0. 8.
Lyra	1.	0.	1.	1.	6.	0.	0. 17.
Cygnus	0.	1.	7.	18.	18.	5.	0. 27
Vulpec. cum Ans.	0.	0.	0.	1.	12.	15.	0. 27
Lacerta f. Stellio	0.	0.	0.	6.	4.	0.	0. 10
Cassiopea	0.	1.	9.	6.	14.	3.	4. 56
Perseus	0.	4.	4.	11.	13.	16.	0. 48
Auriga	1.	1.	6.	13.	17.	5.	0. 40.
Serpentarius	0.	1.	5.	17.	12.	5.	24. 49
Serpens	0.	1.	7.	6.	3.	3.	0. 40.
Camelopardalis	0.	0.	0.	9.	12.	23.	0. 38.
Scutum Sobiescian.	0.	0.	0.	2.	1.	0.	7.
Sagitta	0.	0.	0.	2.	1.	0.	0. 5.
Aquila	0.	1.	2.	5.	4.	11.	0. 23
Antinous	0.	0.	3.	6.	6.	4.	0. 19
Delphinus	0.	0.	4.	1.	3.	7.	0. 14
Equuleus	0.	0.	1.	2.	1.	1.	0. 6.
Pegasus	0.	3.	3.	5.	7.	17.	0. 37
Andromeda	0.	3.	2.	10.	18.	22.	10. 46
Triangulum majus	0.	0.	0.	3.	6.	0.	0. 9.
Triangulum minus	0.	0.	0.	0.	0.	3.	0. 3.
Coma Berenices	0.	0.	0.	2.	8.	5.	1. 21.
Lepus f. Pyris	0.	0.	1.	0.	2.	10.	0. 19.
Leo minor	0.	0.	3.	2.	12.	10.	0. 18.
Summa	4.	26.	64.	196.	250.	78.	10. 856.

OBSERVATORIUM HEVELIANUM.
Hoc Iohañes Hevelius Consul Dantisci circa annum 1650 erigi curavit.

OBSERVATORIUM TYCHONICUM.
Arx hæc Uraniburgium cum ædificiis Stelleburg.

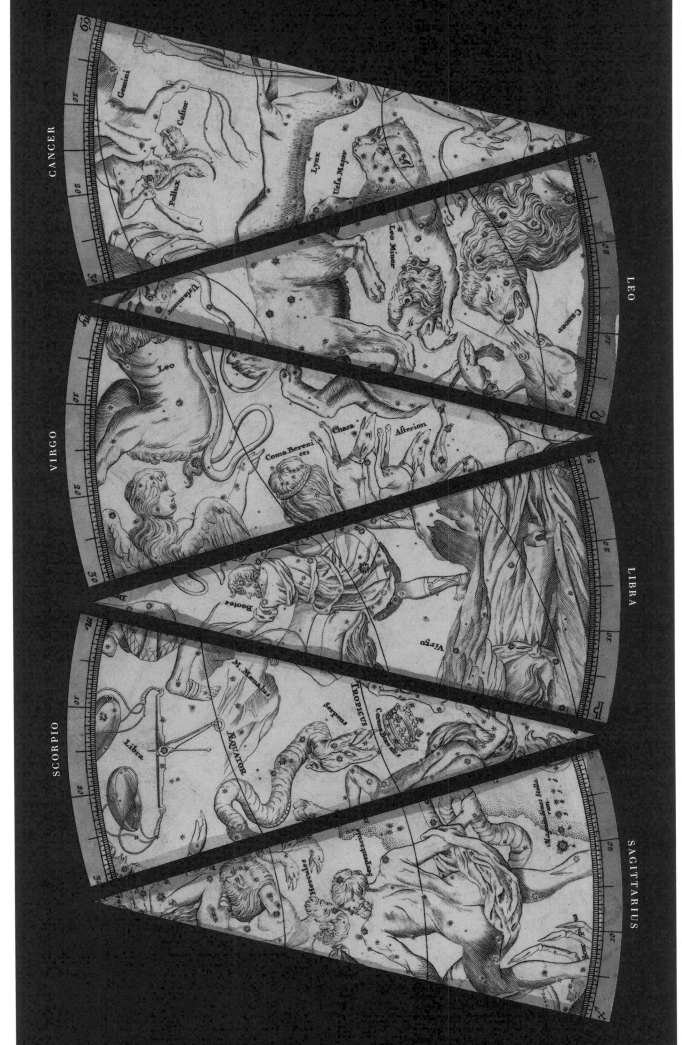

Around the edges of
Plate 18, Doppelmayr
provides tallies of the stars
of different magnitudes
in the constellations lying
entirely to the north of the
ecliptic. There is a separate
list of stars north of the
ecliptic within the zodiac
constellations themselves.
For space reasons, these
equivalent charts cite only
the totals for each and
summarize Doppelmayr's
calculations of the total
number of stars in the sky
by magnitude.

SECTION OF THE SKY.	Total
Stars of the constellations north of the zodiac	854
Stars in the zodiac, north of the ecliptic	242
Total stars in northern ecliptic hemisphere	1096
Stars of the constellations south of the zodiac	574
Stars in the zodiac, south of the ecliptic	200
Total stars in southern ecliptic hemisphere	774
TOTAL STARS IN THE SKY	1870

	1ST MAGNITUDE
	2ND MAGNITUDE
	3RD MAGNITUDE
	4TH MAGNITUDE
	5TH MAGNITUDE
	6TH MAGNITUDE

TYCHO'S OBSERVATORIES ON HVEN.

Doppelmayr depicts Uraniborg
in the foreground, with Stjerneborg
in the background.

PARIS OBSERVATORY.

The Paris Observatory founded
in 1667 under Louis XIV in
Montparnasse is shown with
a long-tubed refracting
telescope in its grounds.

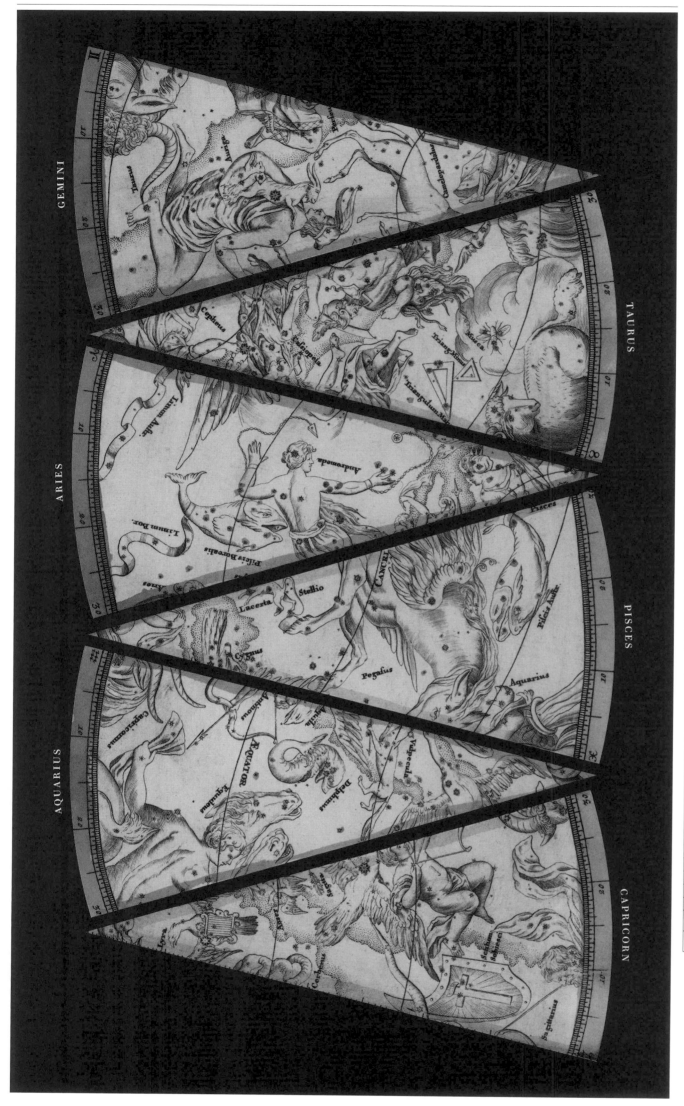

STARS OF THE CONSTELLATIONS, NORTH OF THE ZODIAC.	
	Total stars
Ursa Minor	12
Ursa Major	73
Draco	40
Cepheus	51
Canes Venatici	23
Boötes and Mons Melaus	52
Corona Borealis	8
Hercules	45
Cerberus	4
Lyra	17
Cygnus	47
Vulpecula cum Anser	27
Lacerta	10
Cassiopeia	38
Perseus	46
Auriga	40
Serpentarius	42
Serpens	20
Camelopardalis	32
Scutum Sobiescanum	7
Sagitta	5
Aquila	23
Antinous	19
Delphinus	14
Equuleus	6
Pegasus	37
Andromeda	46
Triangulum Majus	9
Triangulum Minus	3
Coma Berenices	21
Lynx/Tigris	19
Leo Minor	18
TOTAL	854

STARS OF THE ZODIAC CONSTELLATIONS, NORTH OF THE ECLIPTIC.	
	Total stars
Aries	24
Taurus	17
Gemini	22
Cancer	13
Leo	29
Virgo	40
Libra	18
Scorpius	7
Sagittarius	11
Capricornus	13
Aquarius	16
Pisces	32
SUBTOTAL	242
Zodiac stars to the south of the ecliptic	200
TOTAL STARS IN ZODIAC CONSTELLATIONS	442

HEVELIUS'S OBSERVATORY IN DANZIG.

The depiction of the Sternenburg rooftop observatory is borrowed from Hevelius's work *Machina Coelestis* (1673).

NUREMBERG OBSERVATORY.

Doppelmayr's illustration of the observatory founded by Georg Christoph Eimmart shows a view looking towards the north.

PLATE 19.

THE SOUTHERN HEMISPHERE OF THE HEAVENS, DETERMINED IN RELATION TO THE ECLIPTIC.

(HEMISPHÆRIUM COELI AUSTRALE)

Doppelmayr shows the skies centred on the south pole of the ecliptic.

Plate 19 in the *Atlas* is the companion piece to Plate 18: a map of the sky's southern hemisphere defined in ecliptic coordinates. While there is little more to say on this particular topic, this pair of plates has another notable attraction in the form of their marginalia. It is here that Doppelmayr chooses to depict eight of the leading observatories in the story of Renaissance and Enlightenment astronomy.

Temples, tombs and other monuments that align to specific solar and celestial events and objects have been built around the world since prehistoric times. However, the earliest true observatories that we know of (structures designed for the precise observation and measurement of objects in the sky) were built in China, India and across the Islamic world from around the 9th century. Such sites were initially used to measure and time the movements of the Sun and Moon in order to refine solar and lunar calendars, but Islamic astronomers in particular – influenced by the works of Ptolemy (*c.* 100–170 CE) and others – rapidly expanded their interests to the motions of the planets and stars.

Doppelmayr's selection of observatories, however, stays closer to home. Plate 18 features

Tycho Brahe's (1546–1601) pioneering Uraniborg castle on the then-Danish island of Hven, the Paris Observatory in France, Johannes Hevelius's (1611–87) rooftop platform in Danzig (modern Gdańsk, Poland) and the observatory at Nuremberg, Germany, founded in 1678. Plate 19, meanwhile, depicts the Royal Observatory at Greenwich, the Stellaburgis Hafniens in Copenhagen and the Kassel and Berlin observatories in Germany.

Of the locations shown, only Uraniborg (at the upper left on Plate 18) originated in the pre-telescopic era, but its influence was so great that Doppelmayr still found it worthy of inclusion almost two centuries later. Work done at this palatial observatory by Tycho and his army of assistants laid the foundations for the Tychonic model of the solar system, Johannes Kepler's (1571–1630) laws of planetary motion, and much more. Tycho was a wealthy nobleman himself, but the Uraniborg project was so ambitious that it is estimated to have absorbed 1 per cent of Danish crown spending. Doppelmayr depicts Uraniborg in the foreground and its companion underground observatory Stjerneborg, built nearby, in the background. Construction on the island of Hven began around 1576 under the sponsorship of Frederick II (1534–88), and the complex included not only an astronomical observatory but also an alchemical laboratory

FIG. 1.
This coloured engraving (1723) shows the Royal Observatory at Greenwich, designed by Sir Christopher Wren (1632–1723). The main building was constructed between 1675 and 1676.

FIG. 1.

FIG. 2.

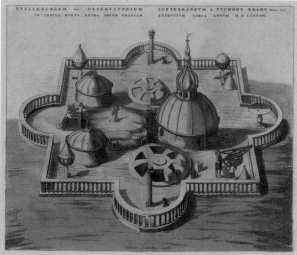

FIG. 3.

Contemporary views of Tycho's twin observatories in the late 16th century: the palatial Uraniborg (far left) and the mostly subterranean Stjerneborg (near left). Although Uraniborg (constructed from 1576) housed many of Tycho's remarkable instruments, he realized that the most accurate measurements would require devices rooted firmly to the ground and sheltered from wind and vibration – hence the construction of the underground crypts of Stjerneborg from 1583.

and workshops for the instrument makers who built some of the ingenious devices depicted on pages 50–1 and page 89. When Tycho fell from favour in the 1690s, he abandoned both the island and his homeland, and both observatories were destroyed soon after his death. The Stellaburgis Hafniens shown at upper right on Plate 19 was a belated successor, completed in Copenhagen in 1642 and used by Christian IV's (1577–1648) astronomer Christen Longomontanus (1562–1647), a former pupil of Tycho who did much to popularize the Tychonic system. The building remains today, and is perhaps best known for the 'equestrian staircase' used by nobles to ride their horses up to its summit.

An equally personal project is depicted in the lower left corner of Plate 18: Hevelius's observatory in Danzig. The wealthy brewer began construction of his rooftop observing platform in 1641 and equipped it with a variety of instruments, including sextants and quadrants for measuring celestial angles, a series of refracting telescopes of increasing ambition and an observing hut that could be used for projecting the image of the Sun.

At lower right on Plate 18, Doppelmayr includes an observatory of particular interest: that of Georg Christoph Eimmart (1638–1705). Eimmart was a Nuremberg engraver, mathematician and instrument maker who in 1678 established a private observatory on the north side of Nuremberg Castle. The open-air observatory functioned as a teaching institution as much as a site for serious research, and after Eimmart's death it was acquired by the city. Doppelmayr himself became its third director (from 1710–50), although sadly his lack of expertise with instruments, coupled with the observatory's exposed outdoor location, led to its decay. It was abandoned in 1751, and attempts at revival foundered in the late 1760s.

The other significant observatories across these pages hint at the increasing state interest in practical astronomy that took hold from the later 17th century (driven in part by the hope that better astronomical knowledge could solve the 'longitude problem', see pages 132–3). The Paris Observatory (at upper right of Plate 18) was the first of these institutions, founded in 1667 by Louis XIV (1638–1715). It was here that Giovanni Domenico Cassini (1625–1712) discovered four satellites of Saturn and the principal division in its rings (which still bears his name). The observatory provided data on the orbits of Jupiter's moons for the *Knowledge of the Times*, an astronomical almanac used for navigation and cartography (see Plate 15), and also became home to the Paris meridian.

The Royal Observatory at Greenwich, at the upper left of Plate 19, was founded by Charles II (1630–85) in 1675, and grew to be equally influential. Appointed as Astronomer Royal, John Flamsteed (1646–1719) was the first observer to work there, using instruments including two pendulum clocks of unprecedented accuracy to re-chart the heavens and finally better the star catalogues of Tycho Brahe. As with Paris, most of Greenwich's work was concerned with solutions to the longitude problem and improved tables for navigation.

At the bottom of Plate 19 are two other German observatories. Doppelmayr may have included the observatory at the new palace of Charles I, Landgrave of Hesse-Kassel (1654–1730), (left) as a nod to Charles's predecessor William IV, who had established a significant pre-telescopic observatory as early as the 1560s. The Berlin Observatory at lower right, meanwhile, was established in 1700, initially to aid calendar reform. However, it would become another major scientific powerhouse in the late 18th and 19th centuries.

HEMISPHÆRIUM

in quo Fixarum loca secundum Eclipticæ ductum ad añum 1730. completum exhibentur à
operâ IOH. BAPTISTÆ HOMANNI

OBSERVATORIUM ANGLICANUM.
Hoc Greenwici prope Londinum Carolus II.
Angliæ Rex (prout ex eius Inscriptione
patet) Astronomicæ et Artis Nau-
ticæ Patronus maximus in utrius-
que comodum año 1676. extrui
curavit.

STELLÆ HEMISPHÆRII AUSTRALIS.
Asterismi australes.

Magnitud	1	2	3	4	5	6	Summa
Cetus	0	3	9	10	12	11	0 45
Orion	2	4	4	9	24	18	1 Sept 62
Eridanus	1	0	3	29	8	2	0 48
Lepus	0	0	9	4	1	2	0 16
Canis maj	1	5	1	5	10	0	0 22
Monoceros	0	0	0	10	7	2	0 19
Canis minor	1	0	1	0	4	7	0 13
Columba	0	4	0	1	8	1	0 13
Argonavis	1	5	12	13	13	4	0 48
Robur Caroli	0	1	2	7	2	0	0 12
Hydra	1	0	8	15	9	8	0 35
Crater	0	0	0	7	1	2	0 10
Corvus	0	0	3	2	2	1	0 8
Centaurus	1	6	7	10	9	1	1 N 35
Sextans Uraniæ	0	0	0	1	5	5	1 Sext 11
Lupus	0	0	3	14	4	0	0 23
Ara	0	0	1	6	1	0	0 9
Corona Austr.	0	0	0	1	8	0	0 12
Piscis notius	0	0	4	10	3	0	0 17
Grus	0	1	4	3	3	0	0 13
Phœnix	0	1	5	5	0	0	0 13
Indus	0	0	1	2	9	0	0 12
Pavo	0	1	3	4	2	0	0 14
Apus Avis Indica	0	0	4	3	4	0	0 11
Musca	0	0	0	3	0	0	0 4
Chamæleon	0	0	0	0	9	1	0 10
Triang Austr.	0	1	2	0	0	0	0 5
Piscis Volans	0	0	0	5	2	0	0 8
Dorado Xiphias	0	0	1	4	1	0	0 6
Toucan	0	0	4	3	0	0	0 9
Hydrus	0	0	4	2	6	1	0 13
Summa	8	32	70	173	181	51	3 514

OBSERVATORIUM CASSELLANUM.
Has Uraniæ, gloriæ Dei munciæ sacras extruxit aedes (ut Inscriptio docet)
Ser. Princeps Carolus I. Hassiæ Landgravius año 1714.

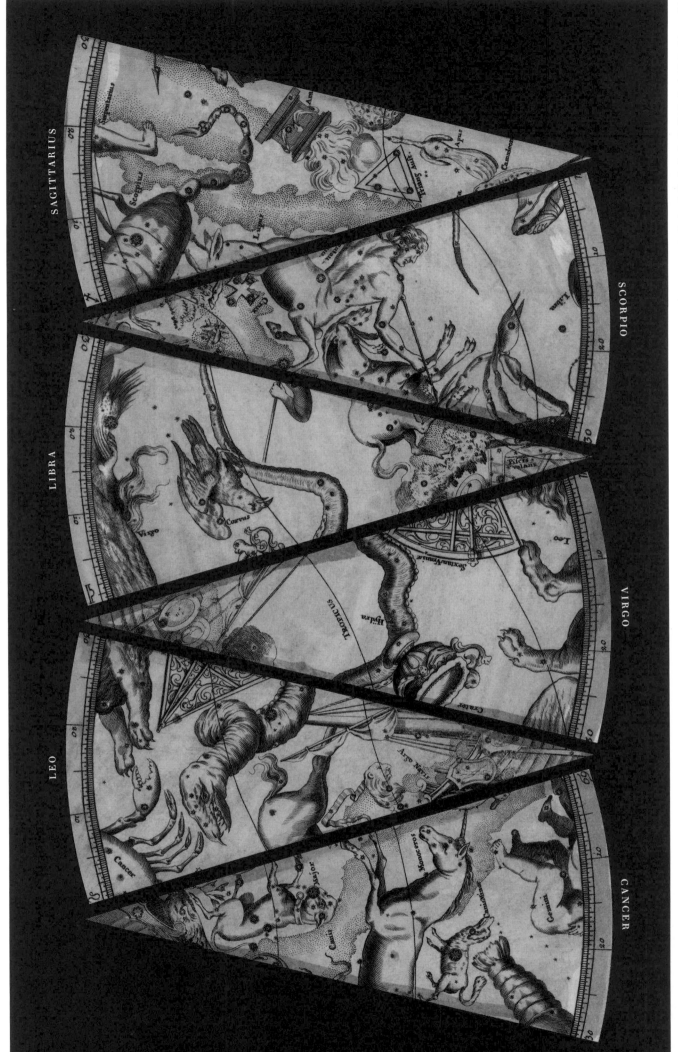

Doppelmayr's chart of the southern ecliptic hemisphere provides counts of the stars with different magnitudes in the constellations lying entirely to the south of the ecliptic, as well as stars in the southern parts of the zodiac constellations. The simplified versions here show only his overall totals. A separate table gives Doppelmayr's count of the number of stars within the complete segment of the sky associated with each zodiac house, revealing variations between crowded and relatively sparse directions in the heavens.

NORTHERN AND SOUTHERN STARS IN THE HOUSES OF THE ZODIAC.

	Northern stars	Southern stars	Total stars
Aries	99	38	137
Taurus	130	62	192
Gemini	107	118	225
Cancer	78	81	159
Leo	80	66	146
Virgo	98	46	144
Libra	71	50	121
Scorpio	91	79	170
Sagittarius	76	61	137
Capricorn	89	51	140
Aquarius	99	68	167
Pisces	77	54	131

1	1ST MAGNITUDE
2	2ND MAGNITUDE
3	3RD MAGNITUDE
4	4TH MAGNITUDE
5	5TH MAGNITUDE
6	6TH MAGNITUDE

ROYAL GREENWICH OBSERVATORY.

The observatory established by Charles II in 1675 is shown with a variety of instruments on its roof. In reality, Flamsteed made many of his positional measurements from outbuildings in the garden.

COPENHAGEN OBSERVATORY.

Christian IV of Denmark's Stellaburgis Hafniens observatory, now better known as the Rundetaarn, was built in the 17th century as a tower on the side of the new Trinitatis Church for the University of Copenhagen.

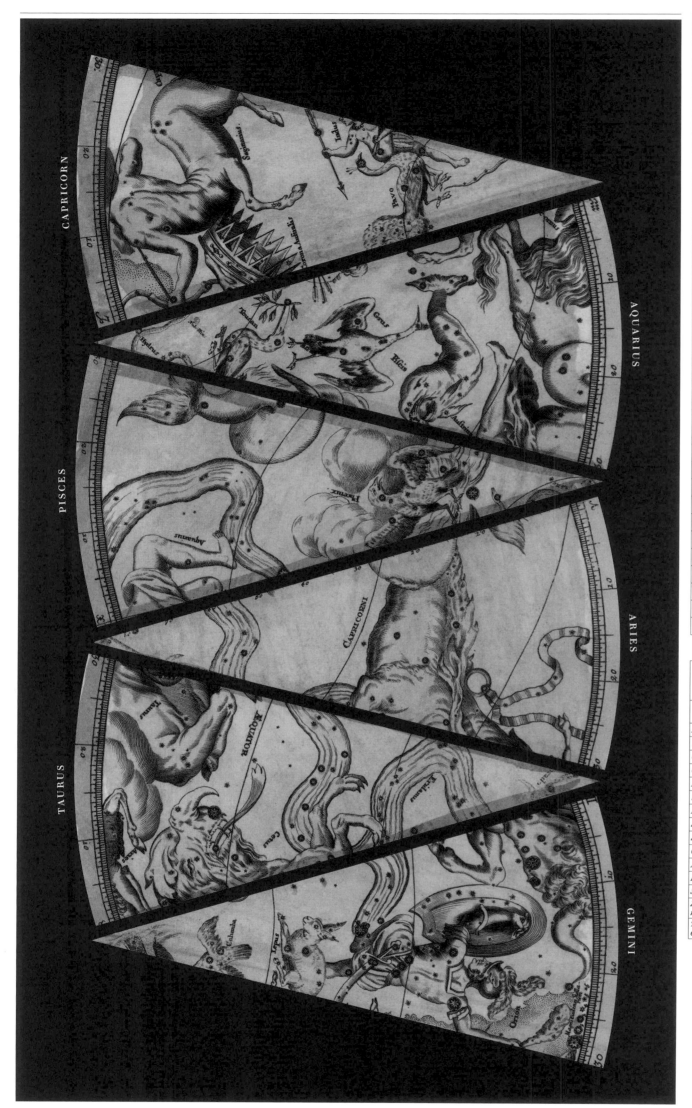

STARS OF THE CONSTELLATIONS, SOUTH OF THE ZODIAC.

	Total stars
Cetus	45
Orion	62
Eridanus	48
Lepus	16
Canis Major	22
Monoceros	19
Canis Minor	13
Columba	10
Argo Navis	48
Robur Caroli	12
Hydra	35
Crater	10
Corvus	8
Centaurus	35
Sextans Uraniae	12
Lupus	23
Ara	9
Corona Australis	12
Piscis Austrinus	17
Grus	13
Phoenix	13
Indus	12
Pavo	14
Apus	11
Musca	4
Chameleon	10
Triangulum Australe	5
Volans	8
Dorado	6
Toucan	9
Hydrus	13
TOTAL	574

STARS OF THE ZODIAC CONSTELLATIONS, SOUTH OF THE ECLIPTIC.

	Total stars
Aries	3
Taurus	34
Gemini	16
Cancer	16
Leo	17
Virgo	10
Libra	2
Scorpius	28
Sagittarius	20
Capricornus	16
Aquarius	31
Pisces	7
SUBTOTAL	200
Zodiac stars to the north of the ecliptic	242
TOTAL STARS IN ZODIAC CONSTELLATIONS	442

KASSEL
OBSERVATORY.

Doppelmayr depicts
Charles I, Landgrave of
Hesse-Kassel's observatory
of 1714 with instruments on
its roof. It is now known as
the Bellevue Palace.

BERLIN
OBSERVATORY.

The 27-m (88-ft) observatory
tower was added to Prussian
King Frederick I's royal stable
complex at Unter den Linden
in the early 1700s.

THE CELESTIAL GLOBE RENDERED AS PLANES, PARTS I–VI.

*Doppelmayr shows the celestial sphere in detail across
a series of maps in a tangential projection.*

In the six large celestial charts that he compiled around 1720, Doppelmayr faces a challenge that dogs all cartographers: the choice of which map projection to use. In its broadest definition, projection is the mathematical process of transferring information about points on one type of surface onto another. However, its most familiar application lies in transferring measurements on curved surfaces (such as the surface of Earth or the inside of the celestial sphere) onto the flat planes of maps suitable for printing.

In modern terms, any map projection involves making a decision about how coordinates of latitude and longitude measured from patterns (of terrain or of stars) on a curving sphere should be plotted on the flat surface. Perhaps the simplest solution (although, curiously, not the oldest) is to simply plot the values of latitude and longitude as if they were the vertical and horizontal axes of a graph, with 1 degree in either direction amounting to a certain distance on either axis. The resulting 'equirectangular' projection is easy to interpret, but exacts a huge cost in terms of accuracy. On a globe, lines of longitude are at their widest at the equator and converge towards the poles, so rendering them as straight parallel lines exaggerates distances and areas at higher latitudes. An equirectangular chart is moderately accurate for a small band around the equator, but becomes wildly inaccurate elsewhere, stretching small objects at high latitudes so they appear horizontally elongated. In general, it renders neither the areas of objects nor the angles between locations accurately.

All map projections involve some trade-off between these properties of angle and area. Maps that preserve angles (for example, accurately representing the angle formed by two intersecting roads) are said to be 'conformal', whereas those that accurately represent scale are termed 'equal-area' projections. Doppelmayr,

like many of his forerunners in mapping the stars, chooses the oldest approach of them all: the 'gnomonic' projection attributed to Thales of Miletus (*c*. 624–*c*. 548 BCE), who is said to have used it for producing star maps in around 580 BCE.

The principal behind the gnomonic projection is simple: the flat map is imagined as a 'tangential plane', one that touches the sphere at just a single point (its centre). The positions of objects on the sphere's surface are then projected onto the map by plotting where a ray from the centre of the sphere, passing through the object, would hit the map. This concept of cast rays explains why the projection shares a name with the gnomon, the shadow-casting central 'pointer' on a sundial.

The gnomonic projection creates a map that shows little distortion for small areas around the centre (where the cast rays diverge at shallow angles), but gets progressively more stretched towards its outer edges. Such a map is fundamentally limited to rendering less than half of a spherical surface (because points at an angle of 90 degrees or more to the tangent point will never make contact with the map plane).

Doppelmayr takes inspiration from French Jesuit physicist Ignace-Gaston Pardies (1636–73), who first used this particular projection to render the sky as the six faces of a cube in maps published posthumously in 1674. Pardie's maps subsequently became well known in Nuremberg. The projections are centred at the north and south celestial poles, and along the celestial equator at 0 hours, 6 hours, 12 hours and 18 hours or right ascension. The style of the constellation figures, meanwhile, owes more than a little debt to Johannes Hevelius's (1611–87) *Firmament of Sobieski, or Map of the Heavens* (1690), while the charts are further decorated by the tracks of notable comets (collated on Plates 27 and 28, where these objects are discussed in more detail).

When it comes to indicating the brightness of stars, Doppelmayr follows a tradition that was popularized by Ptolemy's (*c*. 100–170 CE)

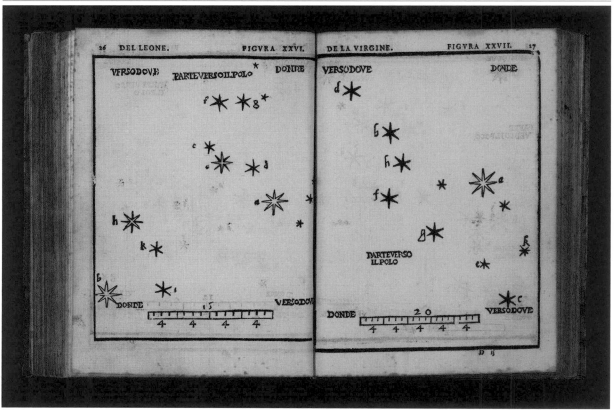

FIG. 1.

FIG. 1.

The maps
of Alessandro
Piccolomini's
(1508–78) book *On
the Fixed Stars* (1540)
are unusual among
early star charts
for abandoning
the pretence of
constellation figures
in favour of simply
showing the stars
themselves at sizes
determined by their
brightness – as shown
on these examples
of Leo (left) and
Virgo (right). A more
enduring shift in the
stellar cartography
of the 16th century,
also highlighted here,
is the abandonment
of the 'external'
viewpoint in plotting
the celestial sphere in
favour of a more useful
internal one, which
showed the stars in the
same orientations as
they are seen in the sky.

Almagest (150 CE), but probably originates earlier in Greek astronomy, and is known today as 'apparent magnitude'. Classical stargazers divided the naked-eye stars into broad categories, with the brightest stars referred to as being of the first magnitude, the next brightest as the second magnitude and so on down to the sixth magnitude stars that were the faintest visible in a clear, dark sky. The size of stars on maps seemed a natural way to represent their magnitude, and a link between a star's brightness and its apparent size seemed to be borne out by Tycho Brahe's (1546–1601) early attempts at measuring the stars (see Plate 9). Tycho listed both positions and magnitudes in his catalogue of 1,004 stars, completed in 1598. This, subsequently, made its way onto the *Uranometria*, the most influential Renaissance atlas of the constellations, published by Johann Bayer (1572–1625) at Augsburg in 1603.

The same atlas invented another influential standard in celestial cartography, which took somewhat longer to become universally accepted. Looking for a handy means of referring to each star in a constellation without resorting to a clumsy description of its position, Bayer hit upon the idea of giving the stars letter designations, mostly in the Greek alphabet. A constellation's brightest star would be alpha, its second brightest beta, and so on. With twenty-four Greek letters to use, this was plenty to account for the visible stars in most constellations.

However, for the most crowded ones, Bayer extended the principle to use most of the letters of the lower-case Latin alphabet (capitalizing 'A' to avoid confusion with alpha).

Other early stargazers followed their own preference. For exanple, in some maps, Hevelius used numbers rather than letters to order a constellation's stars by brightness. Bayer's system only really became the de facto standard through the growing influence of John Flamsteed's (1646–1719) *Historia Coelestis Britannica* – an exhaustive catalogue of some 2,935 stars published in 1725.

As well as popularizing the Greek-letter labels of bright stars, Flamsteed also laid the foundations of a supplementary system that now bears his name. His catalogue attempted to list stars within each constellation by their order from west to east (in other words, by increasing right ascension). A later French translation of 1783, edited by Joseph-Jérôme de Lalande (1732–1807), applied sequential numbers to those stars in Flamsteed's catalogue that did not already have a Bayer designation; these 'Flamsteed numbers' subsequently became the second great pillar of a system that is still used today.

These developments mostly came after Doppelmayr's time, of course, and rather than use Greek letters, the *Atlas* chooses to use Latin capitals for each constellation's brightest stars, with lower-case letters for the fainter ones.

GLOBI COELESTIS IN TABU

in qua Longitudines Stellarum fixarum ad añum Christi c̄

à IOH: GABR: DOPPELMAYR MATH: P.P. Academ: Cæs: I

Operâ IOH: BAPT: HOMAN

Cum Privi

BOOTES

HERCULES

ASTERIO

URSA MINOR

URSA MAJOR

Colurus

LEO MINOR

LYNX ſ. TIGRIS

Magnitudines Stel:

URSA MINOR.

		Mag	Longit	Latit
A	Polaris.			
B	Humerus.			
C	In pectore.			
D	In dorſo.			
E	In lateris ſ. ventre.			
F	Prima cauda.			
G	Media cauda.			
H	Ad humerum proxima.			
	altera.			
I	In capite.			
K	Sub capite ſequens auſtralis.			
L	Sub capite præcedens borealis.			

DRACO.

A	Ad roſtrum.			
B	In vertice.			
C	Ante flexu III. lucida borealis.			
D	Poſt flexum III. ſecunda.			
E	Ante flexum IV. prior.			
F	Ante flexum IV. proxima.			
G	In flexura IV.			
H	Prima cauda.			
I	Penultima cauda.			
K	Ultima cauda.			
L	In lingua.			
M	In ore.			
N	Ad genam.			
O	In flexura I. borealis.			
P	In flexura I. auſtralis.			
Q	In flexura I. media.			
R	Poſt flexuram I.			

CEPHEUS.

A	Humerus ſiniſter.			
B	Cingulum.			
C	In dextro crure.			
D	In tiara prima.			
E	In tiara ſecunda.			
F	In Facie.			
G	In ſiniſtro cubito.			
H	In ſiniſtro lacerto.			
I	In dextro humero.			

URSA MAJOR.

A	In latere.			
B	In ventre.			
C	In dorſo.			
D	In coxa.			
E	Prima cauda.			
F	Media cauda.			
G	Ultima cauda.			

CAMELOPARDALUS.

A	In cervice ſ. ſiniſtra aure.			
B	Ad genam.			
C	In collo prima.			
D	In collo tertia auſtralis.			
E	In collo ſecunda borealis.			

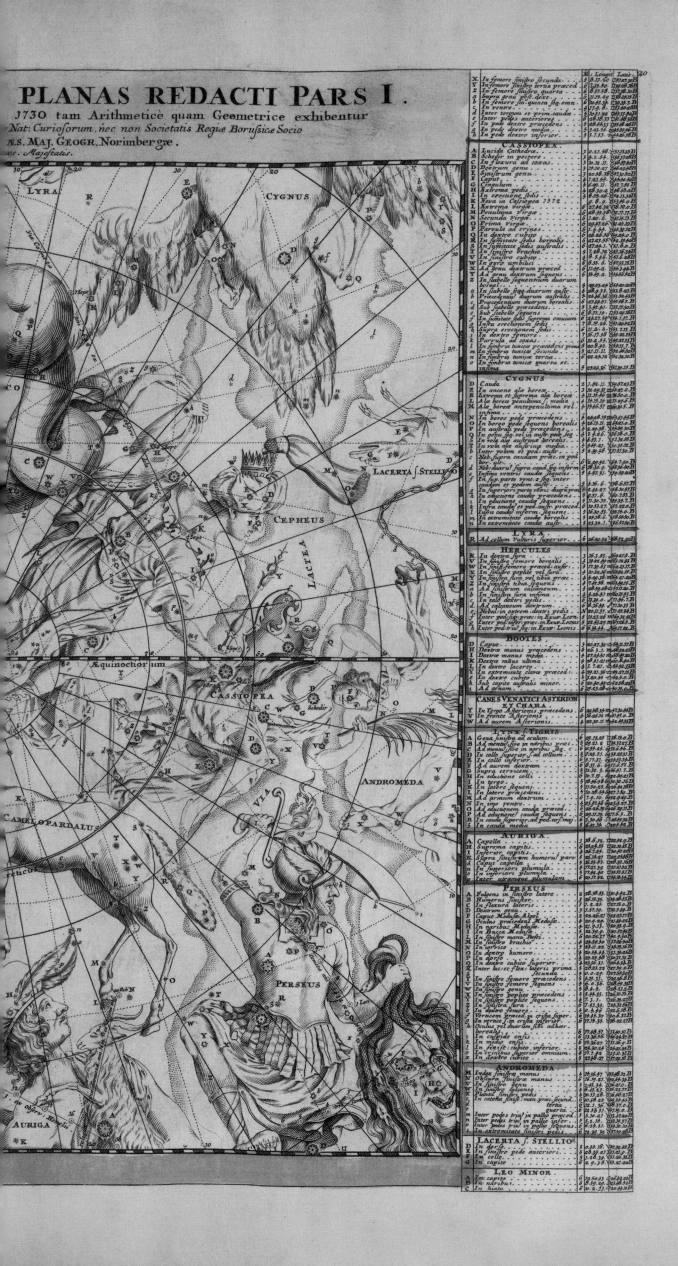

PLANAS REDACTI PARS I.

1730 tam Arithmeticè quam Geometricè exhibentur
Nat: Curioforum, nec non Societatis Regiæ Boruſſicæ Socio
æ.S. Maj. Geogr. Norimbergæ.
us. Majeſtatis.

		Id.	Longit	Latit.
X Y	In femore ſiniſtro ſecunda			
Z	Infemore ſiniſtro tertia praeced.			
a b	In femore ſiniſtro quarta			
	Supra genu poſt. dext.			
c	In femore ſin. quinta ſeq. omn.			
d	In ventre			
e	Inter tergan et prim. caudae			
f	Inter veſtes anteriores			
g	In pede dextro praecedens			
h	In pede dextro media			
i	In pede dextro inferior			

CASSIOPEA.

A	Lucida Cathedrae
B	Schegir in pectore
C	In flexura ad coxas
D	Dextrum genu
E	Siniſtrum genu
F	Caput
G	Cingulum
H	Extrema pedis
I	In erectione ſedis
K	Nova in Caſſiopea 1572
L	Extrema Virgæ
M	Penultima Virgæ
N	Secunda Virgæ
O	Prima Virgæ
P	Parvula ad crines
Q	In dextro cubito
R	In ſumitate ſedis borealis
S	In ſumitate ſedis auſtralis
T	In ſiniſtro brachio
V	In ſiniſtro cubito
W	In gyro umbilici
X	Ad genu dextrum praecel.
Y	Ad genu dextrum ſequens
Z	In ſuabello ſequentium duarum borealis

CYGNUS

D	Cauda
E	In ancone alæ boreæ
F	Extrema et ſuprema alæ boreæ
G	Alæ boreæ penultima f. media
H	Alæ boreæ antepenultima vel infima

LYRA

R	Ad collum Vulturis ſuperior.

HERCULES

K	In dextra ſura

BOOTES

D	Caput
E	Dextrae manus praecedens

CANES VENATICI ASTERION ET CHARA

LYNX ſ. TIGRIS.

AURIGA

A	Capella

PERSEUS

A	Fulgens in ſiniſtro latere

ANDROMEDA

LACERTA ſ. STELLIO

LEO MINOR.

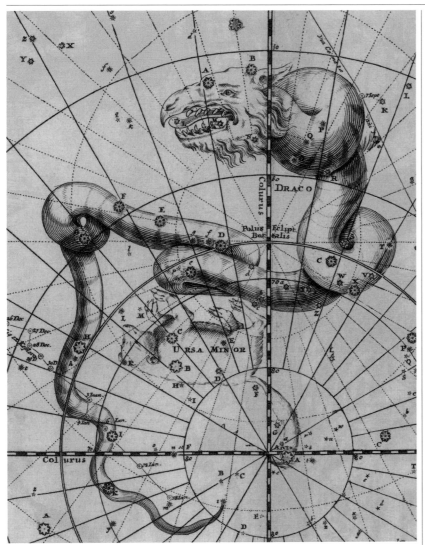

DRACO AND URSA MINOR.

The ancient constellation of the Dragon winds around the north celestial pole, enclosing the Lesser Bear on three sides. Draco was identified with various mythological dragons, most prominently the one that Hercules defeated in the garden of the Hesperides.

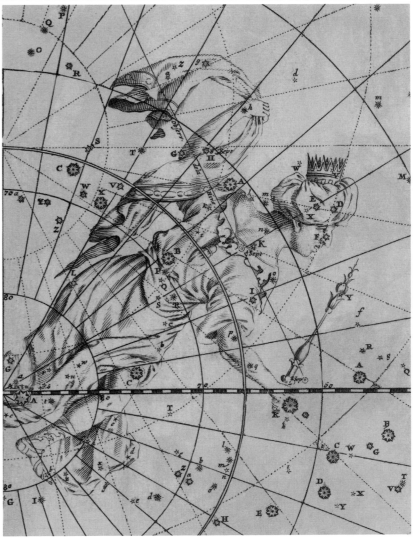

CEPHEUS.

The Ptolemaic constellation Cepheus represents the King of Aethiopia (an ill-defined region of North Africa) in the ancient Perseus myth. Cepheus was the father of Andromeda and husband of Cassiopeia. Doppelmayr shows him beseeching the gods to save his daughter.

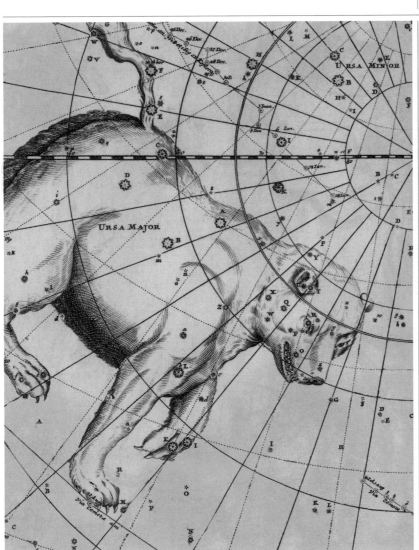

URSA MAJOR.

Seven stars, known variously as the Plough, Big Dipper or Wagon, form the brightest part of the sprawling Ursa Major constellation. The two bears represent Callisto and her son Arcas, transformed by the gods and forever pursued around the north celestial pole by the herdsman Boötes.

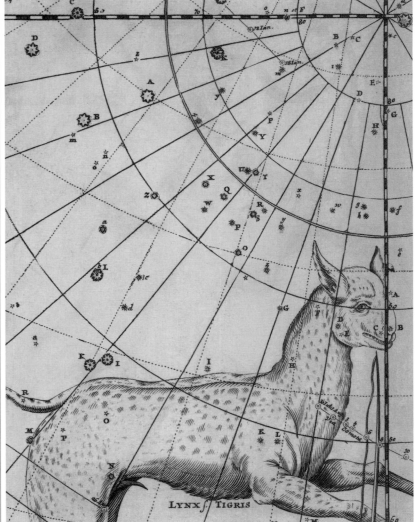

LYNX/TIGRIS.

Hevelius invented this constellation from a chain of faint stars to the south and east of the Great Bear, naming it, in part, because of his somewhat exaggerated claim that only those with the eyes of a lynx could see it. His alternative name of the Tiger had fallen out of use by the late 18th century.

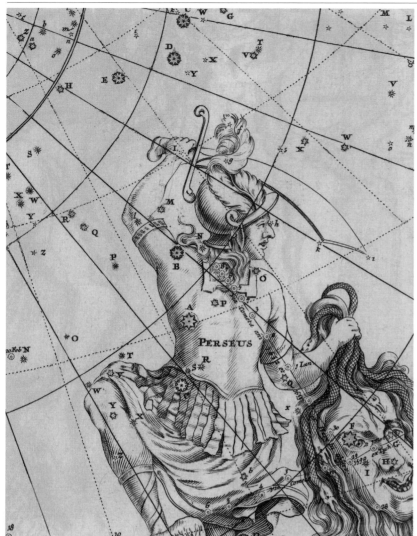

PERSEUS.

The hero Perseus was the slayer of the sea monster Cetus and the rescuer of the princess Andromeda. Here, he is depicted in an ornate plumed helmet wielding a curved sword and the severed head of Medusa, with the famous variable star Algol marking one eye.

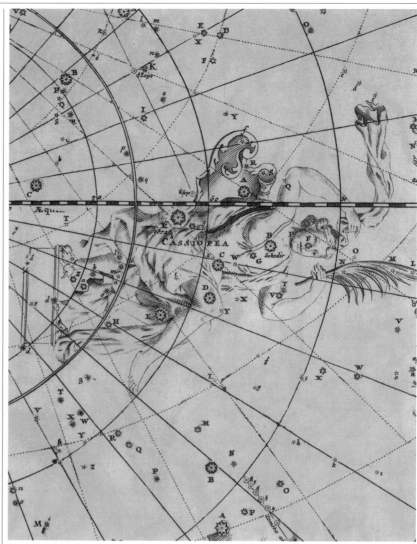

CASSIOPEIA.

Queen Cassiopeia, whose vanity enraged the sea god Poseidon in the ancient Perseus legend, is depicted seated on a throne. In the night sky, the constellation is dominated by a chain of five bright stars forming a distinctive zigzag.

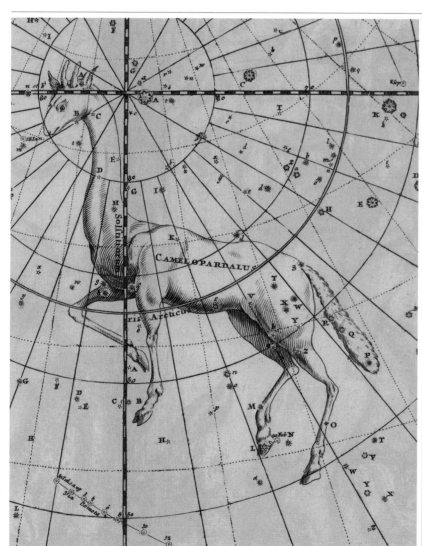

CAMELOPARDALIS.

The large but faint constellation of the Giraffe was invented in or shortly before 1613 by Plancius to fill the void between Cassiopeia and Ursa Major. Its name derives from the Greek word for 'giraffe', which translates as 'camel-leopard', on account of the creature's ungainly size and spotted appearance.

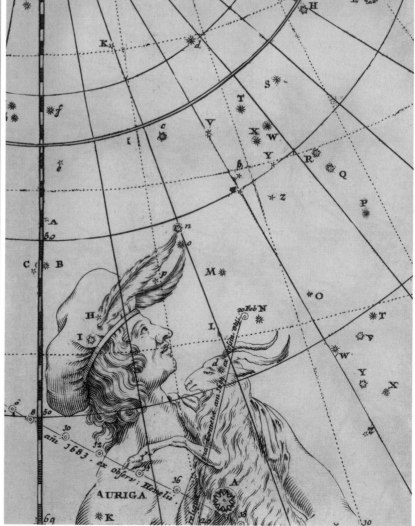

AURIGA.

The northern reaches of Auriga lie on Plate 20, with the bulk represented on Plate 22. Representing a charioteer, the constellation is identified with various mythical figures, including Erichthonius and Myrtilus, none of which explain why he is typically depicted carrying a goat and kids.

PLATE 20.

DÜRER'S HEMISPHERES (1515).

The first star maps to be produced in Europe using the printing press
were by a master of German Renaissance art. Printed from woodcuts,
the twin hemispheres by Albrecht Dürer are based around an ecliptic
projection of the sky that turns the zodiac constellations into a
boundary around the northern hemisphere. The design is thought to
have been inspired by the Vienna star charts of 1440 (see page 155), but
Dürer also worked with two noted astronomers of the time. Johannes
Stabius (c. 1460–1522) of Vienna prepared a relatively simple coordinate
system that divided the perimeter into 360 degrees and the sky into

twelve equal sectors (one for each zodiac sign), while Konrad Heinfogel
(d. 1517) of Nuremberg calculated the present positions of stars from
the *Almagest* (150 CE) and plotted them on the map, numbered to
indicate their place in Ptolemy's star lists. The four corners of the
northern chart show classical authorities: Aratus and Ptolemy at the
top, Marcus Manilius at bottom left and al-Sufi at bottom right. The
southern hemisphere includes the arms of the Archbishop of Salzburg
and a dedication at the top, the names and arms of the contributors
at lower left, and a dedication to Emperor Maximilian I at the bottom.

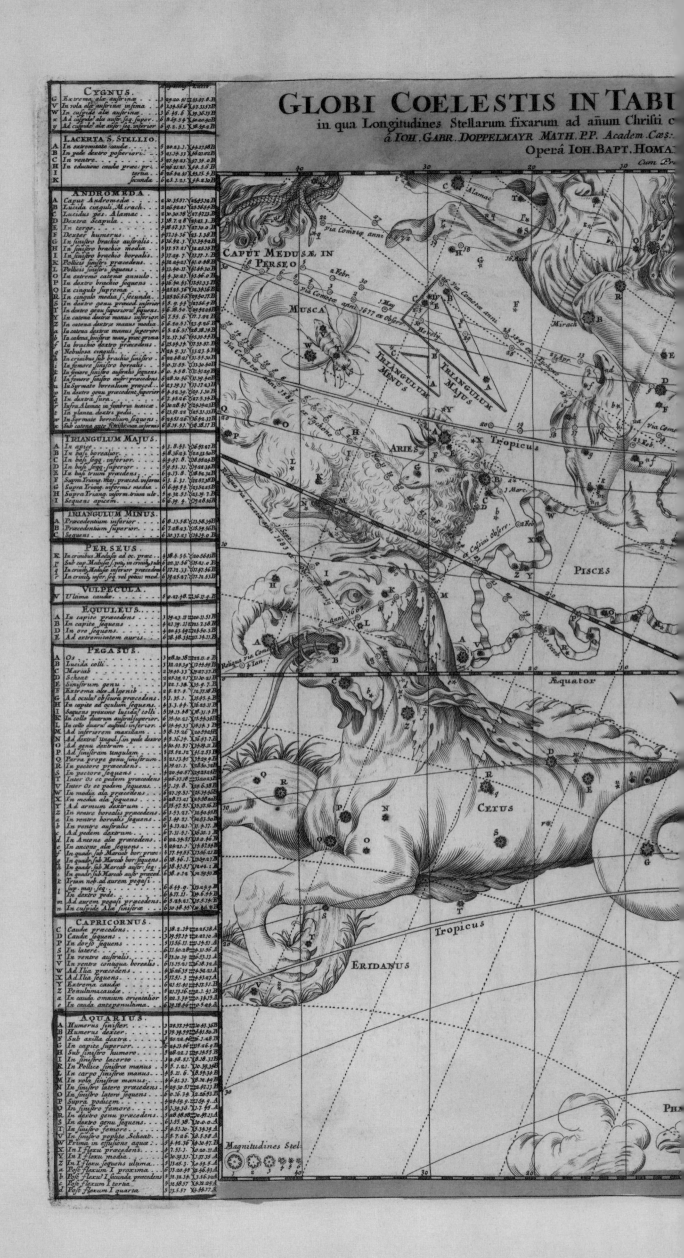

PLANAS REDACTI PARS II.

J730 tam Arithmeticè quam Geometricè exhibentur

dat: Curioforum nec non Societatis Regiæ Boruſſicæ Socio

S. MAJ. GEOGR. Norimbergæ.

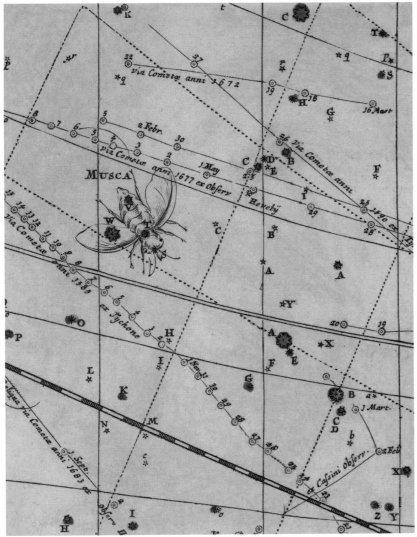

MUSCA.

This now-forgotten constellation was invented by Plancius around 1612 as Apes (the Bees). Hevelius's star catalogue of 1689 transformed it into Musca, the Fly, leading to some confusion with Plancius's own Musca Australis. Today, its stars form part of Aries.

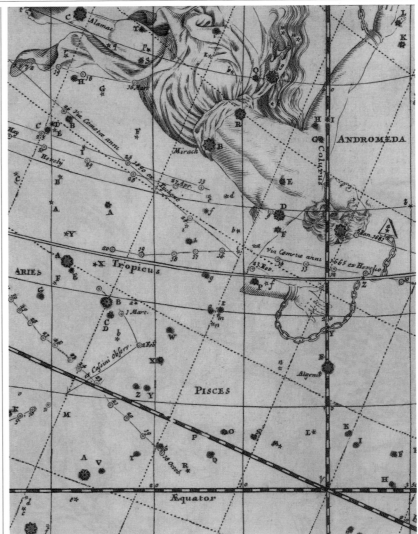

ANDROMEDA.

With her head lying at the northwest corner of the square of Pegasus, Andromeda is the princess of the Perseus story, chained to a rock by her parents, Cepheus and Cassiopeia, in an attempt to assuage the sea monster Cetus sent by Poseidon.

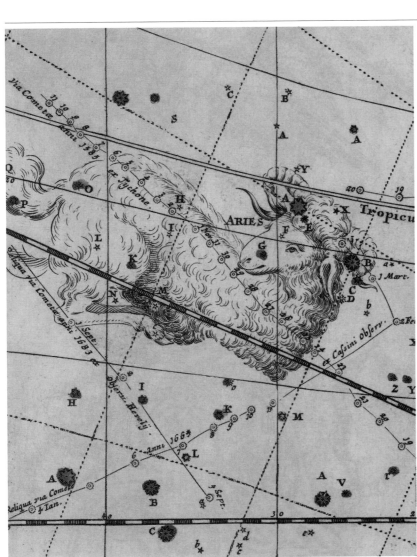

ARIES.

The constellation of Aries is usually associated with the bearer of the golden fleece in Greek mythology. However, its links to a ram (and to fertility) are far older; in ancient times, the Sun entered Aries at the beginning of northern spring.

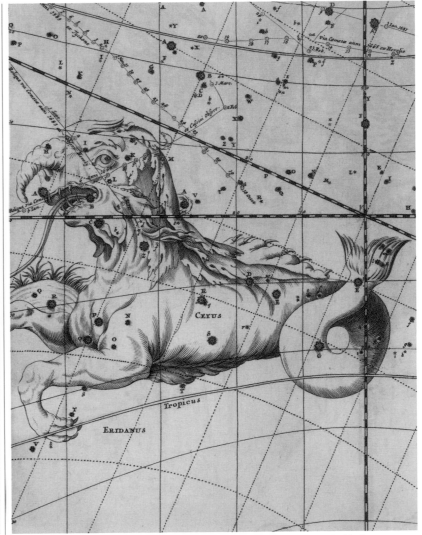

CETUS.

Cetus is usually associated with the monster in the Perseus legend, but the name is, in fact, Latin for 'whale'. Hence, interpretations of the constellation have varied hugely, from naturalistic efforts to the freakish creatures of Hevelius and Doppelmayr.

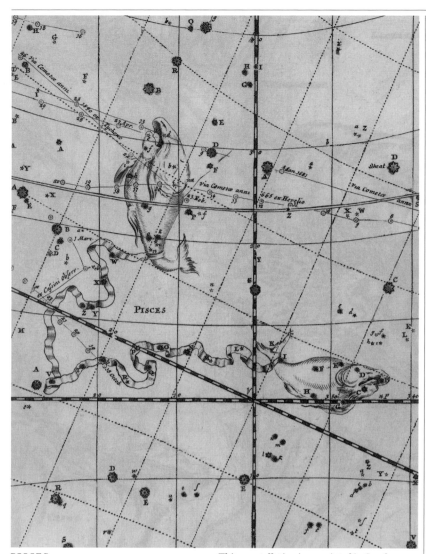

PISCES.

This constellation is associated in Greek mythology with the transformation of Aphrodite and her son Eros in order to escape the sea monster Typhon. The two fishes are usually depicted as a pair fleeing in opposite directions, their tails bound by a connected cord.

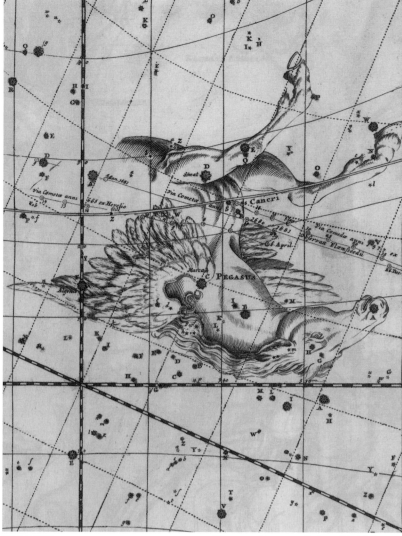

PEGASUS.

The sprawling constellation of the Flying Horse is depicted in the traditional 'upside-down' manner, with the bright stars of the Square of Pegasus forming his chest and body, while three chains of stars trace an outstretched head and forelimbs.

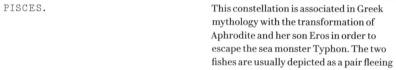

AQUARIUS.

The water carrier Aquarius is among the most ancient of constellations. By classical times, it was associated with Ganymede, a beautiful youth kidnapped by Zeus and taken to Mount Olympus to serve as cup-bearer to the gods.

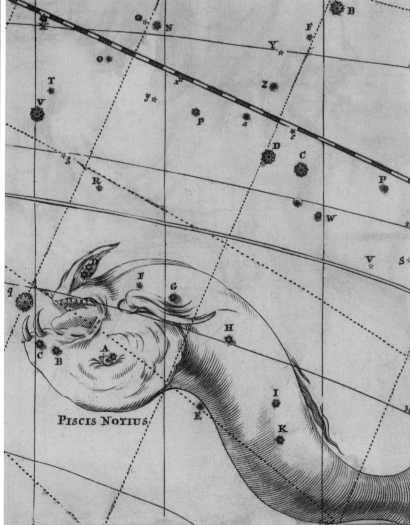

PISCIS NOTIUS.

The constellation of the Southern Fish (now Piscis Austrinus) swims in the waters that pour from Aquarius's jar. The bright star Fomalhaut (once considered part of Aquarius) marks the point where the water enters the fish's mouth.

PLATE 21.

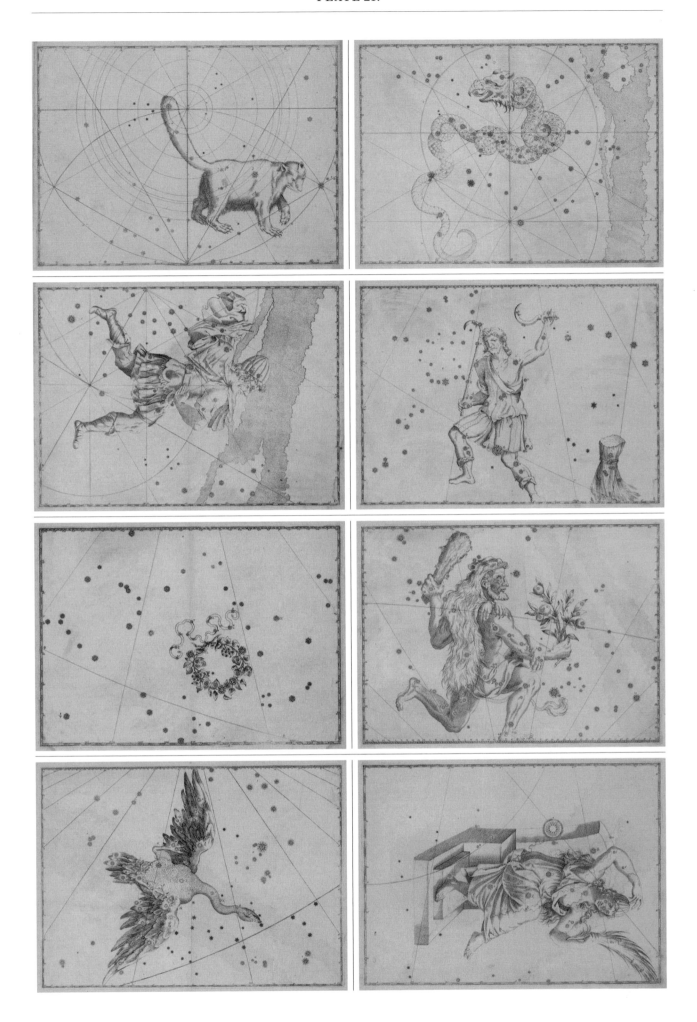

URANOMETRIA (1603).

The first of the great European star atlases was compiled by Johann
Bayer, a lawyer working in Bavaria, and published in 1603. Bayer's
innovative work combined precisely plotted stars (drawn from a list
of 1,005 compiled by Tycho Brahe) with figure illustrations by Alexander
Mair on a series of fifty-one copperplate engravings. In addition to a
pair of celestial hemispheres (centred in the then-traditional way on
the poles of the ecliptic), each of the forty-eight ancient Ptolemaic
constellations was rendered individually, with a single plate of the
twelve recent additions to the far southern sky. Coordinate grids

allowed star positions to be read with high precision, and each
chart was accompanied by a table of its principal stars. Another key
innovation was Bayer's assignment of Greek letters to (roughly) indicate
the order of the brightest stars within each constellation – although
it was more than a century before this became an accepted standard.
The *Uranometria* was not without its quirks. While Bayer plotted the
stellar positions from a geocentric perspective, Mair used an external
perspective in preparing some of his illustrations, resulting in
strange mismatches in the orientation of figures such as Orion.

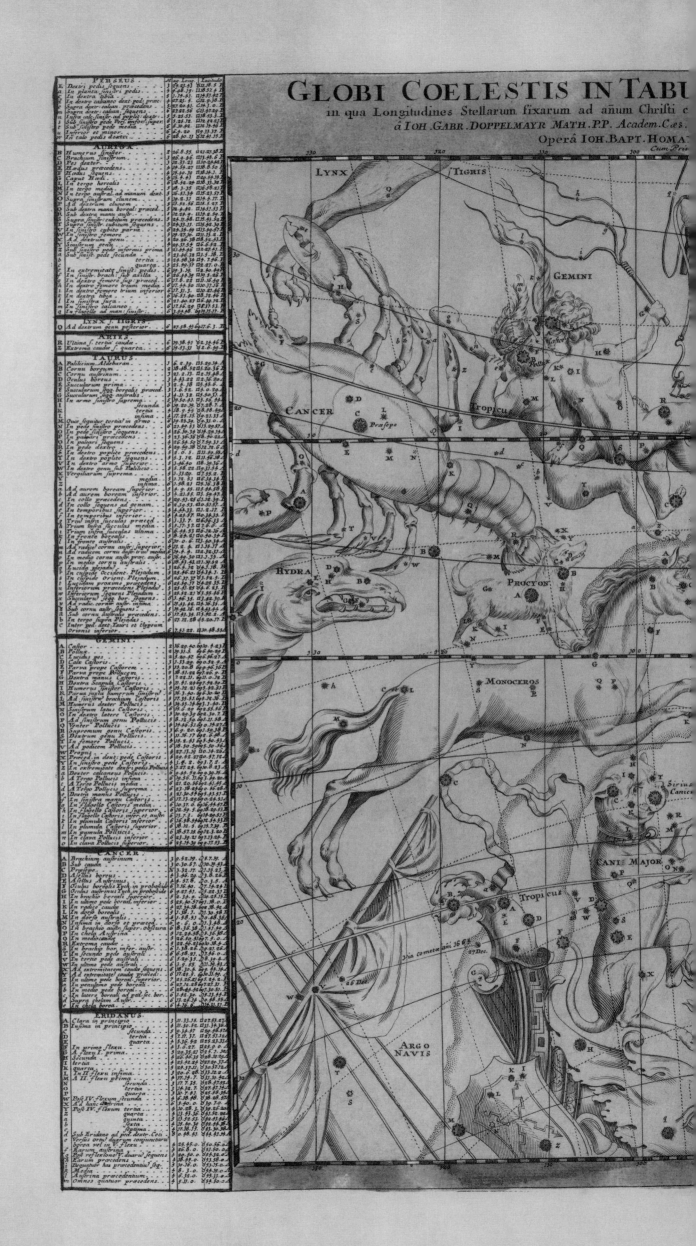

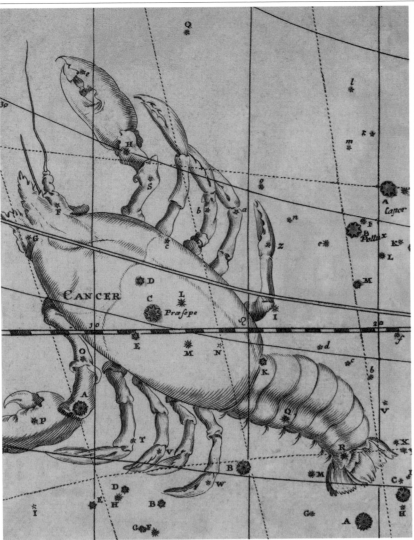

CANCER.

Ptolemy's constellation of the Crab represents a creature that hampered Hercules in his battle with the many-headed Lernean Hydra. Following the style of Hevelius, Doppelmayr depicts the faintest of the zodiac constellations as a rather more lobster-like creature.

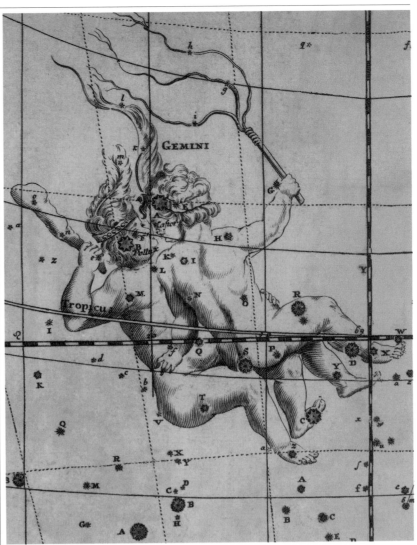

GEMINI.

Distinguished by its pair of bright stars, Gemini has been associated with twin deities from the earliest times. Classical astronomers named these stars after Castor and Pollux, who joined the Argonauts in search of the Golden Fleece.

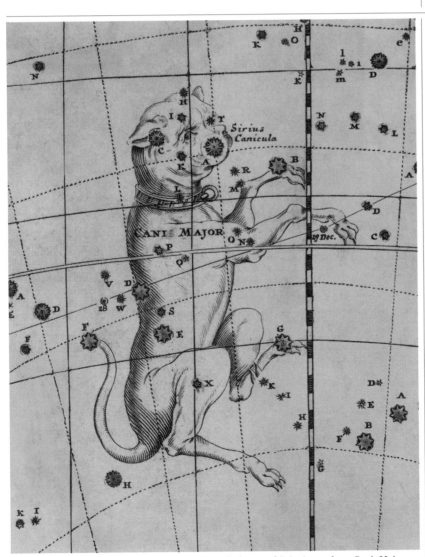

CANIS MAJOR.

The larger of Orion's two dogs, Canis Major, is home to Sirius, the brightest star in the sky. Traditionally, the influence of the Sun and Sirius in alignment was thought to increase the heat during the 'dog days' of summer.

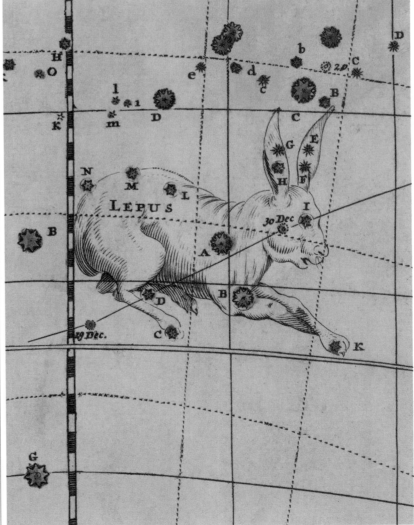

LEPUS.

The constellation of the Hare is one of Ptolemy's original forty-eight, although it does not have any particular stories attached to it. It is usually imagined as a hare fleeing unseen from Canis Major as Orion is distracted by Taurus.

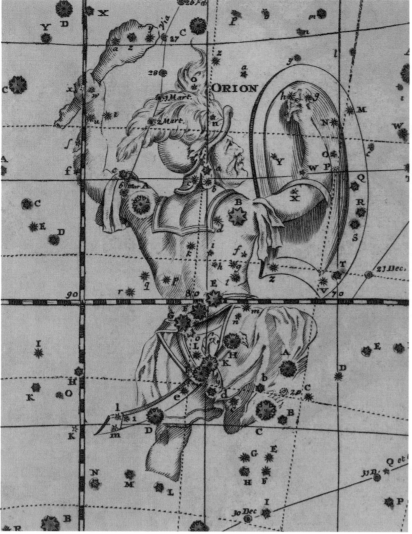

ORION.

The bright constellation of Orion shows the mythical Hunter with a shield, fending off the charging bull Taurus, and a club in his left hand raised to strike. Unusually, Doppelmayr includes the Arabic name of the star at Orion's right foot – Rigel – in the accompanying table.

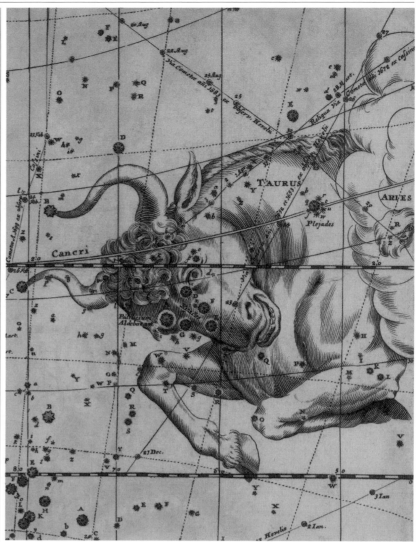

TAURUS.

The charging bull Taurus dates back to prehistoric times. Doppelmayr shows the Pleiades star cluster on its shoulder, the Hyades cluster forming its face and the brilliant red Aldebaran (once also known as Palilicium) marking its eye.

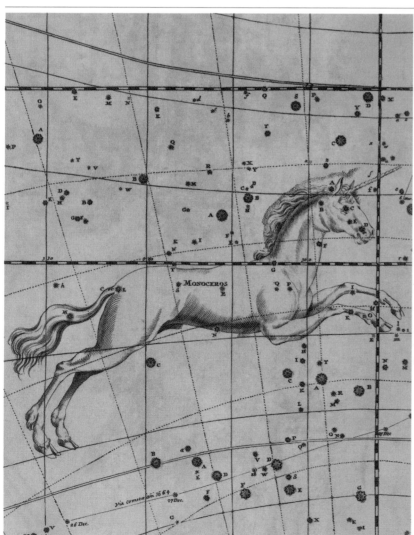

MONOCEROS.

Between Orion's two dogs – Canis Minor and Canis Major – sits Monoceros, the Unicorn. The origins of this star pattern are disputed, but it is first known from Plancius's celestial globe of 1612.

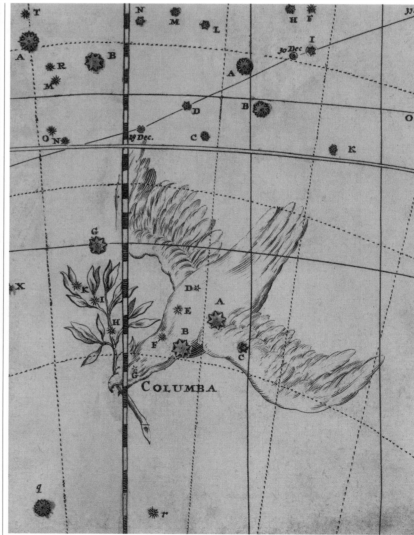

COLUMBA.

This constellation – representing the dove that returned to Noah's Ark bearing a sprig of olive to indicate that the waters of the Great Flood were receding – is first known from a map of 1592 by Plancius.

PLATE 22.

FIRMAMENTUM SOBIESCANUM,
SIVE URANOGRAPHIA (1690).

Johannes Hevelius's last work, known as the *Prodromus Astronomiæ*,
was completed after his death by his widow, Elisabeth. It comprises
three parts: a preface, a stellar catalogue based largely on Hevelius's
observations, and an accompanying atlas, usually known simply as the
Uranographia. This atlas contains some fifty-four plates encompassing
seventy-three constellations (including some of Hevelius's own invention
and some that are no longer used). For the charts of far southern stars,
the *Uranographia* uses stellar positions measured by Edmond Halley

on an expedition to St Helena in 1676. The *Uranographia* is rightly
renowned for the artistry of its engravings – by Charles de la Haye, a
renowned engraver at the Polish court – but many of Hevelius's decisions
hamper its usefulness. For example, the entire atlas is drawn from an
external perspective, rather than representing the constellations as
they appear in the night sky, and the stars are not labelled on the charts
in any way. Furthermore, Hevelius's observations are not as precise as
they might have been due to his insistence that the naked eye was better
than the telescope for positional work.

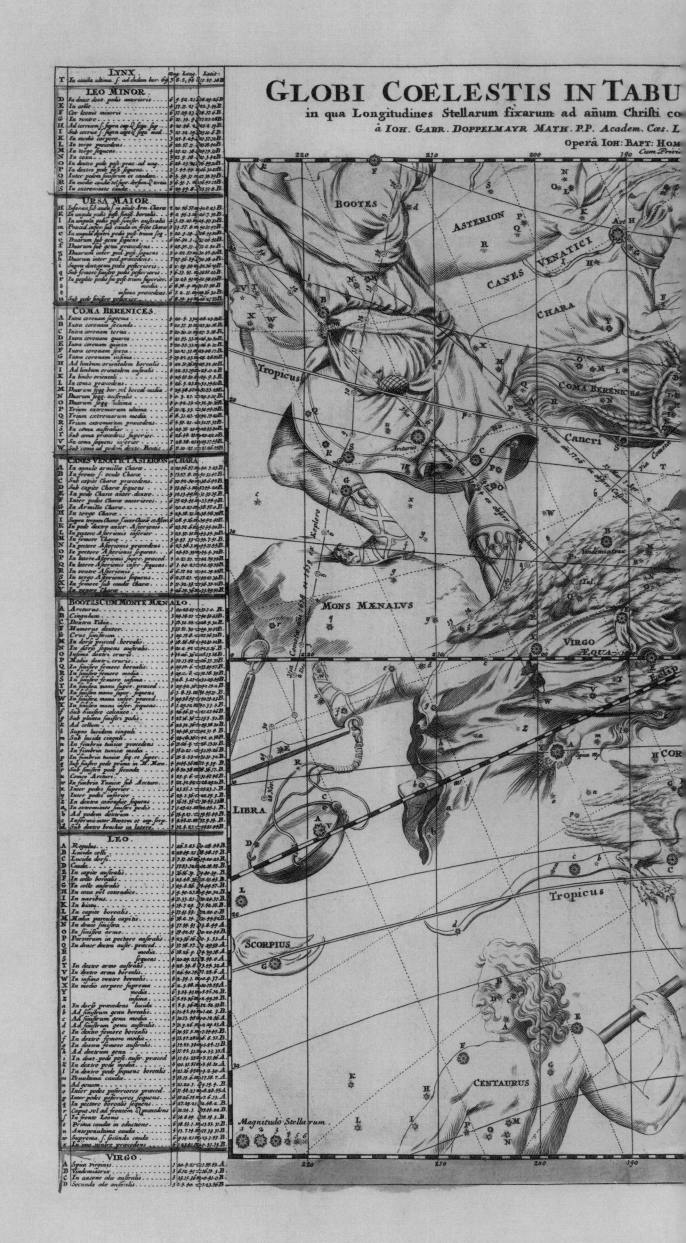

GLOBI COELESTIS IN TABU...

in qua Longitudines Stellarum fixarum ad añum Christi co...

à IOH. GABR. DOPPELMAYR MATH. P.P. Academ. Cæs. L...

Operâ IOH: BAPT: HOM...

Cum Privi...

LYNX

LEO MINOR

URSA MAIOR

COMA BERENICES

CANES VENATICI ASTERION ET CHARA

BOOTES CUM MONTE MÆNALO

LEO

VIRGO

BOOTES

ASTERION

CANES VENATICI

CHARA

COMA BERENICES

Cancri

Tropicus

MONS MÆNALVS

VIRGO
ÆQUA...

Eclip...

LIBRA

H COR...

SCORPIUS

Tropicus

CENTAURUS

Magnitudo Stellarum.

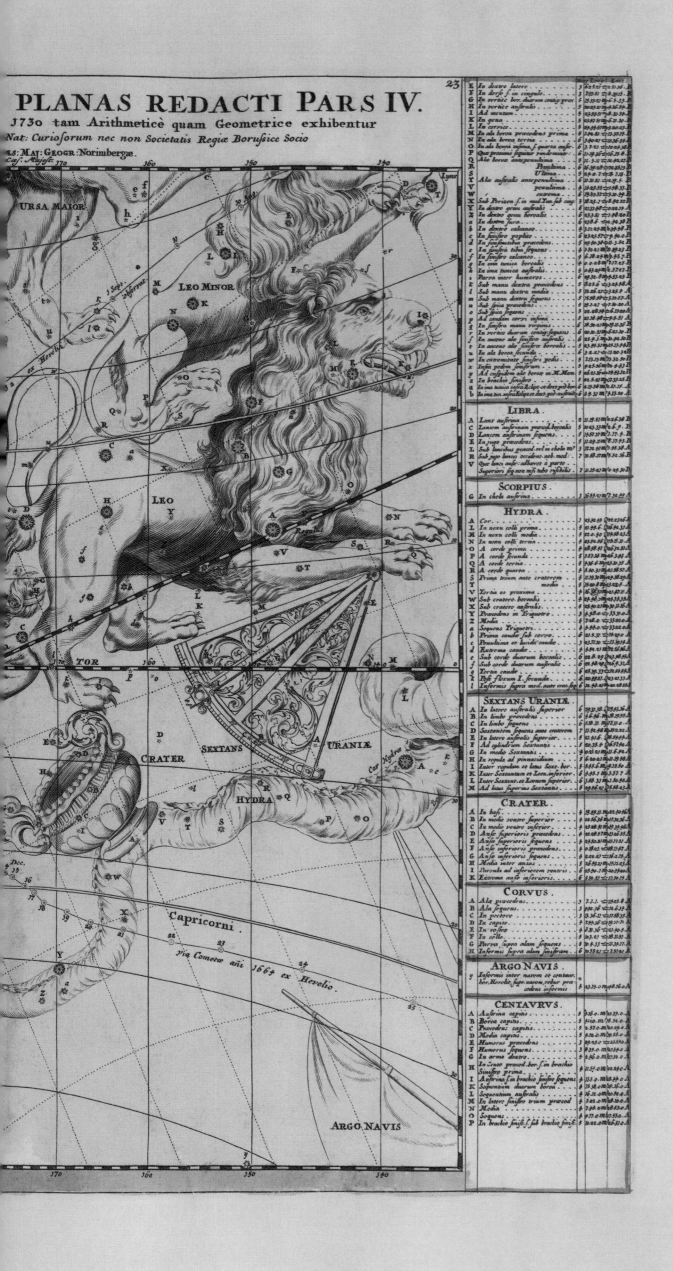

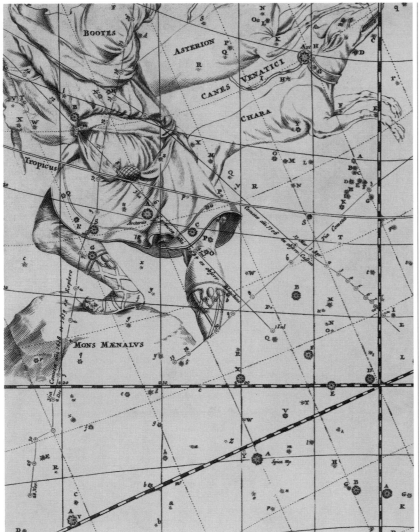

BOÖTES AND CANES VENATICI. The ancient constellation of Boötes, the Herdsman, is truncated by this plate's junction with Plate 20. To his west lie two hunting dogs, hived off into a separate constellation in Hevelius's *Uranographia* and named Asterion and Chara.

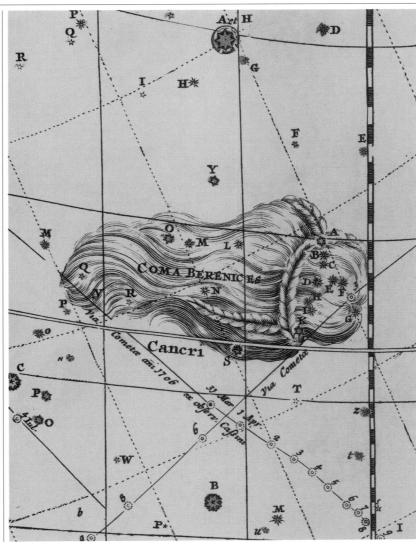

COMA BERENICES. The small constellation of Coma Berenices depicts the hair of Queen Berenice II of Egypt, sacrificed to the gods in exchange for her husband's safe return from war. The region was seen as part of Leo until it gained independence in the 16th century.

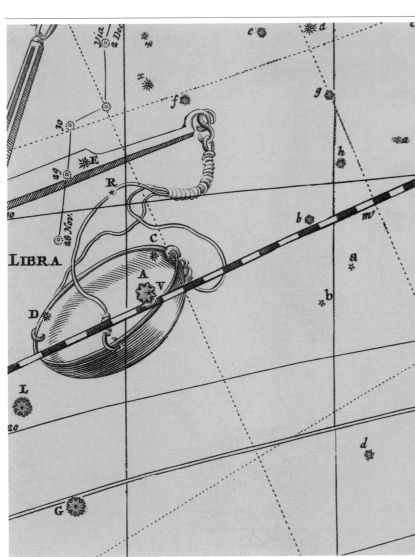

LIBRA. The constellation of the Scales has been recognized for some 3,000 years or more. Although generally seen as the only inanimate object among the zodiac constellations, it has often done 'double duty' as the claws of Scorpius.

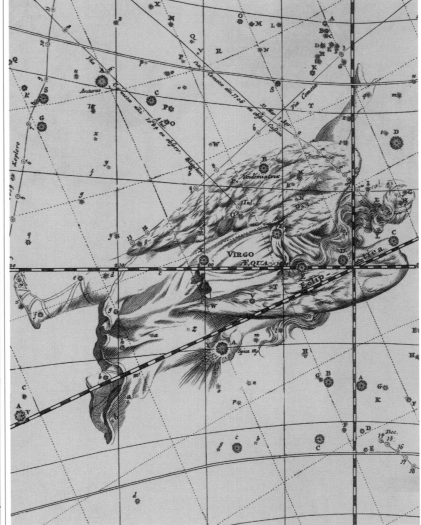

VIRGO. The constellation of the Virgin was first associated with a harvest goddess in ancient Mesopotamia. The Greeks saw her as Demeter or Persephone, and the Romans as Ceres. The bright star Spica marks an ear of wheat in her hand.

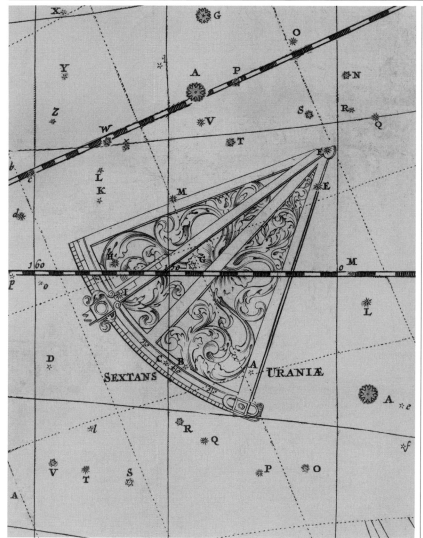

SEXTANS URANIAE.

Sandwiched between Leo and Hydra, the small and faint constellation of the Sextant was introduced by Hevelius to honour a favourite measuring device destroyed in a fire in 1679. Flamsteed was the first to shorten its name simply to Sextans.

LEO.

This group of stars has been seen as a Lion for at least 4,000 years. Greek storytellers imagined it as the Nemean lion fought by Hercules. In 1687, Hevelius added another lion, Leo Minor, to the skies directly north of the Ptolemaic original.

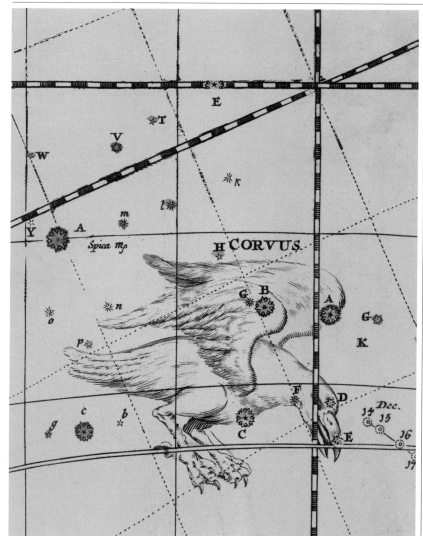

CORVUS.

The small constellation of Corvus, the Crow, was first depicted on the back of the winding water snake Hydra by Babylonian stargazers in c. 1000 BCE. Classical legends associated it with a servant of the god Apollo.

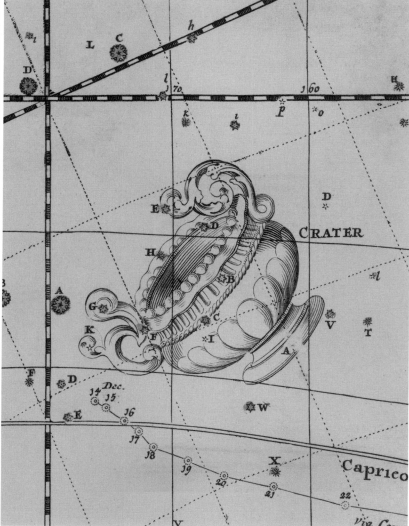

CRATER.

Another small constellation on the back of Hydra, Crater's stars were originally part of the Babylonian raven constellation. Ptolemy, however, treated them as a separate constellation, representing a cup from the same Apollo myth as Corvus.

PLATE 23.

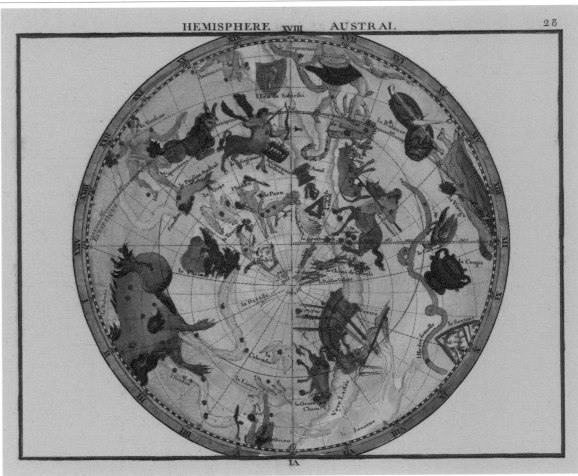

ATLAS COELESTIS (1729).

The first star atlas to truly benefit from the accuracy of telescopic observations was that of John Flamsteed. Notoriously precise and unwilling to commit his observations to posterity without satisfactory verification, Flamsteed fiercely protected his data during his lifetime. As a result, his major works – the star catalogue *Historia Coelestis Britannica* and companion *Atlas Coelestis* – were both published posthumously (in 1725 and 1729, respectively) by his widow, Margaret. Flamsteed's stars were engraved on twenty-five maps, each showing one or more constellations from a geocentric point of view, with rococo

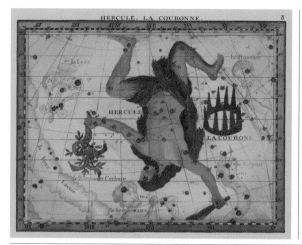

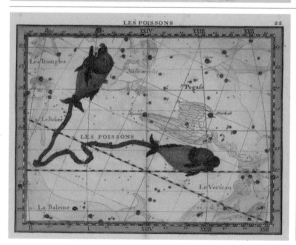

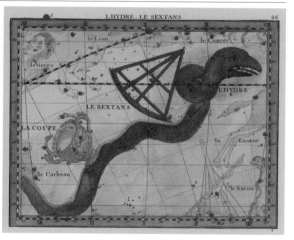

figures by the artist James Thornhill. Flamsteed's assistant, Abraham Sharp (1653–1742), prepared a pair of maps showing the hemispheres of the sky. The maps favour the equatorial coordinate system over the ecliptic one, but this practical advance was negated by the sheer size of the *Atlas* plates. The versions shown here are, in fact, re-engravings from Jean-Nicolas Fortin's *Atlas Céleste de Flamstéed* (1776). This popular French edition added another constellation map and several other plates, and replaced the hemispheres with a pair by French astronomer Charles Monnier (1715–99), shown here on the left.

GLOBI COELESTIS IN TABU

in qua Longitudines Stellarum fixarum ad añum Christi cc

à IOH. GABR. DOPPELMAYR MATH. P.P. Academ. Cæs. I.

Operâ IOH. BAPT. HOMAN

Cum Privile

PLANAS REDACTI PARS V.

1730 tam Arithmetice quam Geometrice exhibentur

Nat. Curiosorum, nec non Societatis Regiæ Borussicæ Socio

es. Maj. Geogr. Norimbergæ .

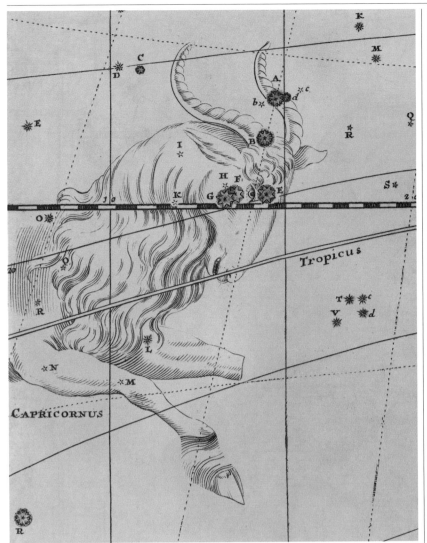

CAPRICORNUS.

One of the oldest known constellations, Capricornus is a hybrid beast, with a goat's body and the tail of a fish. Classical astronomers associated it with the goat Amalthea, who suckled the infant Zeus, and with the goat-like god Pan.

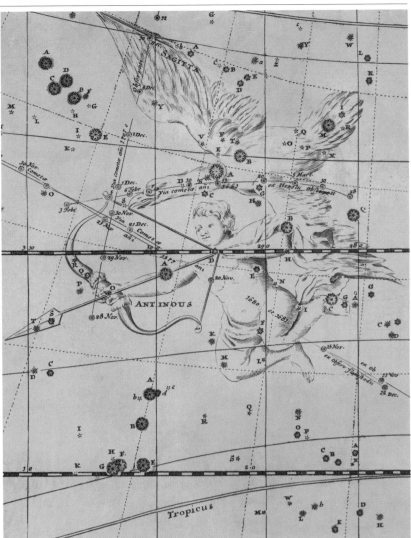

ANTINOUS AND AQUILA.

The obsolete constellation Antinous was invented to honour the Roman Emperor Hadrian's young lover after his death in the River Nile. Ptolemy considered it a subsection of the eagle Aquila, but early star mappers frequently depicted it independently.

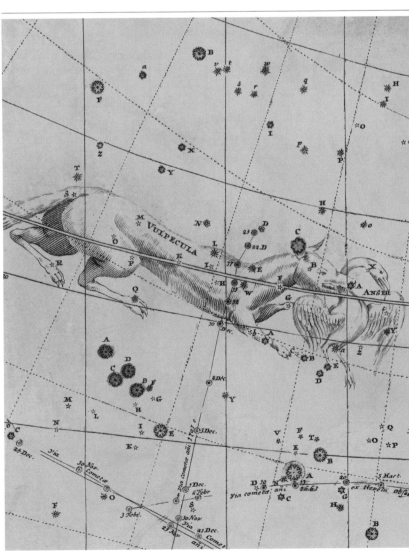

VULPECULA AND ANSER.

Hevelius invented the pattern of a Little Fox clutching a Goose for his star atlas *Uranographia*. Later astronomers differed over whether it should be treated as a single constellation or a pair. Today, only Vulpecula survives.

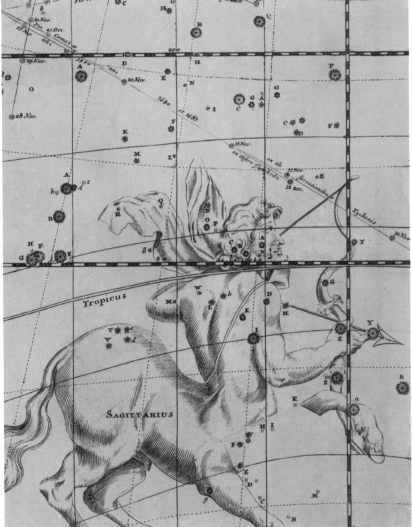

SAGITTARIUS.

The constellation of the Archer has been depicted as a hybrid creature with a horse-like body since Babylonian times. A vertical dashed line marks the right ascension of the Sun at its most southerly point in the sky at northern midwinter.

HERCULES AND CEREBRUS (sic).

Bayer's *Uranometria* (1603) transformed these stars near the ancient hero Hercules into a branch of the Hesperides' apple tree. Hevelius saw them as the three-headed monster Cerberus, although depicted as a multi-headed snake rather than a dog.

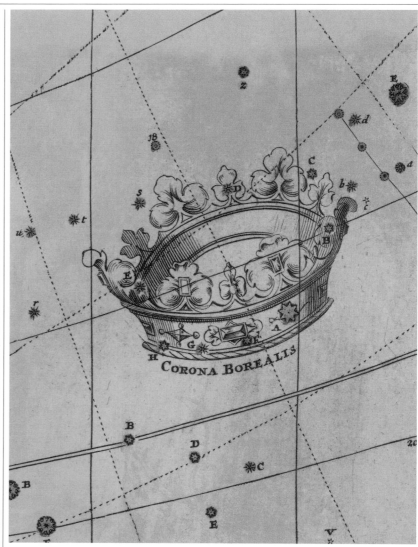

CORONA BOREALIS.

A distinctive arc of stars forms the Northern Crown, one of Ptolemy's forty-eight ancient constellations. Classical astronomers associated it with the crown of princess Ariadne, thrown into the sky in celebration of her wedding to the god Dionysus.

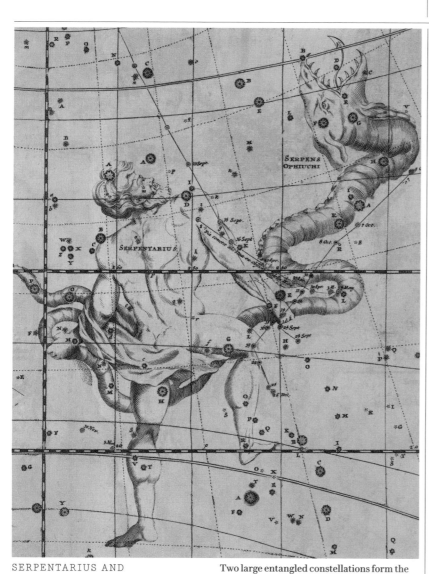

SERPENTARIUS AND SERPENS OPHIUCHI.

Two large entangled constellations form the figure of a giant wrestling a serpent. Today, the component parts are known as Ophiuchus (from the Greek for 'serpent-bearer') and Serpens Caput and Cauda (the snake's head and tail).

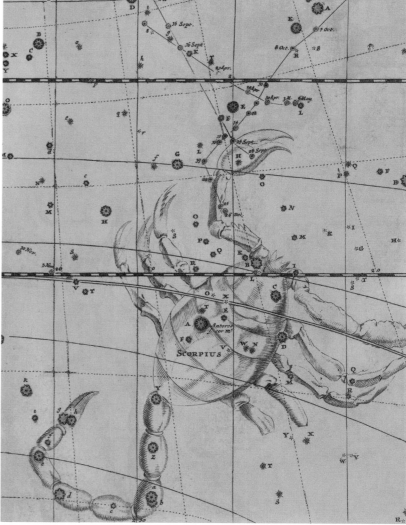

SCORPIUS.

With its heart marked by the red star Antares, the Scorpion constellation has survived unchanged from ancient Mesopotamia. Greek astronomers associated it with a creature sent by the goddesses Artemis and Leto to kill Orion.

PLATE 24.

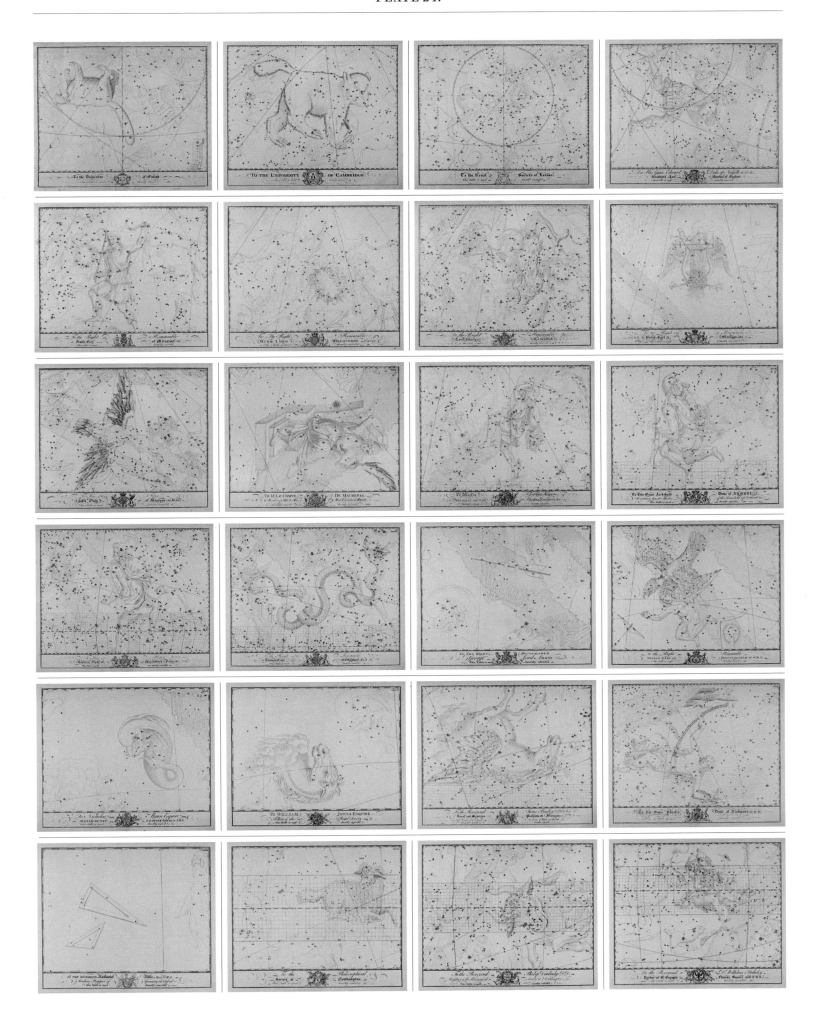

URANOGRAPHIA BRITANNICA (1750).

These plates come from the ambitious but ultimately doomed project of English astronomer John Bevis (1695–1771), best known today as the discoverer of the famous Crab Nebula in Taurus. Bevis's *Uranographia Britannica* was intended to be an updated and more precise version of Bayer's *Uranometria*, adding stars from the intervening observations of Hevelius, Flamsteed and Halley, updating stellar positions to take account of almost 150 years of precession, and (somewhat fittingly) adding a scattering of the nebulae, star clusters and other fuzzy telescopic objects that were then drawing attention among the astronomical

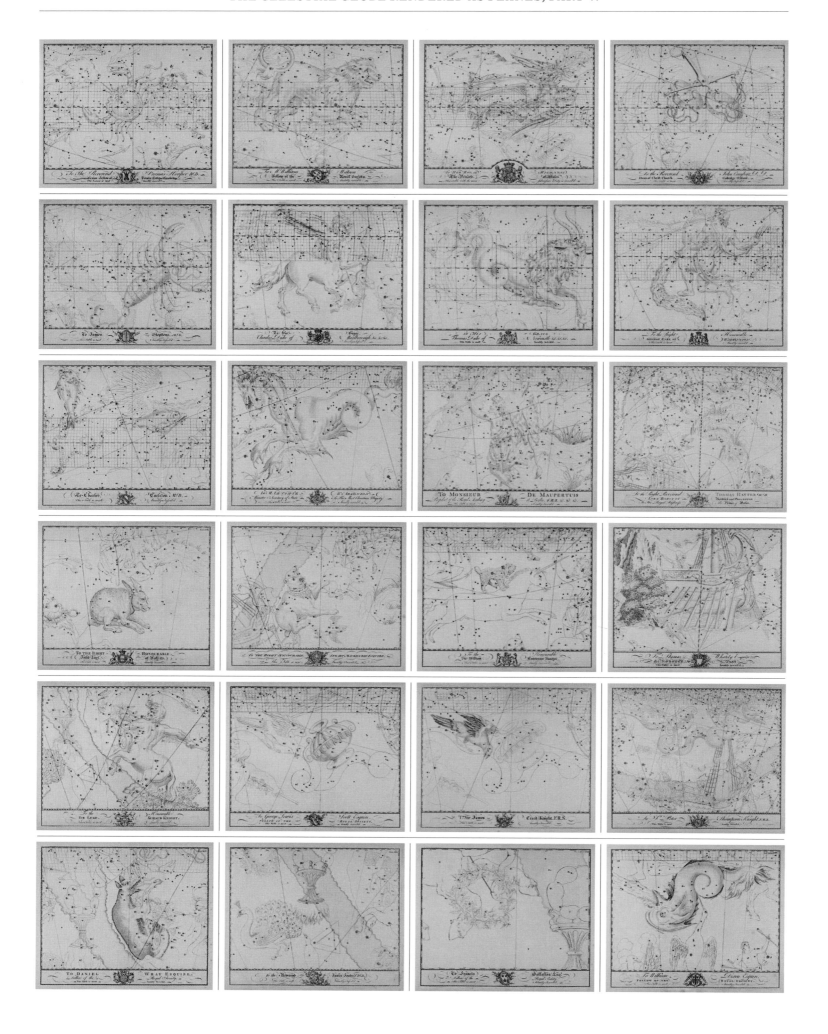

community. In other ways, Bevis's project copied the Bayer atlas.
Its layout was based on ecliptic coordinates (placing the zodiac on
the dividing line between hemispheres), and it even reused Bayer's
illustrations (resulting sometimes in a similar clash between an Earth-
centred mapping of the stars and an external 'mirror-image' view of
some figures). The project foundered when Bevis's sponsor, instrument
maker John Neale, was made bankrupt in 1750, and the original plates
were destroyed. However, bound sets of prints eventually emerged
onto the market under the title *Atlas Celeste* from 1786.

CORONA AUSTRALIS.		Mag.	Long.	Latit.
A	Quæ foris prima ad Auftr. in coron. prox.	5	♐ 20. 0	♑ 22.33. 0. A
	SAGITTARIUS.		Hall.	
f	In insiragine dextro prove conjug. auftr.	4	♐ 17.57. 0	♑ 22.27. 0. A
g	Earum borea	4	♐ 17.59. 0	♑ 22. 5. 0. A

PISCIS NOTIUS.		Expl.		
M	Præcedentium Piscem quæ anteit	5	♑ 1. 0. 0	♒ 20.40. A
N	Media	4	♑ 4.30. 0	♒ 20. 0. A
O	Sequens trium	3	♑ 7. 0. 0	♒ 20.30. A
P	Quæ hanc præcedit obscura	5	♑ 5. 0. 0	♒ 20.30. A

CENTAURUS.		Hall.		
Q	In sinistro cubito	5	♎ 16.45. 0	♍ 25.25. 0. A
R	In extrema manu sinistra	4	♎ 10. 8. 0	♍ 25.37. 0. A
S	In eductione corporis humani	3	♎ 11.13. 0	♍ 24.49. 0. A
T	Duarum obscurarum sequens	3	♎ 11.35. 0	♍ 31. 0. 0. A
V	præcedens	3	♎ 10.38. 0	♍ 10.00. 0. A
W	In dorso equino nebula	N	♎ 6.36. 0	♍ 35. 7. 0. A
X	In humeris duarum sequens	2	♎ 28.34. 0	♍ 30. 3. 0. A
Y	Præcedens ex illis	4	♎ 27.38. 0	♍ 30. 3. 0. A
Z	In sinistro femore duarum borea	3	♎ 23.38. 0	♍ 44.43. 0. A
a	Australis	2	♎ 23.46. 0	♍ 46.48. 0. A
b	E thalur femor sequentibus Austr.	4	♎ 27.38. 0	♍ 47.47. 0. A
c	Borea	4	♎ 27. 3. 0	♍ 48.18. 0. A
d	Clara in alvo	2	♎ 21.56. 0	♍ 49.26. 0. A
e	De quatuor fictis à ſtaula Croſero borea	2	♏ 1. 1. 0	♍ 47.41. 0. A
f	Australis, per crucis	2	♎ 21.13. 0	♍ 48.25. 0. A
g	Præcedens crucis	3	♎ 20. 0. 0	♍ 50.38. 0. A
h	Sequens crucis	2	♏ 2.37. 0	♍ 48.42. 0. A
i	Quæ in pede sinistro	3	♎ 30.30. 0	♍ 48.43. 0. A
k	In genu dextro, ſ. potius ad ung.	4	♏ 9. 3. 0	♍ 44.20. 0. A
l	Quæ inter hanc et præced. Triang.			
	qual. ſ. ad ling. pedis sinistri.	3	♎ 28.36. 0	♍ 46. 2. 0. A

LUPUS.		Hall.		
A	In summo pede posteriori ad brachium			
	centauri borealis	3	♎ 21.36. 0	♍ 29.56. 0. A
B	In poplite pedis ejusdem	3	♎ 19.46. 0	♍ 29.39. 0. A
C	Duarum magis australium borea	3	♎ 15.55. 0	♍ 27. 8. 0. A
D	Australis	3	♎ 15.23. 0	♍ 27.39. 0. A
E	Parvula conjig. in poft. australis	3	♎ 20. 3. 0	♍ 23.26. 0. A
F	In armo duarum præcedens	4	♎ 22.39. 0	♍ 24.37. 0. A
G	In armo duarum sequens	4	♎ 27.34. 0	♍ 29. 3. 0. A
H	In medio corpore	4	♎ 20.17. 0	♍ 25. 72. 0. A
I	In alvo	5	♎ 23.59. 0	♍ 20.40. 0. A
K	In femore	3	♎ 23.46. 0	♍ 23.13. 0. A
L	In eductione femoris borea	4	♎ 20.30. 0	♍ 28.43. 0. A
M	Australis	3	♎ 25.40. 0	♍ 29. 4. 0. A
N	In summo lumbo	4	♎ 26.58. 0	♍ 31.44. 0. A
O	In extrema cauda duarum ausfr:	3	♎ 26. 9. 0	♍ 29. 5. 0. A
P	Borealis	3	♎ 25.44. 0	♍ 28.33. 0. A

TRIANGULUM AUSTRALE.		Hall.		
A	Quæ in cuspide	2	♏ 17. 5. 0	♍ 38. 0. 0. A
B	Borea basis	3	♏ 2. 8. 0	♍ 38.46. 0. A
C	Hanc sequens parva	4	♏ 3. 3. 0	♍ 31.49. 0. A
D	Australis basis	3	♏ 6.37. 0	♍ 37.37. 0. A
E	Parva in media	5	♏ 6.49. 0	♍ 35. 80. 0. A

ARA THURIBULI.		Hall.		
A	In basi duarum borea	6	♏ 31.38. 0	♍ 35.36. 0. A
B	Australis	4	♏ 22.21. 0	♍ 36.31. 0. A
C	In medio Aræ	3	♏ 23. 6. 0	♍ 38.46. 0. A
D	In foco trium borea	3	♏ 23.30. 0	♍ 30.40. 0. A
E	Reliquarum duarum contig. ausfr.	4	♏ 20.30. 0	♍ 39.58. 0. A
F	Borea	4	♏ 20.53. 0	♍ 34.70. 0. A
G	In medio flamæ	4	♏ 16. 5. 0	♍ 33.32. 0. A
H	In fummitate flamæ præced.	4	♏ 28.57. 0	♍ 31.30. 0. A
I	Sequens	4	♏ 21.57. 0	♍ 30.32. 0. A

PAVO.		Hall.		
A	Pavonis oculus	3	♐ 10. 3. 0	♑ 36. 6. 0. A
B	Quæ in pectore	3	♐ 16.58. 0	♑ 36.30. 0. A
C	In radice alæ sinistra	3	♐ 13. 0. 0	♑ 40.47. 0. A
D	In media ala	3	♐ 13.41. 0	♑ 44.23. 0. A
E	In eductione caudæ prima	4	♐ 1.28. 0	♑ 47. 2. 0. A
F	Secunda caudæ	4	♐ 9.34. 0	♑ 44.33. 0. A
G	Tertia caudæ	4	♐ 4.37. 0	♑ 10.48.30. A
H	Quarta caudæ	6	♐ 14.13. 0	♑ 47.13.30. A
I	Quinta caudæ	4	♐ 11.36. 0	♑ 04.49.0. A
K	Sexta caudæ	4	♐ 10.31.50	♑ 51.23.0. A
L	Septima caudæ	5	♐ 27.23. 0	♑ 40. 5. 0. A
M	Ultima caudæ	4	♐ 20.20.50	♑ 45. 5. 0. A
N	In sinist: pede Hevelio, ſub cauda	5	♐ 1.29.66	♑ 49.43. 0. A
O	In dextro pede Hevel. in sinistra	4	♐ 9.46.20	♑ 40.49. 0. A

INDUS.		Expl.		
A	In capite	6	♑ 19.40. 0	♒ 34.30. 0. A
B	In axilla dextra	4	♑ 1.39. 0	♒ 33.48. 0. A
C	In dextra manu sagitta prima	6	♑ 2.16. 0	♒ 37. 0. 0. A
D	Secunda	6	♑ 2.59. 0	♒ 38.33. 0. A
E	Tertia	5	♑ 1.19. 0	♒ 39. 5. 0. A
F	In fuma parte fagittæ man ſin.	3	♑ 2. 6. 0	♒ 37.52. 0. A
G	In ima	5	♑ 15.06. 0	♒ 32.18. 0. A
H	In axilla sinistr: occidente	4	♑ 2.46. 0	♒ 33.45. 0. A
I	Sequens	6	♑ 4.34. 0	♒ 33.53. 0. A
K	Orientalior	6	♑ 4.34. 0	♒ 33.40. 0. A
L	In pectore	5	♑ 4.46. 0	♒ 36.10. 0. A
M	In ventre	4	♑ 4. 6. 0	♒ 39.10. 0. A

GRUS.		Hall.		
B	In medio colli	3	♑ 18. 5. 0	♒ 28.53. 0. A
C	In eductione colli contig: bori:	3	♑ 16.31. 0	♒ 28.45. 0. A
D	Australia	3	♑ 16.36. 0	♒ 28.33. 0. A
E	In dorso duaru' borea, Hevel: in pectore	4	♑ 17.47. 0	♒ 33.18. 0. A
F	Australia	4	♑ 15.32. 0	♒ 32.25. 0. A
G	In sinistra ala borea, Hevel: in dextra	5	♑ 18. 0. 0	♒ 35.17. 0. A
H	Australior	2	♑ 19. 2. 0	♒ 30.58. 0. A
I	In ala dextra Hevelio, sinistra	4	♑ 18.26. 0	♒ 35.22. 0. A
K	In eductione caudæ Hevel in ventre	4	♑ 16.31. 0	♒ 30.43.40. A
L	In cauda trium borea	4	♑ 11.56. 0	♒ 33.52. 0. A
M	Præcedens	3	♑ 14. 8. 0	♒ 34.56. 0. A
N	Sequens			

MUSCA APIS.		Hall.		
A	In capite	4	♏ 15.26. 0	♍ 33. 7. 0. A
B	In Ala dextra	3	♏ 16.51. 0	♍ 36.44. 0. A
C	In Ala sinistra	4	♏ 22.46. 0	♍ 36.20. 0. A
D	In cauda	4	♏ 20.18. 0	♍ 38.43. 0. A

CHAMÆLEON.		Hall.		
A	Ad collum	5	♏ 1.33. 0	♍ 63.38. 0. A
B	In prioribus pedibus	5	♏ 20.30. 0	♍ 63.59. 0. A
C	In posterioribus pedibus	5	♏ 16.37. 0	♍ 68. 7. 0. A
D	In dorso	5	♏ 1.30. 0	♍ 67.50. 0. A
E	In eductione caudæ præcedens	4	♏ 18.33. 0	♍ 72. 8. 0. A
F	Sequens	5	♏ 3.48. 0	♍ 70.30. 0. A
G	In media cauda præcedens	5	♏ 0.25. 0	♍ 73.49. 0. A
H	Sequens	6	♏ 1.37. 0	♍ 73. 3. 0. A
I	In extremitate caudæ borealis	4	♏ 25.30. 0	♍ 73.52. 0. A
K	Australis	5	♏ 17.13. 0	♍ 76. 8. 0. A

APUS AVIS INDICÆ.		Hall.		
A	In capite	6	♐ 21. 7. 0	♑ 44.28. 0. A
B	In collo	4	♐ 13.26. 0	♑ 48.52. 0. A
C	Trium in eductione caudæ bor:	4	♐ 19. 6. 0	♑ 48.23. 0. A
D	Media	5	♐ 17.37. 0	♑ 49.50. 0. A
E	Australior	4	♐ 18.23. 0	♑ 46.80. 0. A
F	In borea parte caudæ contig: prox:	6	♐ 9.52. 0	♑ 48.83. 0. A
G	Sequens	4	♐ 10.51. 0	♑ 51.42. 0. A
H	In media caudæ trium borealior	4	♐ 10.36. 0	♑ 48. 5. 0. A
I	Media	5	♐ 10. 4. 0	♑ 49. 3. 0. A
K	Australior	4	♐ 10.58. 0	♑ 60.25. 0. A
L	Pole vicinior	5	♐ 19.36. 0	♑ 61.27. 0. A

HYDRUS.		Hall.		
A	Caput Hydri	3	♏ 7.59. 0	♍ 63. 3. 0. A
B	Prima colli	5	♏ 23. 3. 0	♍ 67.30. 0. A
C	Secunda	3	♏ 28.13. 0	♍ 67.50. 0. A
D	Tertia	3	♏ 27.67. 0	♍ 73.19. 0. A
E	Quarta	3	♏ 0.26. 0	♍ 71.36. 0. A
F	Inter utramque nubes duaru' ſeq:	3	♏ 16.21. 0	♍ 76.23. 0. A
G	Præcedens	3	♏ 6.36. 0	♍ 77.20. 0. A
H	Quæ adjacet nubeculæ minori	3	♍ 16.50. 0	♍ 76.27. 0. A
I	Duarum collum hydri ſeq. borea	3	♏ 3.38. 0	♍ 79. 6. 0. A
K	Præcedens et australis	3	♏ 17.37. 0	♍ 75. 6. 0. A
L	Antepenultima caudæ	3	♏ 19.41. 0	♍ 80.40. 0. A
M	Penultima caudæ	3	♏ 10. 4. 0	♍ 88.20. 0. A
N	Ultima caudæ	3	♏ 28.19. 0	♍ 88. 0. 0. A

TOUCAN ANSER AMER.		Hall.		
A	In extremo rostro	3	♑ 5.30. 0	♒ 49.42. 0. A
B	In capite	3	♑ 16.31. 0	♒ 46.25. 0. A
C	In ancone alæ dextræ ſuper.	3	♑ 17.47. 0	♒ 40.59. 0. A
D	Inferior	3	♑ 16.42. 0	♒ 53.30. 0. A
E	In India ala	3	♑ 18. 7. 0	♒ 57.13. 0. A
F	Hanc præcedens	3	♑ 19.43. 0	♒ 56.32. 0. A
G	In dorso, vel potius in ala dext: ult:	3	♑ 22.41. 0	♒ 55.13. 0. A
H	In cauda vel potius ad pedes	3	♑ 20.30. 0	♒ 57.33. 0. A
I	In rami folio, ſ. nuce myristica	3	♑ 1.18. 0	♒ 59. 9. 0. A

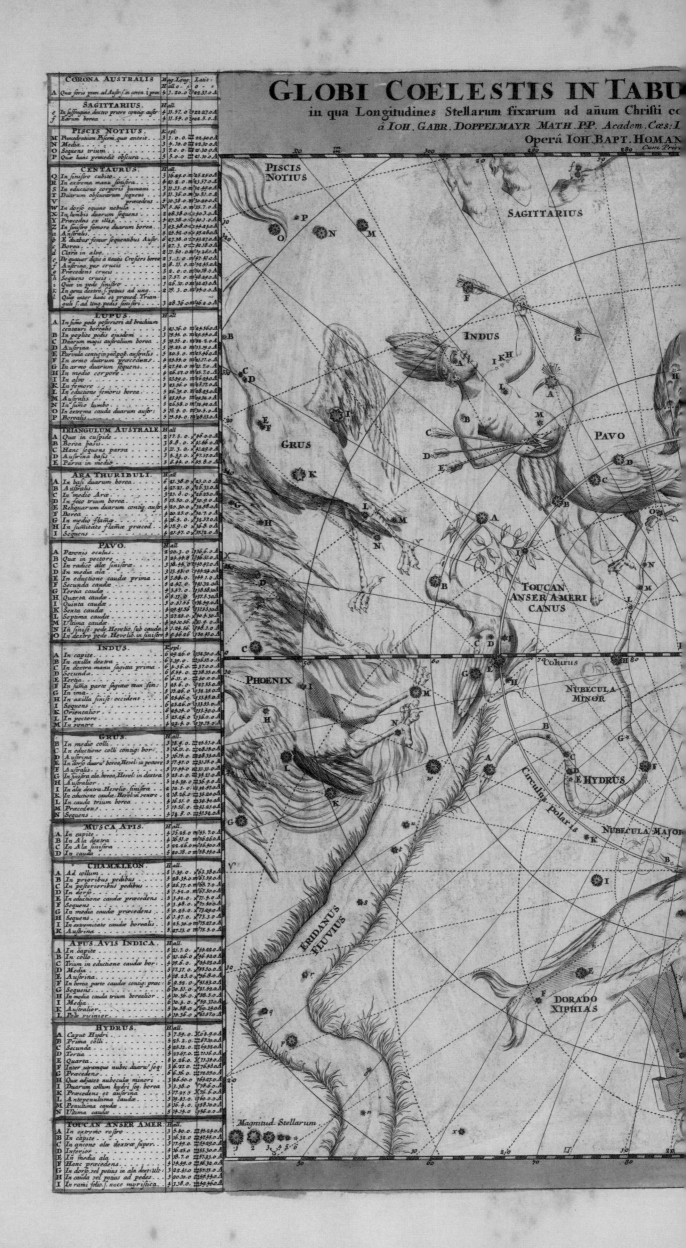

GLOBI COELESTIS IN TABU

in qua Longitudines Stellarum fixarum ad añum Christi co...

à IOH. GABR. DOPPELMAYR MATH. P.P. Academ. Cæs. L...

Operâ IOH. BAPT. HOMAN...

Cum Priv...

PISCIS NOTIUS

SAGITTARIUS

INDUS

PAVO

GRUS

TOUCAN ANSER AMERICANUS

PHOENIX

Cetus

NUBECULA MINOR

HYDRUS

Circulus Polaris

NUBECULA MAJOR

ERIDANUS FLUVIUS

DORADO XIPHIAS

Magnitud. Stellarum

PLANAS REDACTI PARS VI.

1730. tam Arithmetice quam Geometrice exhibentur

Nat:Curiosorum, nec non Societatis Regiæ Borussicæ Socio

...S. MAJ. GEOGR. Norimbergæ.

PHOENIX.

		H. all.	L. all.
C	In ancone alæ sinistræ		
D	In ala sinistra trium australi.		
G	In extrema ala dextra		
H	Ejusdem alæ educho.		
I	Ad ped. sinstr. Hevelio in pectore.		
K	In foco sub ala dextra australis.		
L	Borealis Hevelio in ala dextra		
L	In foco sub ala sinistr. prec. sin ped. fin.		
M	Sequens. Hevelio in pede dextro		

ROBUR CAROLI

		H. all.	
A	Quæ ad radicem		
B	In fumo trunce		
C	In ramis præced. de quatuor bor.		
D	Sequens		
E	Præcedens		
F	Media		
G	In ramis seq. duarum borea		
H	Australis		
I	In fuma arbore duarum præced.		
K	Sequens		
L	Informum ad truncum præced.		
M	Sequens		

PISCIS VOLANS.

		H. all.	
A	In capite.		
B	In medio corpore		
C	In cauda		
D	In ala dextra superior		
E	Inferior.		
F	In ala sinistra superior		
G	Inferior		
H	Superimus inter Chamæl. et pisc. vol.		

DORADO, XIPHIAS.

		H. all.	
A	In capite		
B	In Branchiis.		
C	Circa caput		
D	Quæ supra dorsum.		
E	Quæ in cauda		
F	In extrema cauda		

ERIDANUS.

		H. all.	
a	Omnes quatuor præced. duarū seq.		
b	Earum præcedens.		
c	Ultima flumunis		
d	Hanc præcedens.		
e	Trium in recta descendentium bor.		
s	Earum media		
t	Trium austrina.		
u	Hanc præcedens		
v	Ultima Eridani. Achernar.		
x	Informis inter fluvi. et caud. Xiph.		

ARGO NAVIS.

		H. all.	
O	In scamine mali borealis sequens		
P	Hevelio. in malo inferiorum sequens		
P	Præcedens. Hevelio in malo super. præc.		
Q	Australis sequens Hevelio in tertio		
	scuto trium superiorum media		
R	Præcedens. Hevelio in tertio scuto triu'		
	superiorum præcedens		
X	In scut. transfer. Hevelio sub velo in fundo		
Y	In ead. sectione inseralver. Hevelio in		
	tertio scuto triu' superioru' sequens		

I.

Tabula pro definiendis Longitudinibus Stellarum fixarum per singulos años ab anno 1700 ad añum 1760.

Subduc. Differentia Add. Dif.				Subduc. Differentia Add. Dif.			
An. compl.	Min. Sec. Tert.	An. compl	An. compl.	Min. Sec. Tert.	An. compl		
1700	25. 26. 0	1760	1735	12. 43. 0	1725		
1701	24. 35. 8	1759	1736	11. 52. 8	1724		
1702	23. 44. 16	1758	1737	11. 1. 16	1723		
1703	22. 53. 24	1757	1738	10. 10. 24	1722		
1704	22. 2. 32	1756	1739	9. 19. 32	1721		
1705	21. 11. 40	1755	1740	8. 28. 40	1720		
1706	20. 20. 48	1754	1741	7. 37. 48	1739		
1707	19. 29. 56	1753	1742	6. 46. 56	1738		
1708	18. 39. 4	1752	1743	5. 56. 4	1737		
1709	17. 48. 12	1751	1744	5. 5. 12	1736		
1710	16. 57. 20	1750	1745	4. 14. 20	1735		
1711	16. 6. 48	1749	1746	3. 43. 28	1734		
1712	15. 15. 36	1748	1747	2. 32. 36	1733		
1713	14. 24. 44	1747	1748	1. 41. 44	1732		
1714	13. 33. 52	1746	1749	0. 50. 52	1731		

II.
Motus Stellarum fixarum pro singulis mensibus.

Mens. compl.	Sec. Tert.	Mens. compl.	Sec. Tert.	Mens. compl.	Sec. Tert.
Ianuarius	4. 14.	Majus	21. 12.	Septembris	38. 9.
Februarius	8. 29.	Iulius	25. 26.	Octobris	42. 23.
Martius	12. 43.	Iulius	29. 40.	Novembr.	46. 38.
Aprilus	16. 57.	Augustus	33. 55.	Decembr.	50. 92.

III.
Motus Stellarum fixarum diurnus.

Dies	1.	2.	3.	4.	5.	6.	7.	8.	9.	10.
Mot. diurn.	S. T.	S. T.	S. T.	S. T.	S. T.	S. T.	S. T.	S. T.	S. T.	S. T.

Dies	11.	12.	13.	14.	15.	16.	17.	18.	19.	20.
Mot. diurn.	S. T.	S. T.	S. T.	S. T.	S. T.	S. T.	S. T.	S. T.	S. T.	S. T.

Dies	21.	22.	23.	24.	25.	26.	27.	28.	29.	30.
Mot. diurn.	S. T.	S. T.	S. T.	S. T.	S. T.	S. T.	S. T.	S. T.	S. T.	S. T.

Usus præcedentium Tabularum.

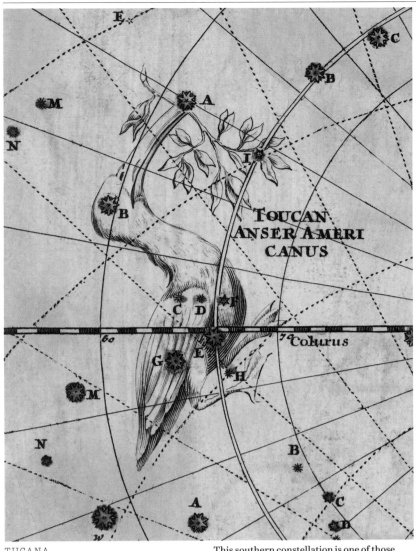

TUCANA.

This southern constellation is one of those invented by Pieter Dirkszoon Keyser and Frederick de Houtman on the first Dutch trading voyage to the East Indies in the 1590s. Doppelmayr also offers an alternative name, the American Goose.

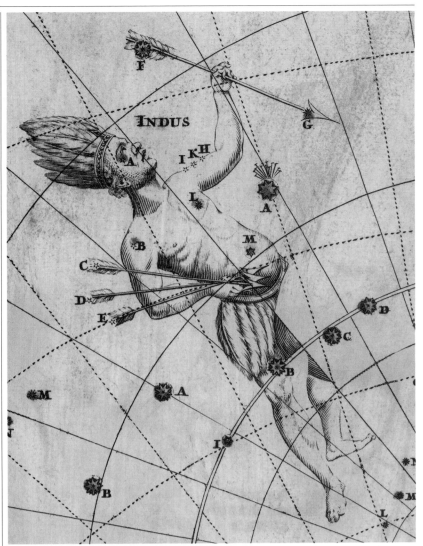

INDUS.

The constellation of the Indian is another invention of Keyser and de Houtman that first appears on Plancius's globe of 1598. Most early charts give the figure a generic loincloth and arrows; Doppelmayr's feathered headdress is an unusual departure.

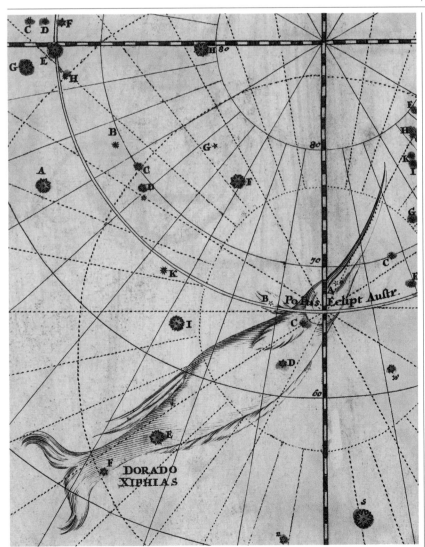

DORADO.

Keyser and de Houtman invented this constellation with the intention of depicting the mahi-mahi or dolphinfish. Later astronomers frequently preferred to see it as a swordfish (hence the long-lasting alternative name Xiphias).

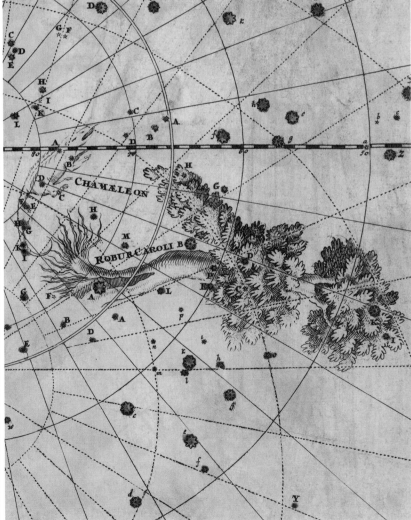

CHAMELEON AND ROBUR CAROLI.

The constellation of the Chameleon is another Dutch invention of the late 16th century. Nearby, Doppelmayr shows the now-obsolete constellation of King Charles's Oak, sequestered from the stars of Argo by Edmond Halley in 1678.

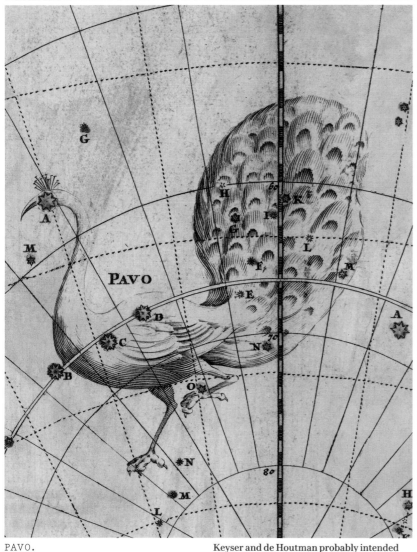

PAVO.

Keyser and de Houtman probably intended this constellation to represent the Javan green peacock. A generation after Doppelmayr, its tail was abruptly truncated when its stars were co-opted by Nicolas Louis de Lacaille for his new constellation Telescopium.

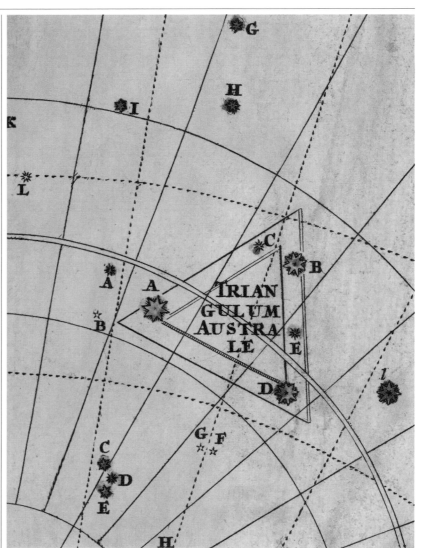

TRIANGULUM AUSTRALE.

The presence of a distinctive triangle of stars in southern skies was first reported in the early 1500s by Amerigo Vespucci. The modern constellation, renamed Triangulum Australe, first appears on Plancius's globe of 1598, in an incorrect position.

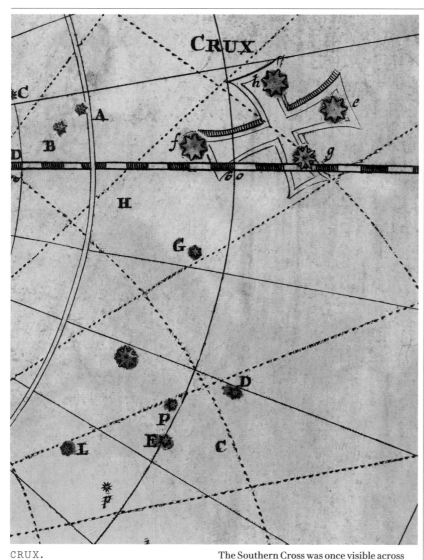
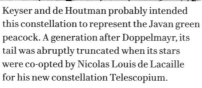

CRUX.

The Southern Cross was once visible across the ancient Mediterranean, and seen as part of Centaurus. Rediscovered by Europeans around 1500, it was first charted as a separate constellation by Plancius and Emery Molyneux (d. 1598) in 1592.

ARGO NAVIS.

The vast ship Argo may have originated as the boat of Egyptian god Osiris. In 1756, Lacaille split the ungainly constellation into three areas (Vela, the Sails; Carina, the Keel; and Puppis, the Poop Deck), which are now independent constellations.

PLATE 25.

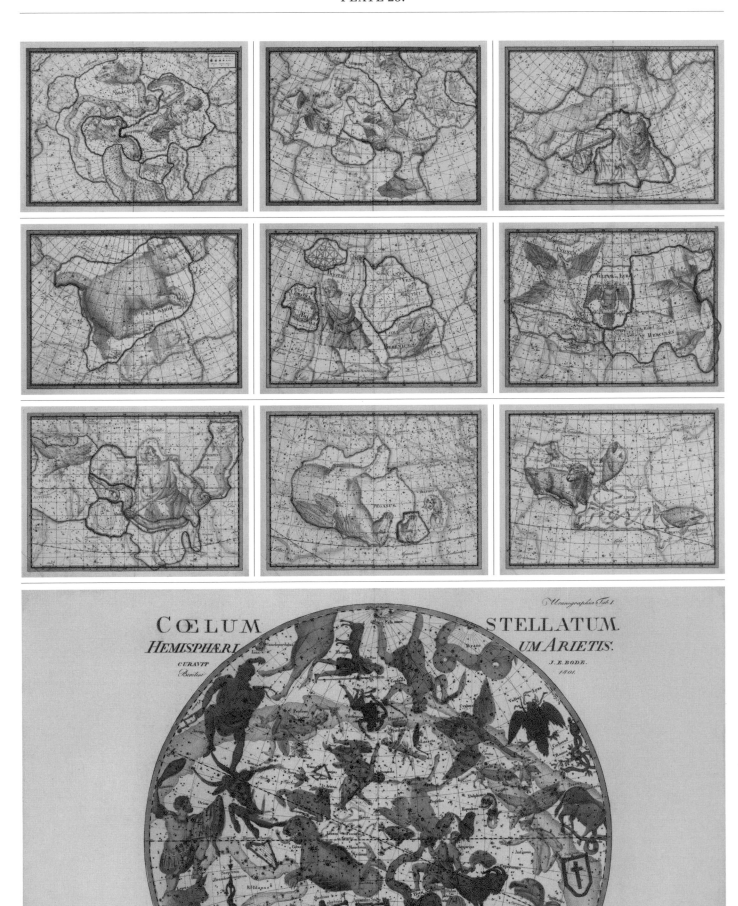

URANOGRAPHIA SIVE ASTRORUM DESCRIPTIO (1801).

Johann Elert Bode's grand project of the early 19th century marks the culmination of the star atlas tradition that began with Bayer two centuries before. Bode published a German equivalent of the *Atlas Céleste de Flamstéed* in 1782, including some extra stars and innovative drawn boundaries between constellations. However, after becoming director of the Berlin Observatory in 1786, he began a far more ambitious work.

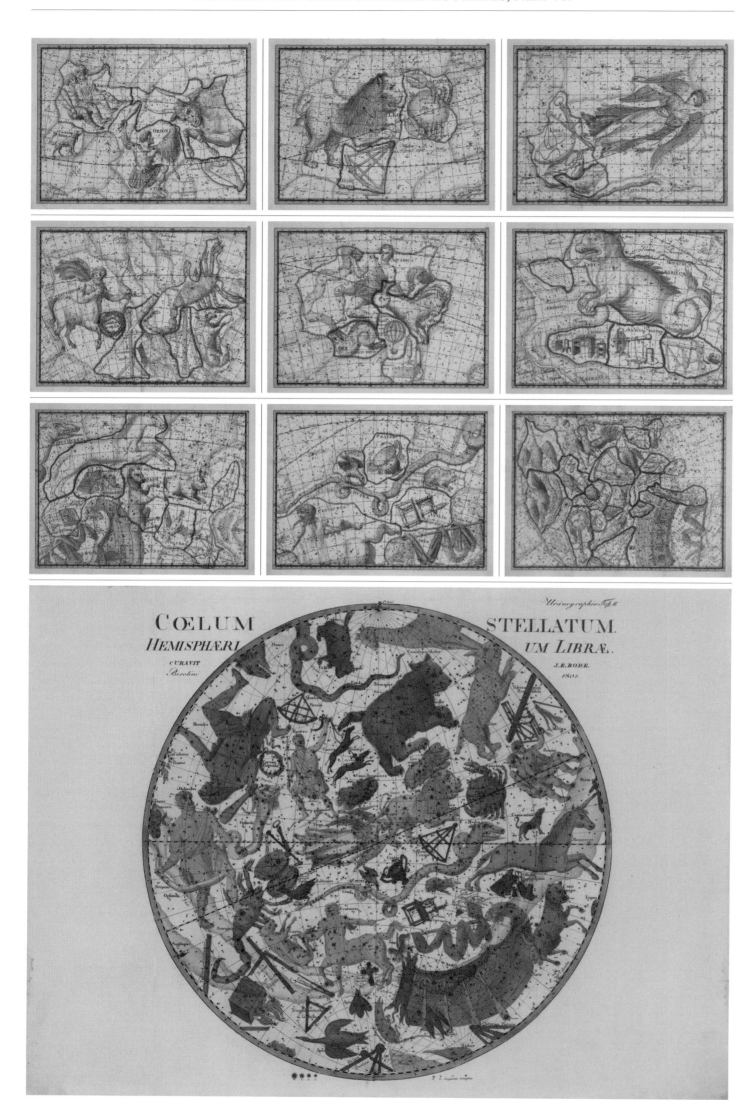

Bode's *Uranographia* was based on a catalogue of some 17,240 stars –
far more than any previous atlas and going well below naked-eye visibility
– as well as around 2,500 nebulae and deep-sky objects catalogued by
William and Caroline Herschel. Two hemispheres showed opposite
sides of the entire sky (centred on the celestial equator at 0 hours and
12 hours right ascension) while eighteen plates encompassed no fewer
than one hundred constellations and other patterns.

PLATE 26.

THEORY
OF COMETS.

*Doppelmayr describes changing ideas about the orbits of comets
from the beginning of the 17th century to his own time.*

T he nature of comets was a cause of much debate from ancient times through to Doppelmayr's own age, when they would ultimately deliver some of the most startling proofs of both the heliocentric solar system and Isaac Newton's (1643–1727) laws of motion and gravitation.

Early stargazers interpreted comets as phenomena of the upper atmosphere. Believing that they were fiery in nature, they mostly placed them in the Aristotelean sphere of fire that lay above the air but below the lunar sphere (and naturally drew fiery objects towards it). The fact that they were not restricted to the zodiac, but could instead appear in any part of the sky, was taken as a strong argument against the idea that they might be related to planets. Of course, their rapid shifts in appearance also confined them to the changeable sublunary realm. It was only the appearance of the Great Comet of 1577 that allowed Tycho Brahe (1546–1601) and others to put a maximum value on the parallax of a comet, and determine that this one, at least, lay far beyond the Moon.

Once comets were accepted as moving in the realm of the planets, the most obvious question was what sort of orbits they followed. Tycho's answer was to place them on circular paths around the Sun, similar to those taken by the planets (although, naturally, in Tycho's model, the Sun itself still orbited Earth). Doppelmayr, however, illustrates four alternative hypotheses in the corners of Plate 26: those of Johannes Kepler (1571–1630), Johannes Hevelius (1611–87), Pierre Petit (1598–1677) and Giovanni Domenico Cassini (1625–1712).

Any theory of comets had to account for three factors in their appearance: their largely rectilinear (straight-line) motion across the sky, the random orientation of their tracks with respect to the other planetary bodies and their remarkable variations in appearance (both individually in a single apparition, and from one comet to the next). Typically, a comet would first become visible while faint and relatively slow moving; if it already displayed a tail, it would appear fairly long. Over a matter of weeks, it would increase in brightness and pick up speed, perhaps shifting its position by several degrees each day, although its movements could vary unpredictably. Its tail might shorten and then lengthen again, and in the case of the brightest comets more than one tail might develop, perhaps even pointing in different directions. Eventually, its motion would slow and it would fade away.

In this context, perhaps the early assumption that comets were a type of weather phenomenon in the upper atmosphere does not seem so wrong-headed. Aristotle (384–322 BCE) himself pondered whether they might be a special form of shooting star, objects that similarly can cross the sky in any direction and develop long tails, despite having only brief lives of a second or two.

Based on his observations in the early 1600s, Kepler concluded that comets travelled in straight lines and that their appearances were somehow triggered by conjunctions (planetary close encounters) in certain signs of the zodiac. As shown in the diagram at top left of Doppelmayr's plate, these lines were related to the orbits of successive planets. The changing speed and appearance of comets in the sky was attributed to the relative motion of Earth and comet, and the perspective of observers on Earth.

Hevelius, meanwhile, came to a different and perceptive conclusion in his *Cometographia* (1668). He realized that while the paths of comets might generally appear straight, they often curved towards the Sun when at their closest approach. He pondered whether a comet's path might be a parabola – a type of open-ended curve that comes very close to being a straight line for most of it length, but bends sharply as it changes direction. He attributed the comet's changing appearance

FIG. 1.

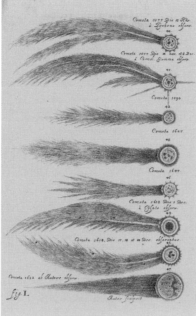

FIG. 2.

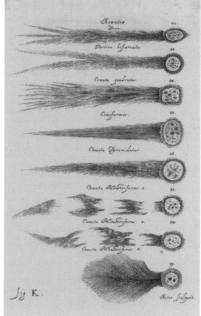

FIG. 3.

At far left is the
frontispiece
to Hevelius's
Cometographia (1668).
It shows astronomers
discussing three
models of comets:
the atmospheric
ideas of Aristotle, the
Keplerian theory of
straight-line motion,
and Hevelius's own
model of parabolic
orbits. In the centre
is Hevelius's attempt
to put comet tails in a
chronological order
of development. At
near left, he classifies
different forms of
comet tail.

to rotation of an elongated form that caused
different parts to be illuminated by the Sun,
and visible from Earth.

French astronomer Petit, ordered by Louis XIV
(1638–1715) to report on the bright comet of 1664,
was among the first to argue that comets might
return on a regular basis. In fact, he (incorrectly)
argued that the 1664 comet was identical to one
seen in 1618, due to their similar paths across
the sky, and predicted that it would return in
1710. Petit thought that comets followed regular
orbits that lay entirely beyond the orbit of
Saturn (explaining the long intervals between
their supposed reappearances), but he correctly
suggested that they only became visible when
at their closest to Earth and the Sun, and that
their tails were permanently directed away
from the Sun.

At bottom right, Doppelmayr reproduces a
small part of a much larger diagram by Cassini,
attempting to explain the motions of the brilliant
comet of 1680 in terms of his own unique theory.
Cassini, at this stage still a late supporter of
an Earth-centred universe, proposed that the
orbits of comets were vast epicycles, centred on
the orbits of stars far beyond Earth, and never
approaching closer than the orbit of Saturn.

Finally, we come to the twin centrepieces of
Plate 26, which display how the true nature of
cometary orbits was finally resolved. The left
circle depicts the orbit of that same brilliant
comet of 1680, showing how it relates to the
comet's changing position in Earth's skies.
Discovered by Gottfried Kirch (1639–1710) of
Coburg, Germany, on 14 November 1680, the
comet developed a spectacularly long tail as

it appeared to head straight towards the Sun
in early December. However, on 22 December,
both Cassini and English Astronomer Royal
John Flamsteed (1646–1719) spotted a comet
re-emerging from the Sun's rays in the
opposite direction. As this comet brightened
spectacularly, to become visible in daylight,
most astronomers came to the conclusion
that this was the same object that had been
discovered by Kirch.

The comet's sudden change in direction
was conclusive proof that it must, indeed,
follow a sharply bending parabolic path.
This proof inspired Isaac Newton (1643–1727)
to demonstrate how the influence of the
Sun's gravity could create such paths in his
masterwork the *Mathematical Principles of
Natural Philosophy* (1687). It was Newton's
friend Edmond Halley (1656–1742), however, who
famously brought matters to their conclusion.
Collecting a large number of comet apparitions
between 1337 and 1698, he used Newton's model
to analyse their paths through space, and
discovered that three comets seen in 1531, 1607
and 1682 followed nearly identical orbits. Halley,
therefore, theorized that the visible parabolic
curve followed by these comets was just the
Sunward end of a closed elliptical orbit, which
a single object took around seventy-six years to
complete. Halley correctly predicted that this
particular comet (which now bears his name)
would return in 1758. The right-hand side of the
plate is devoted to a diagram showing Halley's
calculated orbits, from a 1715 original by William
Whiston (1667–1752), an early popularizer of
Newtonian physics.

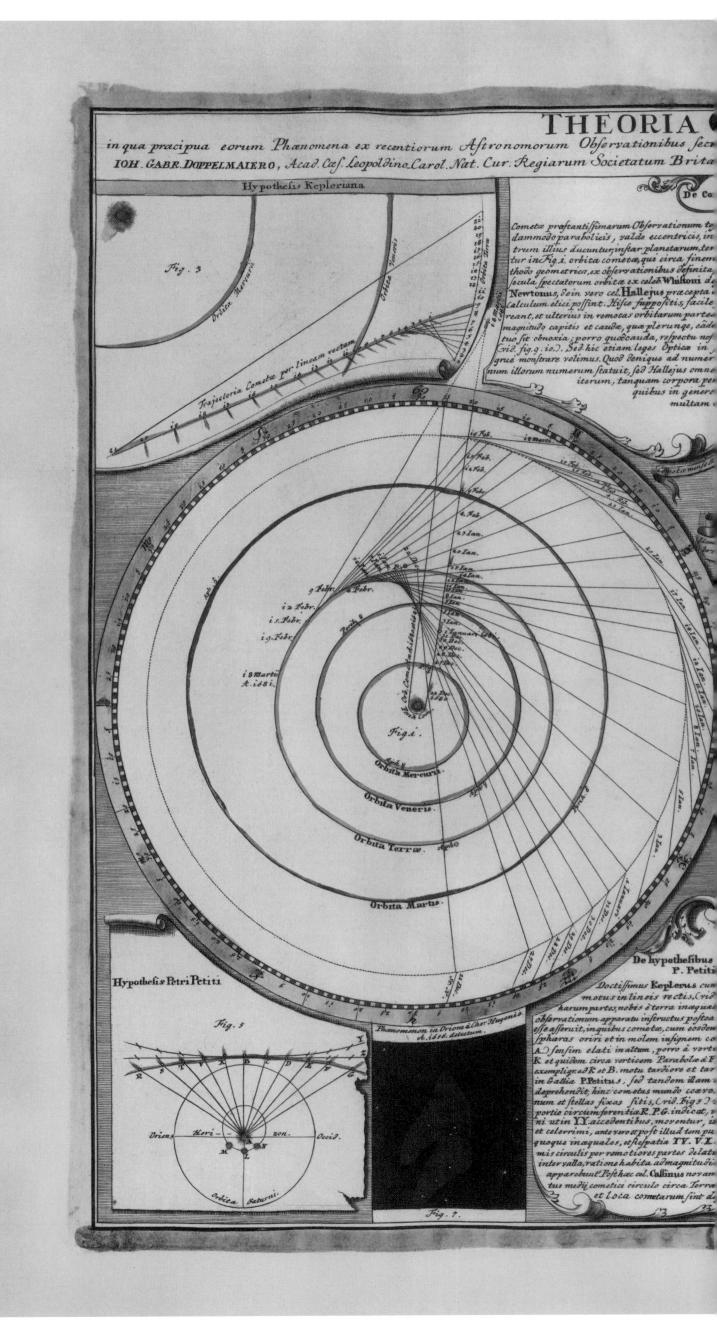

in qua pracipua eorum Phænomena ex recentiorum Astronomorum Observationibus secr
IOH.GABR.DOPPELMAIERO, Acad.Cæs.Leopoldino.Carol.Nat.Cur.Regiarum Societatum Brita

Hypothesis Kepleriana

De Co

Fig. 3

Orbita Mercurii

Orbita Veneris

Orbita Terra

Trajectoria Cometa per lineam rectam.

Cometæ præstantissimarum Observationum to
dammodo parabolicis, valde eccentricis, in
trum illius ducuntur; instar planetarum, ter
tur in Fig. i. orbita cometæ, qui circa finem
thodo geometrica, ex observationibus definita,
secula spectatorum orbitæ ex celeb.Whistoni d
Newtonus, dein vero cel.Hallejus præcepta,
calculum elici possint. Hisce suppositis, facile
reant, et ulterius in remotas orbitarum partes
magnitudo capitis et caudæ, quæ plerunque, eâde
tuo sit obnoxia; porro quoad caudâ, respectu nos
crid. fig.g.io.]. Sed hic etiam leges Opticæ in
grue monstrare velimus. Quod denique ad numer
num illorum numerum statuit, sed Hallejus omne
iterum, tanquam corpora per
quibus in genere
multam

Fig.i.

Orbita Mercurii.

Orbita Veneris.

Orbita Terræ.

Orbita Martis.

De hypothesibus
P. Petiti

Hypothesis Petri Petiti.

Fig. 5

Phænomenon in Orione à Chr. Hugenio.
A.1656. detectum.

Oriens   Hori - - - zon.   Occid.

Orbita Saturni.

Fig. 7.

Doctissimus Keplerus cum
motus in lineis rectis.[crid.
harum partes, nobis è terra inæquat
observationum apparatu instructus postea
esse asseruit, in quibus cometæ, cum eosdem
spheras oriri et in molem insignem co
A.] sensim elati in altum, porro à vorte
K et quidem circa verticem Parabolæ à F.
exempligr. ad R et B. motu tardiore et tar
in G adlia P.Petitus, sed tandem illam m
deprehendit, hinc cometas mundo coævo.
num et stellas fixas sitis,[crid.Fig 3.]
portio circumferentiæ R.P.G.indicat, v
ni ut in ♓ accedentibus, moventur, is
et celerrimi, ante vero et post illud tem pu
quoque inæquales, et spatia IV.V.L.
mis circulis per remotiores partes delata
intervalla, ratione habita ad magnitudi
apparebunt Posthæc cel.Cassinus nov an
tus medij cometici circulo circa Terra
et loca cometarum sint de

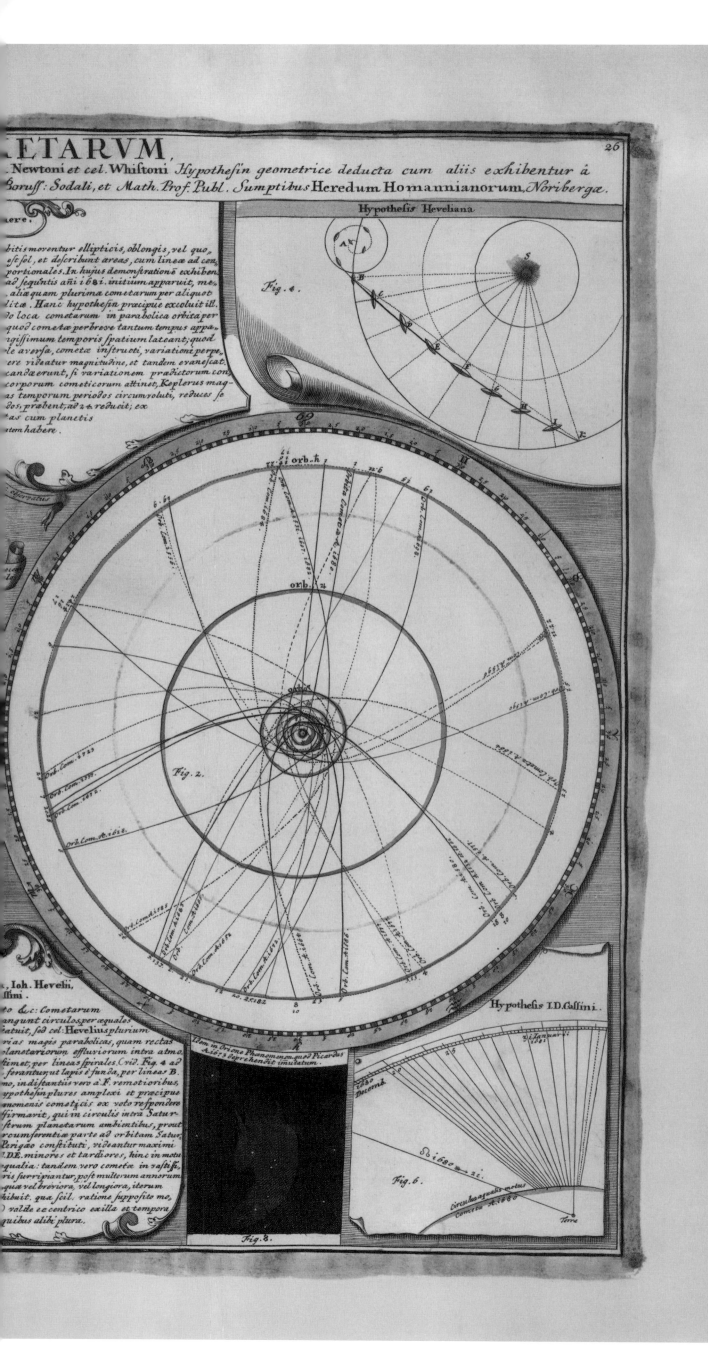

ETARVM,
. Newtoni et cel. Whiftoni Hypothefin geometrice deducta cum aliis exhibentur à
Boruff: Sodali, et Math. Prof. Publ. Sumptibus Heredum Homannianorum, Noribergæ.

ære.

bitis moventur ellipticis, oblongis, vel quo,
est sol, et describunt æreas, cum lineæ ad cen,
portionales. In hujus demonstratione exhiben.
ad sequentis añi 1682. initium apparuit, me,
aliæ quam plurima cometarum per aliquot
dita. Hanc hypothefin præcipue excoluit ill.
o loca cometarum in parabolica orbita per
quod cometæ perbreve tantum tempus appa,
ngissimum temporis spatium lateant; quod
le aversa, cometæ instructi, variationi perpe,
re videatur magnitudine, et tandem evanescat.
candæ erunt, si variationem prædictorum con,
corporum cometicorum attinet, Keplerus mag-
as temporum periodos circumvoluti, reduces se
dos, præbent, ad 24 reducit; ex
as cum planetis
tem habere.

Hypothefis Heveliana

Fig. 4.

orb. ♄
Orbita Cometæ A 1661.
orb. 24

Fig. 2.

Orb. Com. 1723
Orb. Com. 1577
Orb. Com. 1672
Orb. Com. A. 1618

, Ioh. Hevelii,
ſſini .

o &c: Cometarum
angunt circulos per æquales
tatuit, sed cel. Hevelius plurium
rias magis parabolicas, quam rectas
planetariorum effluviorum intra atmo.
stimet, per lineas spirales. Crid. Fig. 4 ad
ferantur; ut lapis è funda, per lineas B.
mo, in distantius vero à F. remotioribus,
hypothefin plures amplexi et præcipue
momenis cometicis ex voto respondere
firmavit, qui in circulis intra Satur-
sterum planetarum ambientibus, prout
rcumferentiæ parte ad orbitam Satur,
Perigæo constituti, videantur maximi
DE. minores et tardiores, hinc in motu
æqualia: tandem vero cometæ in vastissi,
ris surripiantur, post multorum annorum
quæ vel breviora, vel longiora, iterum
hibuit. quæ, scil. ratione supposito me,
valde excentrico exilla et tempora,
quibus alibi plura.

Item in Orione Phænomenon, quod Picardus
A 1673 deprehendit imutatum.

Hypothefis I.D. Caſſini.

Janquarii 1681

Orbo Deionbe

☿ 1680 = 22.

Fig. 6.

circulus æqualis motus
Cometæ A. 1680

Terra

Fig. 8.

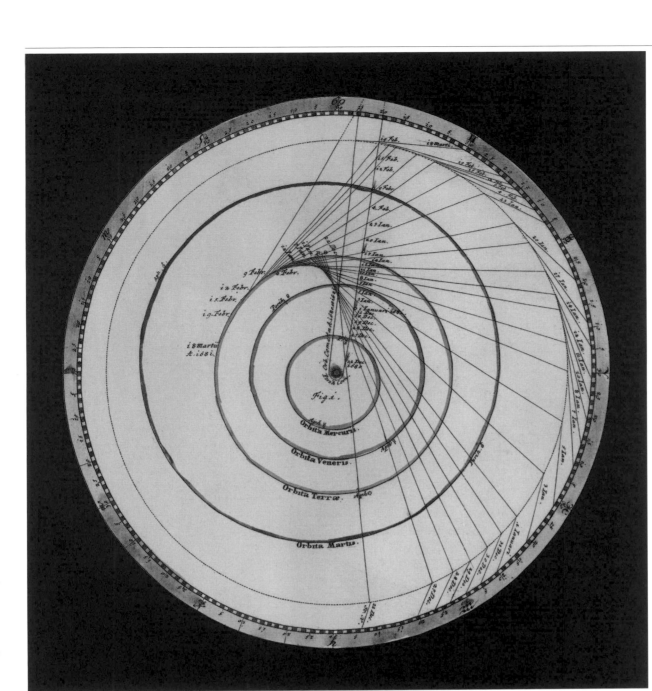

Doppelmayr plots the path of the Great
Comet of 1680 using measurements taken
after its reappearance on 22 December
1680 through to March 1681. By comparing
the comet's direction in Earth's sky with
Earth's own changing position on its orbit,
he reveals a path that is very close to a straight
line. It was only prior observations of the
comet heading towards the Sun in November
that led to the suggestion (first made by
Flamsteed) that it had changed direction –
and eventually to Newton's proof that the
comet was actually following an extreme
example of a parabolic curve.

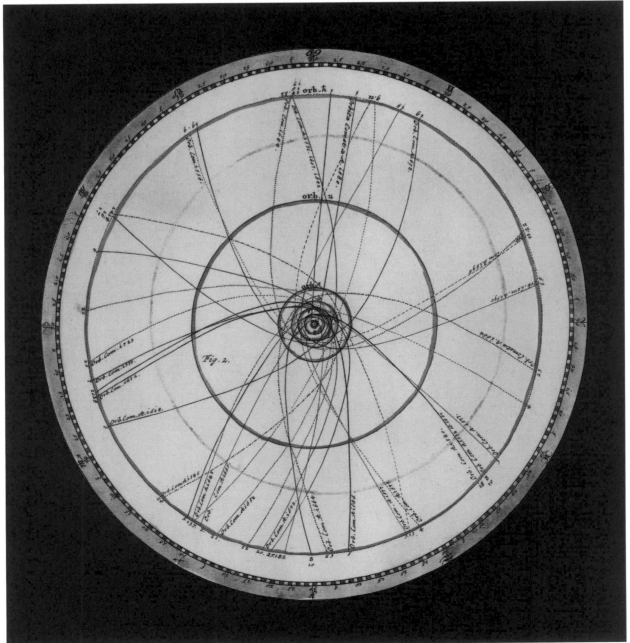

ELLIPTICAL ORBITS OF COMETS
AFTER NEWTON AND WHISTON.
(FIG. 2.)

This chart copies Whiston's 1712 chart of
cometary orbits, itself drawn from Halley's
1705 work on the subject. The Great Comet
of 1680 follows the same orientation as in the
left-hand chart, entering and departing the
inner solar system just to the right of the 12
o'clock position. To its left is the path of the
comet known today as Halley's Comet.

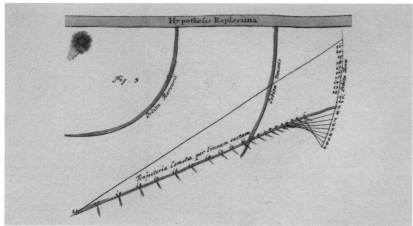

KEPLER'S HYPOTHESIS.
(FIG. 3.)

This diagram demonstrates Kepler's idea that the erratic motions of comets through Earth's skies could be explained by our own planet's motion affecting the apparent direction of an object following a straight-line path through space.

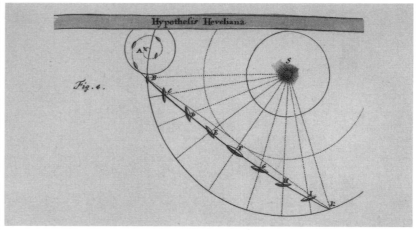

HEVELIUS'S HYPOTHESIS.
(FIG. 4.)

Hevelius speculated that comets originated in the atmospheres of planets from which they escaped on spiral vortices, growing over time. Eventually, their paths became almost straight lines, curving only slightly.

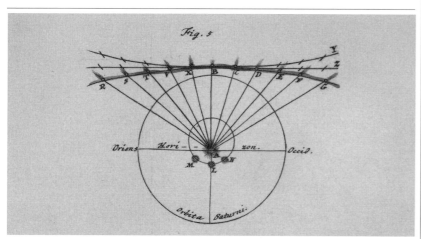

PETIT'S HYPOTHESIS.
(FIG. 5.)

According to Petit's model of 1665, comets followed circular or elliptical orbits beyond the orbit of Saturn. They could, therefore, return to Earth's skies, becoming visible only around the time of their closest approach.

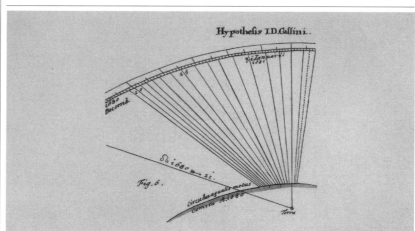

CASSINI'S HYPOTHESIS.
(FIG. 6.)

The ingenious theory of Cassini placed comets on vast circular orbits, with a centre in the direction of Sirius and encompassing Earth and the orbit of the Sun in such a way as to account for at least some of their unusual motion.

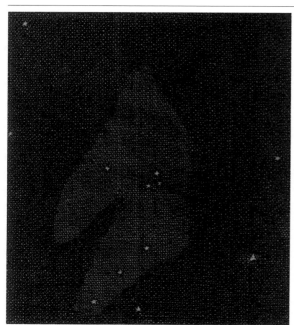

ORION NEBULA OBSERVED BY HUYGENS. (FIG. 7.)

Presumably on account of the resemblance between gaseous nebulae and comet tails, Doppelmayr chooses this plate to include a reproduction of Christiaan Huygens's pioneering sketch of the Orion Nebula.

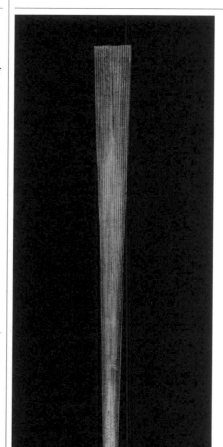

KIRCH'S COMET IN DECEMBER 1680. (FIG. 9.)

This sketch depicts the comet at its most spectacular, around the time of its perihelion passage (closest point to the Sun). There are reports of the comet being bright enough to see in daylight at around this time. Its tail, then displayed almost side-on as seen from Earth, stretched across up to 70 degrees of the sky.

ORION NEBULA OBSERVED BY PICARD. (FIG. 8.)

A second sketch of the Orion Nebula, made by Jean Picard in 1673, shows more structure. See pages 242–5 for more on the subsequent evolution of astronomical illustration in general, and renditions of this object in particular.

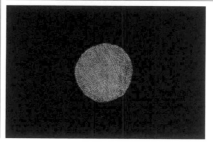

KIRCH'S COMET IN MARCH 1681. (FIG. 10.)

A small sketch shows the comet of 1680–81 shortly before its disappearance. By this time, its tail had disappeared, due to a combination of the comet's dwindling activity and fore-shortening, as its angle changed with respect to Earth.

PLATE 27.

MOVEMENTS OF COMETS IN THE SKY'S NORTHERN HEMISPHERE.

(MOTUS COMETARUM IN HEMISPHÆRIO BOREALI)

Doppelmayr charts the paths of comets and the apparitions of novae – evidence that the heavens are not static, but subject to change.

Doppelmayr's cometary plates were among the last to be compiled for the *Atlas*. Their depiction of events in the sky up to 1742 shows that they were probably completed shortly before publication. In addition to the introductory Plate 26, describing cometary theories, a pair of celestial hemispheres divided along the ecliptic (the Sun's path through the zodiac and de facto plane of the solar system) highlight the paths of comets and the locations of other unusual objects in the sky.

Changes in the sky have been known and observed since ancient times: not only comets, but also showers of shooting stars or meteors that were correctly associated with Earth's upper atmosphere and rare stars that flared into brilliant life before fading back to nothing (discussed alongside Plate 28).

With the previous plate, we reviewed the debate over the paths of comets, but there was also a similar discussion about their physical nature. For early observers, the only comparison that could be made from everyday experience was with fire (hence, the Aristotelean assumption that they belonged in the outermost fiery sphere of the sublunary world). Their shifting appearance and unpredictable nature led to an astrological association with imbalance, and they were generally seen as omens of change (Chinese astronomers developed a similar theory, although their record-keeping was more assiduous).

Some researchers have argued that the association between comets and doom is deeply rooted in human prehistory and may reflect the terror unleashed when a comet passed perilously close to Earth. However, a more likely reason for the link may be the brilliant comet that appeared in 44 BCE, shortly after the assassination of the Roman general Julius Caesar (100–44 BCE). This comet appears to have been a one-off visitor from the depths of the solar system, but many of the other particularly bright comets recorded in pre-telescopic times are actually returns of a single object: the famous Comet Halley. Orbiting the Sun roughly every seventy-six years, this consistently bright comet is first reliably recorded in a Chinese chronicle of 240 BCE, although records may go back two centuries further. Classical, Chinese and medieval European sources capture most of its appearances through the first millennium, including a spectacular close approach in

FIGS. 1–2. This pair of spectacular comets are from the *Augsburg Book of Miracle*s (mid-1550s). Such visitations were often depicted as signs of ill omen. At near right is a comet with a 'peacock tail' that appeared in 1401 before a plague in Swabia; at far right is a comet of 1527 that appeared to wield a sword.

FIG. 1.

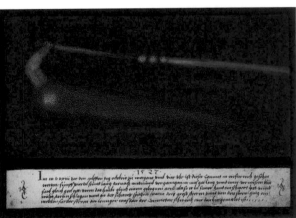

FIG. 2.

FIG. 3.

837 CE, before its well-known appearance on the Bayeux Tapestry, which chronicles the Norman Conquest of England in 1066.

Doppelmayr's chart – inspired by similar ones in *The Theatre of Comets* (1668) by Polish astronomer Stanisław Lubieniecki (1623–75) – captures three appearances of this comet: the 1531 sighting recorded by German humanist Peter Apian (1495–1552), the 1607 apparition tracked by Johannes Kepler (1571–1630) and the 1682 one measured by British Astronomer Royal John Flamsteed (1646–1719). Their near-parallel tracks are all shown here to the right of the celestial pole and give some hint to the similarity in orbits that led Edmond Halley (1656–1742) to recognize them as a single object and predict a further return for 1758.

While the remaining comets on the chart are all rare visitors in much longer orbits around the Sun, they are not without interest. Four of them are bright enough to be considered 'great comets' – those of 1556, 1577, 1618 and 1680. We have already seen how the 1680 comet played a vital role not only in resolving arguments about the shape of cometary orbits, but also as evidence for Isaac Newton's (1643–1727) theory of gravity. Of the others, however, the comet of 1577 (whose track is marked by a long untinted line passing close to Pegasus) is the most historically significant. Visible across all of Europe and much further afield over the course of several months, it reportedly reached a peak brightness that rivalled the Moon, and developed a long tail.

The great Danish observer Tycho Brahe (1546–1601) noted that this tail shifted its orientation but always pointed away from the Sun. His discovery hinted at the reality of comets as small icy bodies, which develop huge atmospheres and tails as they are heated by sunlight during their passage through the inner solar system. More importantly in the short term, however,

Tycho was able to use the instruments on his observatory island of Hven to track the comet's path across the sky with high precision. In an attempt to estimate the comet's distance, he collaborated with Tadeáš Hájek (1525–1600) in Prague, some 805 km (500 miles) to the south, to take simultaneous measurements from both locations. The two astronomers found that while the Moon's position in the sky appeared to shift slightly between Hven and Prague compared to the background stars, no such shift could be detected for the comet. This convinced Tycho that the comet must lie at least three times further away than the Moon, and was, therefore, a phenomenon of the supposedly inviolable heavens, rather than the changeable sublunary world.

Half a century later, the Great Comet of 1618 proved almost as influential. It was the last of three naked-eye comets to appear in that year, visible from November until January 1619, and was the first comet to be observed in some detail with a telescope. Known as the Angry Star on account of its distinctive reddish colour and ominous sword-like tail, its appearance inspired mentions in works of literature, including in a poem by James I of England (James VI of Scotland) (1566–1625) admonishing his subjects for their superstitious fears that it was a bad omen.

Alongside the comets, Doppelmayr represents the locations of six 'new stars', traditionally called novae, recorded in the two centuries before his publication. The earliest of these, in 1572, reached a brightness equal to that of Venus in November 1572. It was the most intense stellar explosion seen on Earth for centuries, and the long memory of the event inspired astronomers to keep a closer eye on the sky. A similar stellar eruption that took place in the constellation of Ophiuchus in 1604 became almost as bright as its predecessor.

MOTVS COMETARUM IN

qui intra 210 años ab añ 1530 usque ad añ: 1740, cum sex stellis
geometrice exhibiti à IOH. GABR: DOPPELMAIERO Acad. Imp. Leopol –
Sumtibus Heredum Homannianorum

Hypothesis Ieremiæ Horroccii.

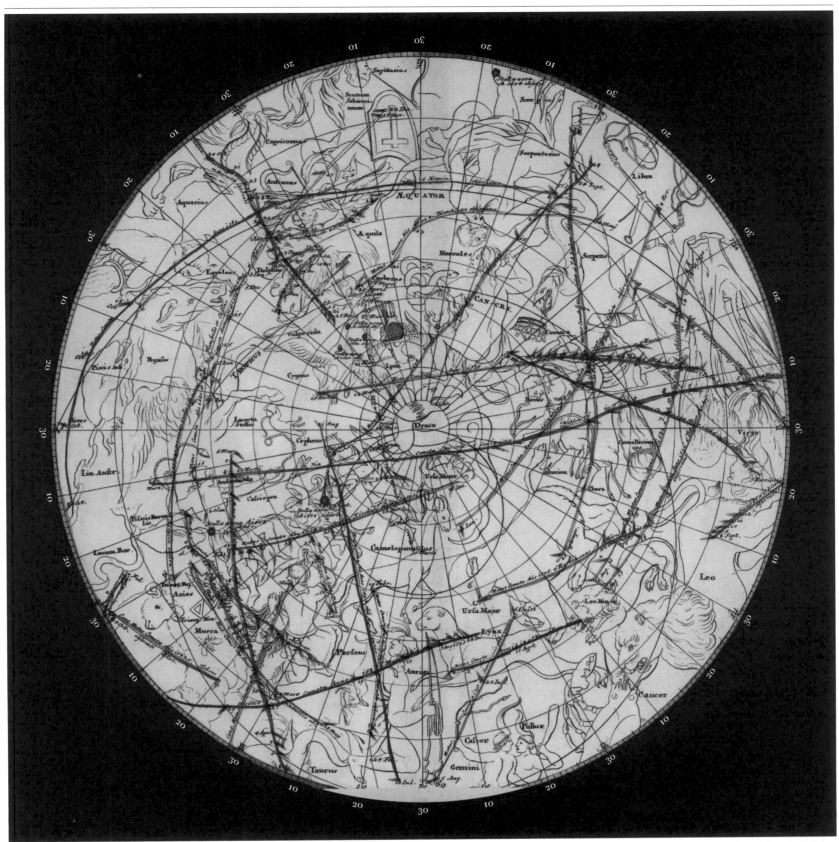

MOVING COMETS OBSERVED OVER
THE COURSE OF TWO LIFETIMES.

Plate 27's main chart plots the paths of comets observed since 1530 onto the northern half of the sky. In order to better interpret the motions of solar system objects, Doppelmayr plots this map in ecliptic coordinates, so that the line of the ecliptic (plane of the solar system) divides his maps of the northern and southern sky. Positions of six novae or 'new stars' are also marked and dated.

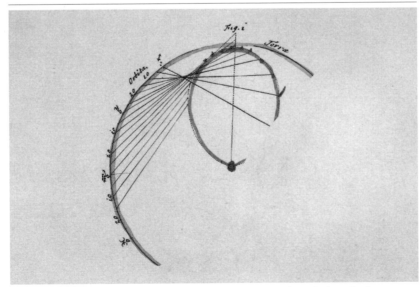

THEORY OF JEREMIAH
HORROCKS. (FIG. 1.)

Doppelmayr outlines the hypothesis of Jeremiah Horrocks that comets follow elliptical orbits around the Sun (later proved correct). He demonstrates how this can explain dramatic changes in the speed of comets through the sky.

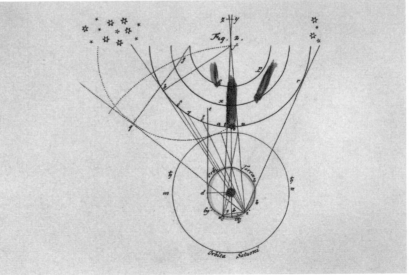

HYPOTHESIS OF JACOB
BERNOULLI. (FIG. 2.)

Doppelmayr also illustrates an intriguing but incorrect cometary hypothesis of Jacob Bernoulli (1655-1705), which proposed that comets were satellites orbiting a distant unseen planet lying beyond Saturn.

This pair of diagrams illustrates
how a curious faint glow along the
zodiac constellations is centred on
the Sun, and most easily seen when
the ecliptic is steeply tilted to the
horizon in spring.

THE ZODIACAL LIGHT
SEEN IN OCTOBER.
(FIGS. 4 AND 5.)

This second pair of diagrams
shows the zodiacal light in October.
Both pairs of diagrams (Figs. 3 and
6, and 4 and 5) depict the glow as
a lenticular extension of the Sun's
atmosphere. However, it is now
known to be caused by sunlight
reflecting from dust grains in the
plane of the solar system.

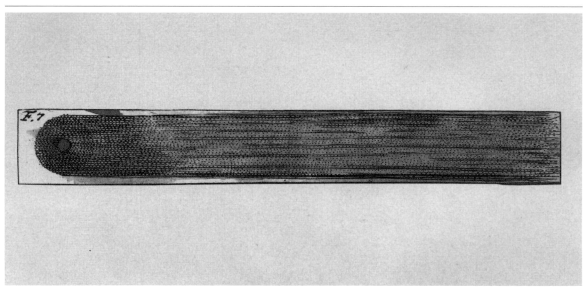

STRUCTURE OF A COMET.
(FIG. 7.)

Doppelmayr illustrates the typical structure of a comet, with a bright
central nucleus, extensive atmosphere or coma and elongated tail.
While early philosophers viewed comets as a phenomenon of the upper
atmosphere, Newton deduced that they were compact solid bodies
that emitted vapour when heated by their passage around the Sun.

PLATE 27.

KOMETENBUCH (c. 1587).

Created by an anonymous author in Flanders around 1587, the
Kometenbuch is an extraordinary illuminated manuscript, describing
the astrological interpretation of comets. Drawing from classical,
medieval and Arabic sources, the book's illustrations elucidate the
sometimes fanciful descriptions of the appearance of historic comets,
depicting them as lances, tumbling wheels and even faces. With its
roots in a philosophy that saw comets as phenomena of the upper
atmosphere – appearing in the spheres of air and fire – it is little wonder

that the book is mostly concerned with the possible consequence of these apparitions for people on Earth. This is perhaps unsurprising when the *Kometenbuch* also includes records of stones falling from the heavens; the superficial similarity between comets and meteors (shooting stars) appeared to confirm that comet-like objects could sometimes have a physical effect on Earth. Despite their fantastical elements, the *Kometenbuch* illustrations hint at the wide variety in the appearance of physical comets, created by interactions of their gas and dust tails and central comas, and the way these reflect sunlight.

PLATE 28.

MOVEMENTS OF COMETS IN THE SKY'S SOUTHERN HEMISPHERE.

(MOTUS COMETARUM IN HEMISPHÆRIO AUSTRALI)

*Movements of comets in the sky's southern hemisphere,
between 1530 and 1740, with two new stars seen in Doppelmayr's time.*

In comparison with his crowded chart of comets in the sky's northern hemisphere, Doppelmayr's companion map for the celestial hemisphere to the south of the ecliptic is relatively empty, mostly showing extensions to the paths of comets preceding or following their appearance in northern skies. The main exception is the track of the great comet of 1664–65, which hung low on the winter horizon throughout its apparition. The reason for the relative paucity of southern comets is simply the lack of methodical observers who had visited the southern hemisphere before Doppelmayr's time. Alongside the comets, however, Doppelmayr marks the locations of two 'new stars', or novae, accompanying the six shown on Plate 27.

The idea that the stars were fixed and unchanging was a mainstay of European cosmology from classical times until the mid-16th century. The constancy of stars provided one of the key justifications for viewing them as qualitatively different from either the earthly realm, planets or comets, and for imagining them as fixed objects on a celestial sphere. Nevertheless, occasional novae, caused by various forms of stellar explosion in distant parts of our Milky Way galaxy, were certainly occurring throughout this period, as evidenced in the records of 'guest stars' seen by Chinese astronomers, and other historical observations from around the world.

In an age before the telescope, only the brightest explosions would have been noticed by naked-eye observers, and these events (now called 'supernovae' to distinguish them from 'novae', which have a different physical cause) are the rarest of all. Apparent European ignorance of them can sometimes be explained by accidents of timing or location that put these novae out of sight of Western stargazers, but in some cases (most notably the famous Crab Supernova in Taurus, which outshone every star in the sky for several months in 1054), the lack of European reports is harder to explain and has even led to the idea of records being deliberately suppressed. In autumn 1572, however, a super-nova eruption in the constellation of Cassiopeia thrust itself onto European consciousness in a way that could not be ignored.

Shining as brightly as Venus, this was the most intense stellar explosion seen on Earth for centuries. Today, it is known as Tycho's Supernova, since the Danish astronomer Tycho Brahe (1546–1601) not only observed it, but also used the same method that he would later employ for the Great Comet of 1577 to show that it lay well beyond the orbit of the Moon and was a genuine celestial phenomenon. Until it faded back to invisibility some sixteen months later, the new star remained an awkwardly undeniable sight for

FIG. 1.
An illustration from Tycho's *Concerning the New Star* (1573) shows the 'nova stella' of 1572 outshining all the stars of Cassiopeia. Tycho's observations secured his reputation among the European scientific community of the time.

FIG. 1.

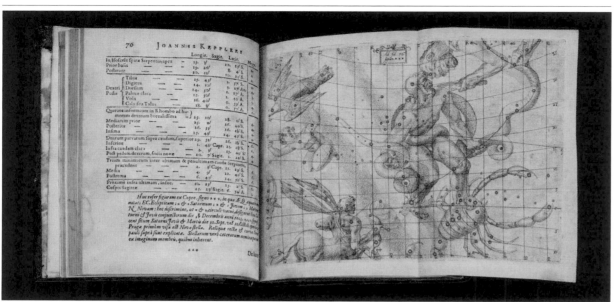

FIG. 2.

FIG. 2.
Kepler's chart from
*On the New Star in
the Foot of the Serpent
Handler* (1606) marks
the location of the
1604 supernova within
the constellation now
known as Ophiuchus.

those who believed the heavens should remain fixed. It appears on Doppelmayr's Plate 27.

The eruption of a second brilliant supernova in 1604 (also shown on Plate 27) had an even greater effect on the Western view of the heavens. This explosion in the constellation of Ophiuchus, the Serpent-Bearer (Doppelmayr uses the alternative name Serpentarius) did not quite reach the brightness of the nova of 1572, but it was even more widely observed and speculated upon, not least because it coincided with a rare planetary conjunction between Mars, Jupiter and Saturn. Johannes Kepler (1571–1630) was among those summoned to prognosticate upon the new star's meaning, and the event remains known today as Kepler's Supernova.

This 1604 supernova remains the most recent to have been observed in the Milky Way galaxy, but within a few years the invention of the telescope and the beginning of more methodical stellar surveys would allow somewhat fainter and more common novae to be spotted with increasing frequency. Technically speaking, Doppelmayr includes two of these on his Plate 27 map: the stars marked as being seen in Andromeda in 1612 and Lyra in 1670. However, neither marked location corresponds to the correct site of each year's nova. That of 1670 was seen to the south of the head of Cygnus, the Swan (within the present-day boundaries of Vulpecula, the Fox), while that of 1612 was on the other side of the sky from Andromeda, in Leo.

The other four stars marked as 'novae' on Plates 27 and 28, meanwhile, are in fact objects that astronomers today would class as variable stars. The 1600 outburst in the body of Cygnus (Plate 27) was one of many unpredictable changes in the light of a star called P Cygni – after shining at third magnitude for several years around

1600, it slowly faded to invisibility before further brightening from the 1650s onwards. In Doppelmayr's time, it showed at fifth magnitude and it has remained more or less unchanged ever since.

The remaining three 'novae' are all examples of a different type of object known as a 'long-period variable'. Mira, in the throat of Cetus, the Sea Monster (Plate 28), was the first of these stars to be discovered, after German astronomer David Fabricius (1564–1617) noted its changing brightness when attempting to use it as a reference for the motions of planets. After initially seeing it brighten and then fade from visibility, Fabricius assumed it was a nova – until he found, in 1609, that it had returned to its original brightness. The stars Chi Cygni in the neck of Cygnus (Plate 27) and R Hydrae in the tail of Hydra, the Water Snake (Plate 28) were similarly identified as unusual in 1686 and 1704, respectively. All three are red stars that pulsate in brightness, varying with somewhat erratic periods that average out at 332, 409 and 389 days.

While the idea that novae and associated stars were atmospheric phenomena was dismissed almost from the outset, thanks to Tycho's convincing measurements, astronomers were left struggling for plausible explanations for what they really were. Early theories included the idea that they were genuine stellar births (raising questions about why the stars in question did not last) or that they were objects whose orbits brought them briefly closer to Earth and caused them to brighten before they retreated back into the depths of space. Neither explanation would, ultimately, stand up to scrutiny, but it was not until the latter half of the 19th century that the true nature of these changing stars began to become clearer.

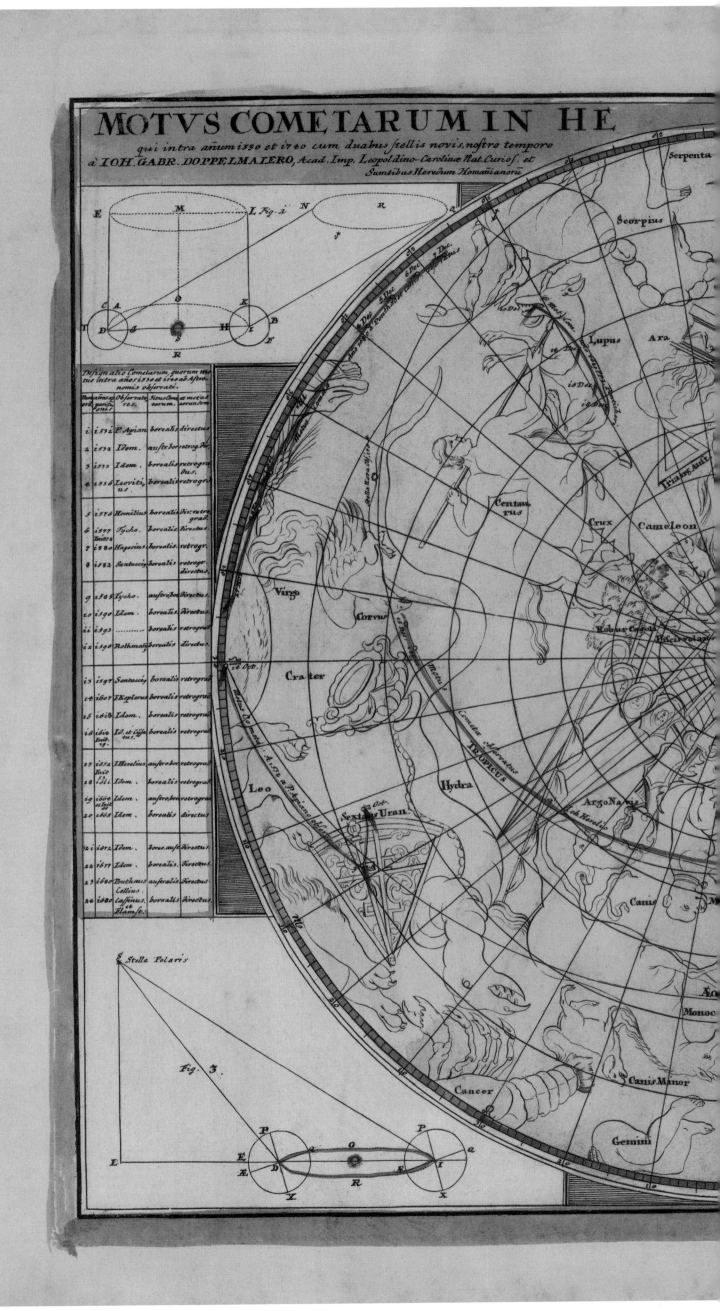

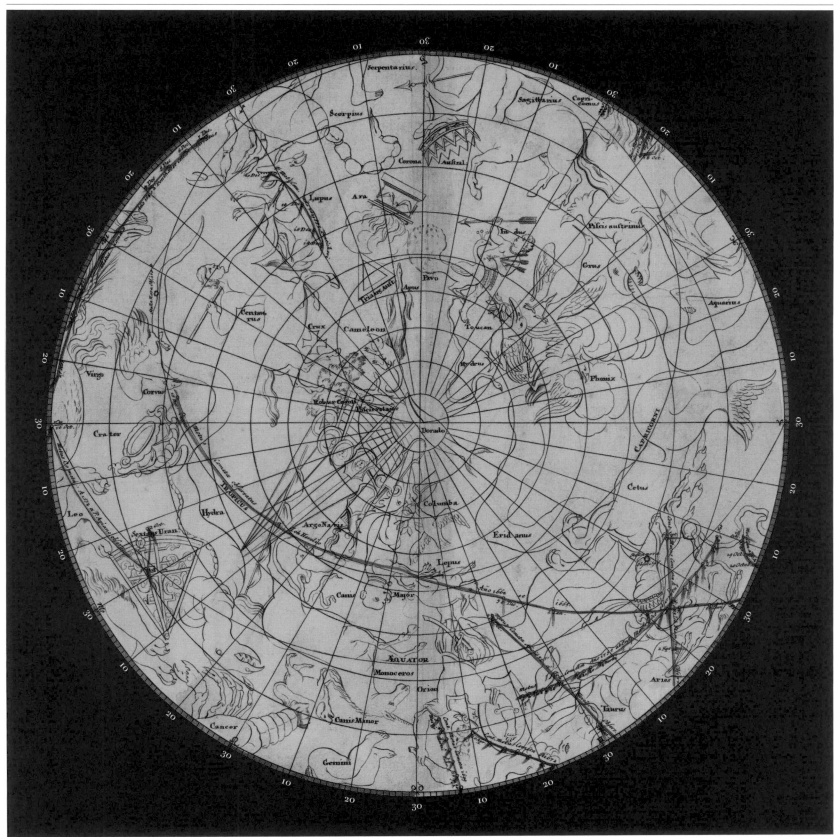

PATHS OF COMETS OBSERVED IN THE
SOUTHERN ECLIPTIC HEMISPHERE.

The main chart of Plate 28 complements the previous chart, showing the paths of comets since 1530 to the south of the ecliptic plane. Inevitably, the chart is somewhat sparse compared to its northern counterpart, due to the lack of European observers suitably located to see comets in the deep southern sky.

TABLE OF COMETS OBSERVED BY ASTRONOMERS BETWEEN 1530 AND 1740.			
Year	Observed by	Location in sky	Motion
1531	Apian	northern	direct
1532	Apian	southern northern	retrograde direct
1533	Apian	northern	retrograde
1556	Lvovický	northern	retrograde
1576	Hommel	northern	direct retrograde
1577–78	Tycho	northern	direct
1580	Hájek	northern	retrograde
1582	Santucci	northern	retrograde direct
1585	Tycho	southern northern	direct
1590	Tycho	northern	direct
1593	....	northern	retrograde
1596	Rothmann	northern	direct
1597	Santucci	northern	retrograde
1607	Kepler	northern	retrograde
1618	Kepler	northern	retrograde
1618–19	Kepler & Cysat	northern	retrograde

CONTINUED			
Year	Observed by	Location in sky	Motion
1652–53	Hevelius	southern northern	retrograde
1661	Hevelius	northern	retrograde
1664–65	Hevelius	southern northern	retrograde
1665	Hevelius	northern	direct
1672	Hevelius	northern southern	direct
1677	Hevelius	northern	direct
1680	Pontio & Cellio	southern	direct
1680	Cassini & Flamsteed	northern	direct
1682	Flamsteed	northern	direct
1683	Flamsteed	northern southern	retrograde
1684	Bianchini & Ciampino	northern	direct
1686	G. Kirch	northern	direct
1689	Richaud	southern	direct
1698	de la Hire	northern	retrograde
1699	Maraldi	northern southern	direct
1702	Bianchini & G. Kirch	northern	retrograde

CONTINUED			
Year	Observed by	Location in sky	Motion
1706	Cassini	northern	retrograde
1707	Maraldi	northern	direct retrograde
1718	C. Kirch	northern	retrograde
1723	Bradley, Maraldi & Bianchini	southern northern	retrograde direct
1729	Maraldi	northern	retrograde
1737	Manfredi	southern	direct

NEW STARS SEEN BY CELEBRATED OBSERVERS FROM TYCHO BRAHE TO GIACOMO MARALDI.			
Location	Date of detection	Discoverer	Modern name
In the throne of Cassiopeia	1572	Tycho Brahe	SN 1572, Tycho's Supernova
In the neck of Cetus	1596	David Fabricius	Mira, Omicron (o) Ceti
In the chest of Cygnus	1600	Willem Janszoon Blaeu	P Cygni
In the right foot of Serpentarius	1604	Johannes Kepler	SN 1604, Kepler's Supernova
In the belt of Andromeda	1612	Simon Marius	Messier 31, Andromeda Galaxy
Below the head of Cygnus	1670	Don Anthelme of Dijon/ Hevelius	Nova Vulpeculae 1670
In the neck of Cygnus	1686	Gottfried Kirch	Chi (χ) Cygni
In the tail of Hydra	1704	Giacomo Maraldi	R Hydrae

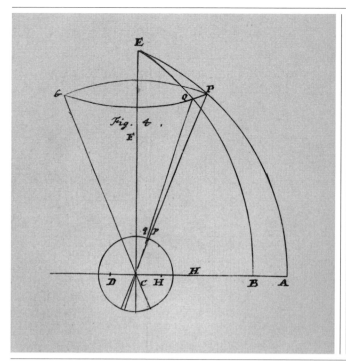

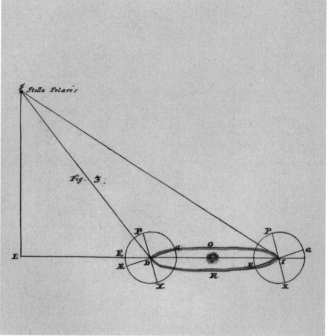

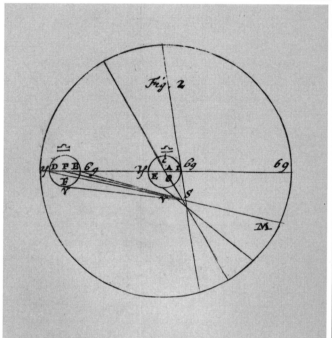

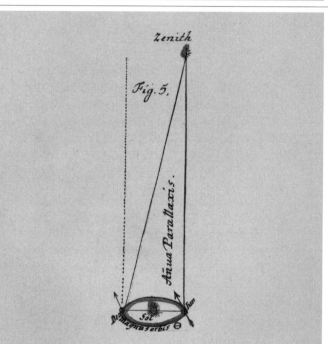

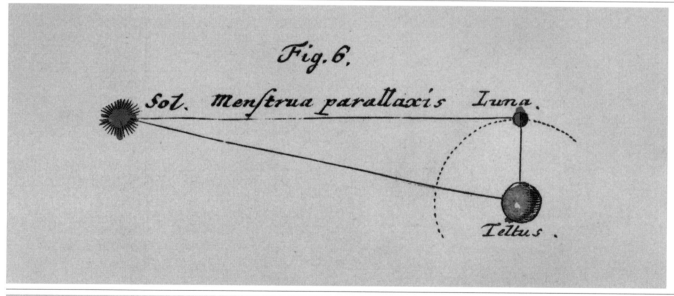

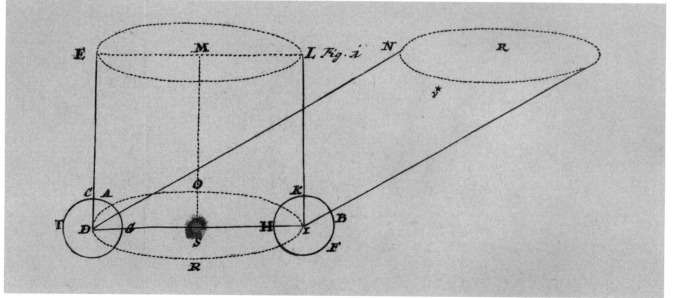

PRECESSION OF THE
EQUINOXES (FIG. 4.)

Described in text on Plate 27, this
diagram demonstrates the slow
change in the orientation of the
sky known as precession, created
as Earth's axis of rotation 'wobbles'
in space over 25,800 years.

FLAMSTEED'S 'PARALLAX'
(FIG. 3.)

Doppelmayr copies a diagram
used by Flamsteed to support his
claimed observations of parallax
in the pole star, also described
on Plate 27.

CASSINI'S REBUTTAL
(FIG. 2.)

Here, Doppelmayr copies a 1702
diagram by Cassini, demonstrating
that the true cycle of motion arising
from parallax should be different
from that observed by Flamsteed.

PARALLAX AT THE
ZENITH (FIG. 5.)

A small diagram shows how
the principle of annual parallax
applies to a star at the observer's
zenith point (directly overhead).

SOLAR PARALLAX
(FIG. 6.)

This small illustration
(unreferenced in the text) depicts
the determination of solar parallax:
a method of obtaining the relative
distances of Earth and the Moon
by measuring the Moon's position
at exactly first-quarter phase.
This subjective method formed
the foundation for scales in the
solar system, until the 1761 transit
of Venus permitted more accurate
calculations.

ABERRATION OF
STARLIGHT? (FIG. 1.)

This diagram appears to be an
illustration of the phenomenon
known as 'aberration of starlight',
discovered by James Bradley in
1728. It was the conclusive proof
that Earth is in motion around
the Sun, and yet it is not explained
on either this or the preceding
plate. For more on this topic,
see pages 236–8.

COMPARATIVE ASTRONOMY, I–II.

(ASTRONOMIA COMPARATIVA)

*Doppelmayr compares the phenomena of the solar system
as they would appear from its other worlds.*

The final two plates in the original *Atlas* constitute a remarkable feat of both imagination and calculation for Doppelmayr. They attempt nothing less than an ambitious comparison between views and experiences of the solar system, as seen from its other worlds.

Such a leap in perspective would have seemed nonsensical to astronomers in the pre-Copernican world, and the earliest known attempt at this sort of 'comparative astronomy' is usually attributed to none other than Johannes Kepler (1571–1630). In his curious novel *The Dream* (1608), Kepler uses a fantastical tale of a trip to the Moon as a framing device for discussing lunar astronomy – how the heavens might appear from Earth's satellite. Kepler is thought to have begun work on *The Dream* as a treatise in defence of Copernican astronomy in the 1590s, then later added a narrative framing device and copious footnotes. The complete work was published posthumously in 1634 and has been described as the first work of 'hard' science fiction – earlier writers who dabbled with the concept of a trip to the Moon used it as an opportunity for allegory and satire, glossing over the physical realities.

Doppelmayr divides the treatment of his expansive subject logically enough, focusing on the Moon, Mercury and Venus in Plate 29, and on Mars, Jupiter and Saturn in Plate 30. Therefore, from our Earth-centred perspective, he divides the inferior planets from the superior ones. However, it is worth revisiting just what these concepts mean in the context of Doppelmayr's current purpose.

According to the classical astronomy of Ptolemy (*c.* 100–170 CE) and his successors, the inferior planets were those whose movements on their deferent circles were locked to that of the Sun. As a result, their paths through the sky remained centred on the Sun, appearing as loops that swung out to the east and west, reached a certain maximum separation or elongation and were then inevitably drawn back towards the Sun. Superior planets, in contrast, moved more slowly and freely on their deferents, emerging from a 'conjunction' in which they passed behind the Sun, appearing to drift westwards in relation to it until they were on the opposite side of the sky (at 'opposition', when they would execute a retrograde loop in the heavens), before approaching the Sun again from the east to return to the next conjunction.

While planetary motions themselves did not change, the gradual recognition of the Sun-centred solar system saw the notions of inferior and superior planets take on a more straightforward meaning – now they were simply those closer to the Sun than Earth (and whose orbits could only reach a certain angular

FIG. 1.
Kepler took account of the best understanding of how the heavens and Earth would appear from a lunar perspective as a basis for his fantasy novel *The Dream* (1608). Here, a diagram shows conditions governing the phases of the lunar cycle.

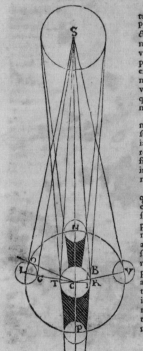

FIG. 1.

separation from the Sun) and those whose paths encircled both Earth and the Sun. It follows, therefore, that any planet can be taken as the reference point for inferiority and superiority. Earth itself will appear as an inferior planet from Mars, Jupiter and Saturn, with its movements seemingly anchored to the Sun, while from the point of view of Mercury, even Venus becomes a superior planet roaming all the way around the sky.

The complex diagrams at the centre of each plate place the planets in various positions within a Sun-centred solar system, and then (around the outer rings) demonstrate how these give rise to the range of apparent planetary motions of each world as seen from the others. For example, on Plate 29, the inner ring demonstrates the apparent paths of the other planets through Mercury's sky over the course of one Mercury year, while the outer ring demonstrates the tracks of the planets as they would be seen from Venus. Other potentially useful information is provided in tables – for example, the maximum elongations of each planet (the greatest angular separation they reach from the Sun) as seen from each of its outer neighbours on Plate 29, and the duration of periods of retrograde motion displayed by each planet's outer neighbours on Plate 30.

Around the corners of each plate, Doppelmayr provides a range of smaller diagrams. The lower half of Plate 29 recalls Kepler's *The Dream*, describing Earth's changing appearance from the Moon as it goes through its own sequence of phases. Two hemispheres attempt to render the appearance of Earth from the Moon, centred on the 'Old World' continents and the 'New World' of the Americas. A diagram at the lower right of Plate 30, meanwhile, shows the comparative size of the Sun in the skies of each of the planets.

Somewhat less practical, at first glance, are the alternative 'Tychonic' models of the solar system that fill the upper corners of Plate 29 and three slots on Plate 30. Each of these imagines a system in which a specific planet is fixed at the centre of the universe and orbited by the Sun, which is, in turn, circled by all of the other planets. This might seem like little more than a curious *jeu d'esprit*, but recall how the Tychonic model offers a powerful tool for understanding planetary motions seen from Earth (discussed on Plate 10). In the same spirit, a similar shift of perspective provides a useful way of thinking about the patterns of planetary movement, approach and retreat in the skies of any arbitrarily chosen world. The small tables showing the angular inclinations of each planet's orbit as seen from the other

FIG. 2.

FIG. 2.
The 19th century saw the first artistic depictions of views from other celestial bodies. It was the birth of the genre that would later be known as 'space art'. This early example from *The Moon* (1868) by French science writer Amédée Guillemin (1826–93) shows the lunar near side illuminated only by the light from a Full Earth.

worlds (that is, how much each planet would appear to diverge from the Sun's ecliptic path through that world's skies) hint at this more practical application.

The last plate of the original *Atlas* ends with a fitting epigram, taken from Roman philosopher Seneca's (*c.* 4 BCE–65 CE) *Investigations into Nature*, written around 64 CE. This curious and somewhat mystical work of natural philosophy is largely focused on meteorology but therefore incorporates some phenomena that we today recognize as within the province of astronomy. Doppelmayr quotes a passage from Book 7, on the subject of comets, to acknowledge that however comprehensive his work might attempt to be, it will inevitably be superseded by new knowledge and discoveries:

Nature does not reveal all her secrets at once: her arcana are not shared with all indiscriminately, they are withdrawn and shut up in the inner shrine. For many ages to come, when our memory will have faded, are reserved those things for which those of this age will see one thing, and those that follow us another. When, then, will these things be brought to our full knowledge? Great results proceed slowly, especially when labour ceases.

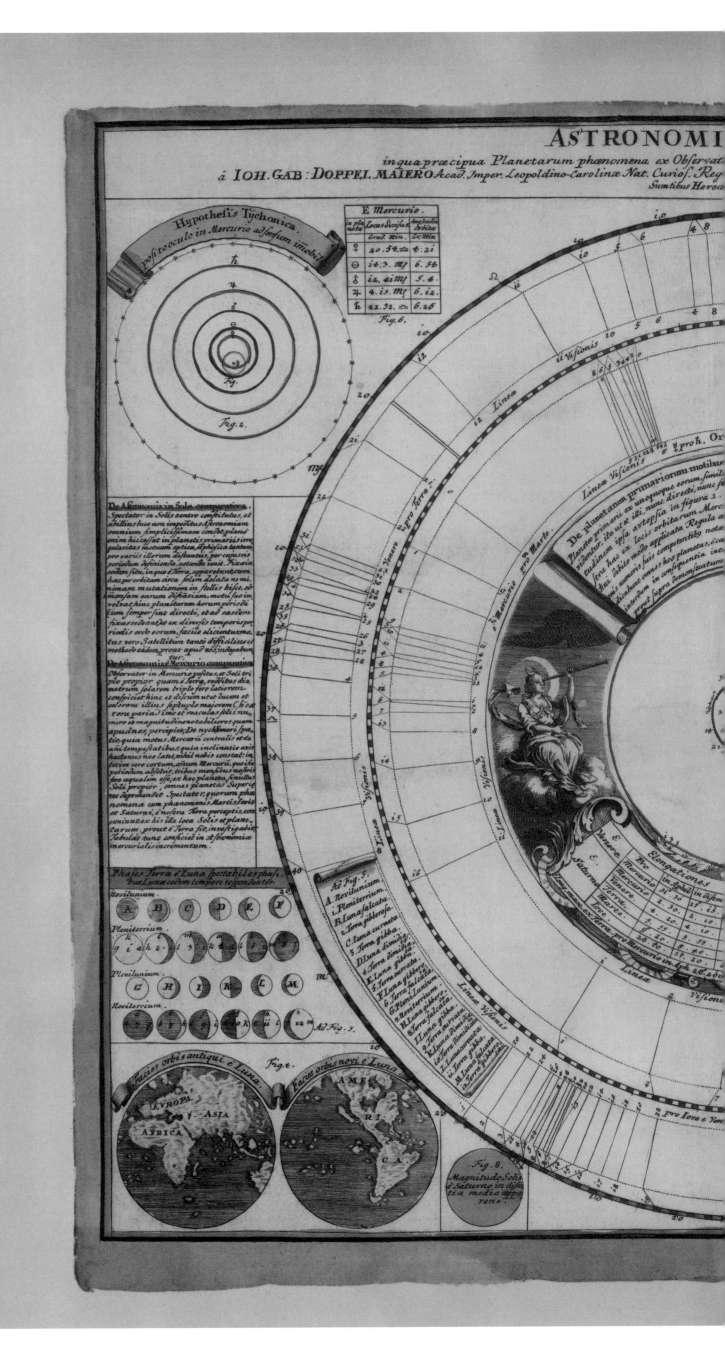

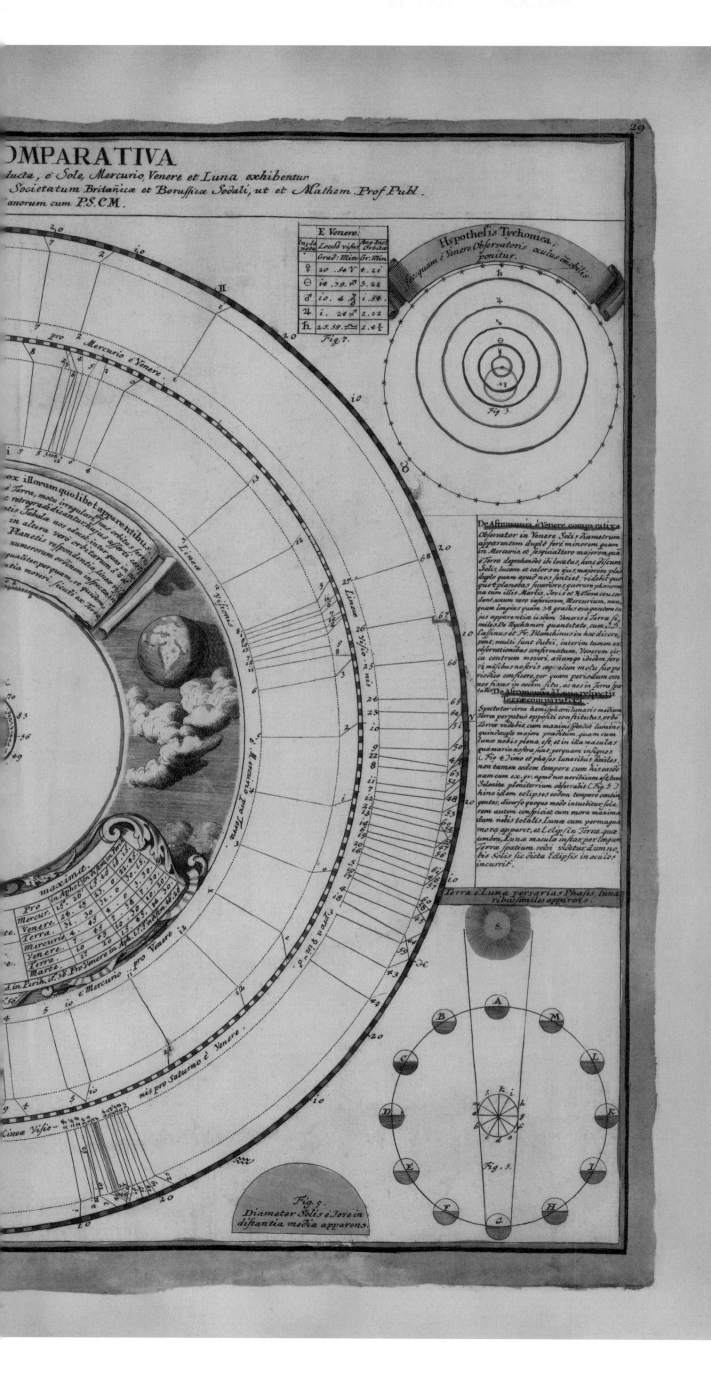

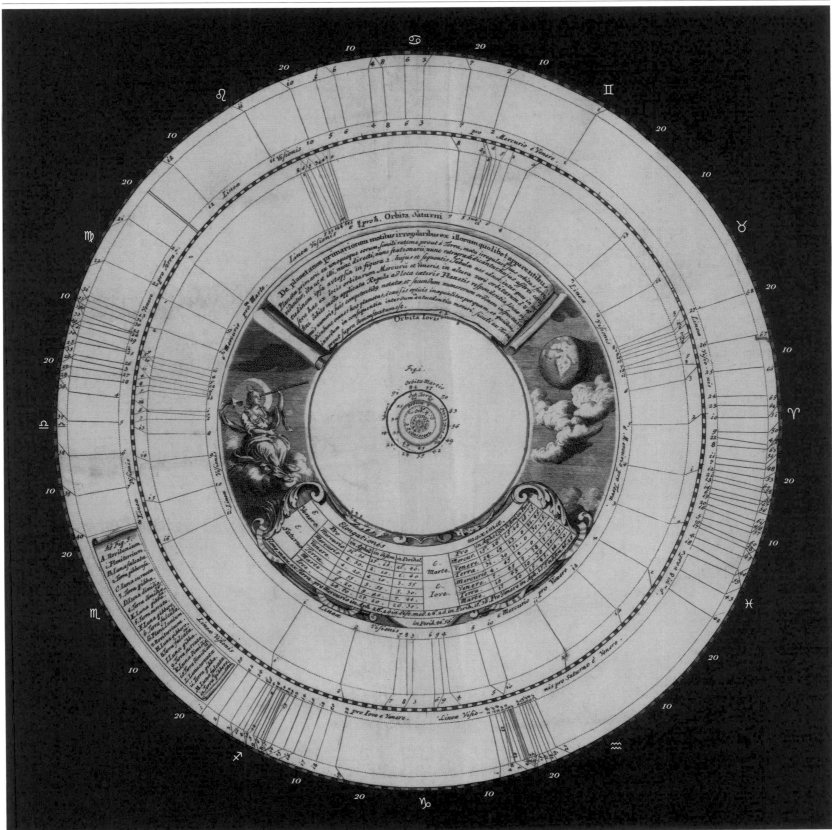

MOTIONS OF THE OTHER PLANETS SEEN
FROM MERCURY AND VENUS. (FIG. 1.)

This central diagram shows the orbits of the planets in relation to one another. Around the outside, the lines of sight that arise from these are constructed to reveal the phenomena of each planet's motion, as seen from Mercury and Venus. The table at the bottom indicates the elongations of the planets (their greatest angular distance from the Sun) when they are inferior (closer to the Sun) in comparison to an observer's location.

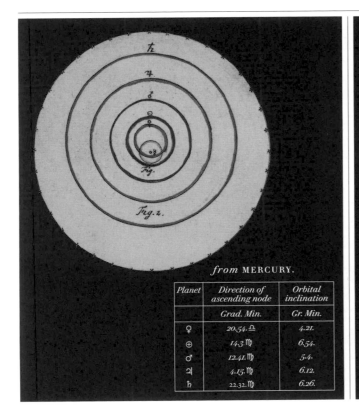

from MERCURY.

Planet	Direction of ascending node	Orbital inclination
	Grad. Min.	Gr. Min.
♀	20.54. ♎	4.21.
⊕	14.3 ♍	6.54.
♂	12.41. ♍	5.4.
♃	4.15. ♍	6.12.
♄	22.32. ♍	6.26.

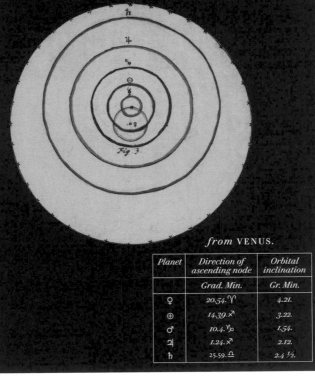

from VENUS.

Planet	Direction of ascending node	Orbital inclination
	Grad. Min.	Gr. Min.
♀	20.54. ♈	4.21.
⊕	14.39. ♐	3.22.
♂	10.4. ♑	1.54.
♃	1.24. ♐	2.12.
♄	25.59. ♎	2.4 ½.

TYCHONIC SYSTEM AROUND A FIXED MERCURY. (FIGS. 2 AND 6.)

Doppelmayr imagines a model of the solar system similar to Tycho's, with all the other planets circling the Sun, which itself circles a static Mercury. The table shows the orientation of the other planets' orbits as measured from Mercury.

TYCHONIC SYSTEM AROUND A FIXED VENUS. (FIGS. 3 AND 7.)

This diagram depicts a Tychonic solar system in which Venus is static, orbited by the Sun, which is, in turn, circled by all the other planets. The table gives the orientation of the orbits of the other planets as seen from Venus.

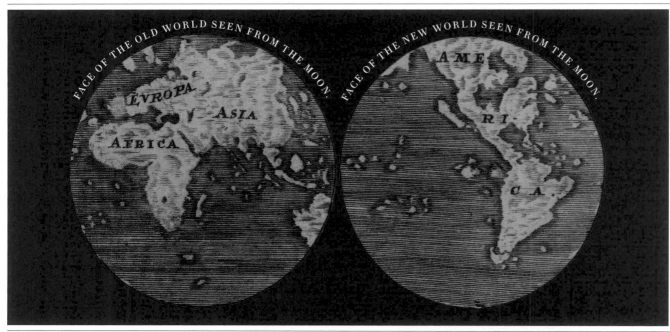

Face of the Old World seen from the Moon.

EVROPA
ASIA
AFRICA

Face of the New World seen from the Moon.

AME
RI-
CA

Here, Doppelmayr considers what
our planet might look like from
the Moon, shining five times more
brightly than the Moon does from
Earth. The oceans appear as
prominent dark areas.

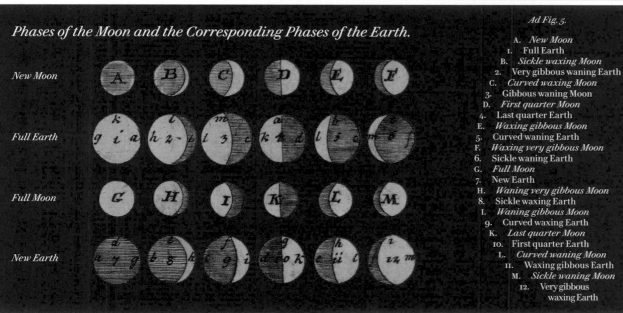

Phases of the Moon and the Corresponding Phases of the Earth.

New Moon

Full Earth

Full Moon

New Earth

Ad Fig. 5.

A. *New Moon*
1. Full Earth
B. *Sickle waxing Moon*
2. Very gibbous waning Earth
C. *Curved waxing Moon*
3. Gibbous waning Moon
D. *First quarter Moon*
4. Last quarter Earth
E. *Waxing gibbous Moon*
5. Curved waning Earth
F. *Waxing very gibbous Moon*
6. Sickle waning Earth
G. *Full Moon*
7. New Earth
H. *Waning very gibbous Moon*
8. Sickle waxing Earth
I. *Waning gibbous Moon*
9. Curved waxing Earth
K. *Last quarter Moon*
10. First quarter Earth
L. *Curved waning Moon*
11. Waxing gibbous Earth
M. *Sickle waning Moon*
12. Very gibbous
waxing Earth

PHASES OF THE MOON
AND EARTH. (FIG. 5. [*sic*])

Doppelmayr depicts the changing
phases of the Moon as seen from
Earth, and the corresponding
phases of the Earth simultaneously
seen from the Moon. Note that while
Earth goes through a 29.5-day cycle
of phases similar to the Moon, its
daily rotation causes the continents
to slide on and off the Moon-facing
hemisphere.

Fig. 8.
Magnitudo Solis
è Saturno in diſtan-
tia media appa-
rens.

SIZE OF THE SUN FROM
SATURN. (FIG. 8.)

A diagram depicting the size
of the Sun, as seen from Saturn.

Fig. 9.
Diameter Solis è Jove in
diſtantia media apparens.

SIZE OF THE SUN FROM
JUPITER. (FIG. 9.)

A diagram depicting the size
of the Sun, as seen from Jupiter.

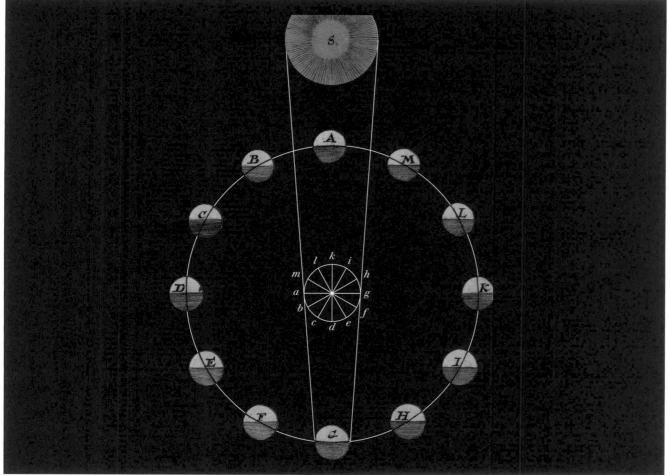

RELATIONSHIP OF THE
PHASES. (FIG. 5. [*sic*])

This diagram demonstrates how the
phases of Earth, labelled elsewhere,
can be identified in relation to the
location and phase of the Moon.

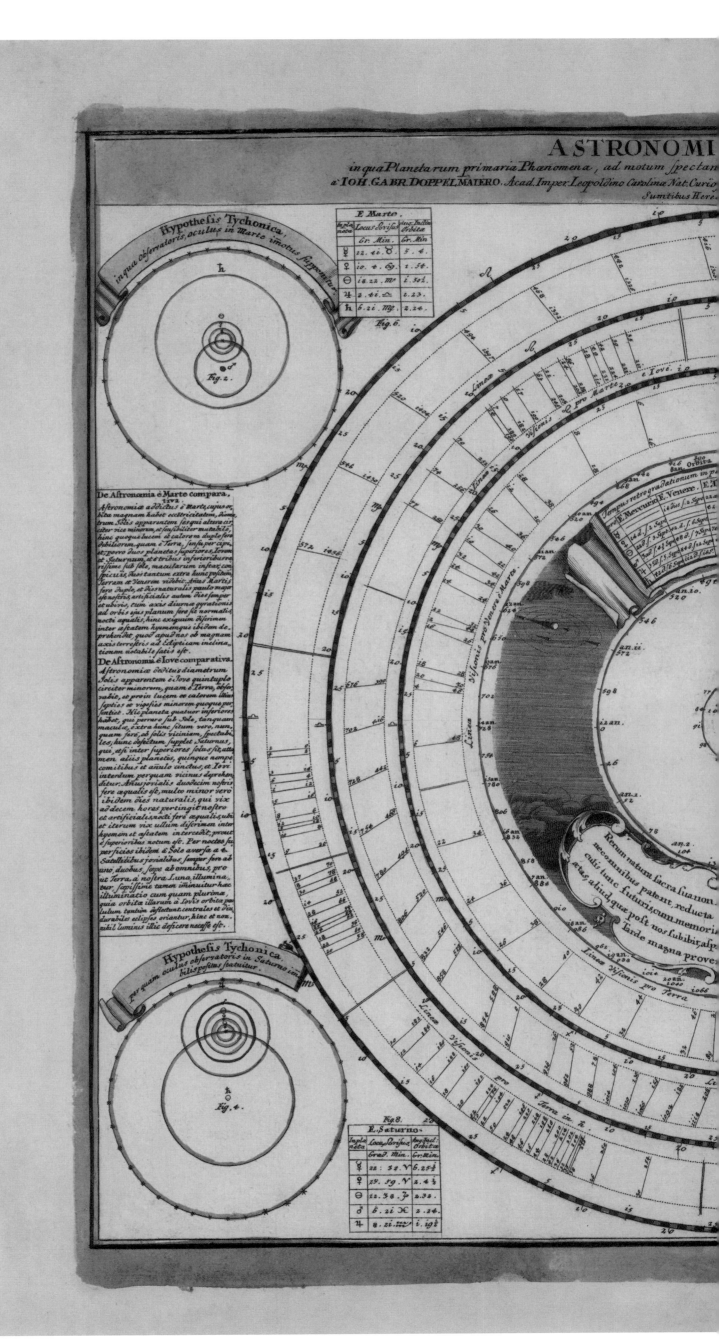

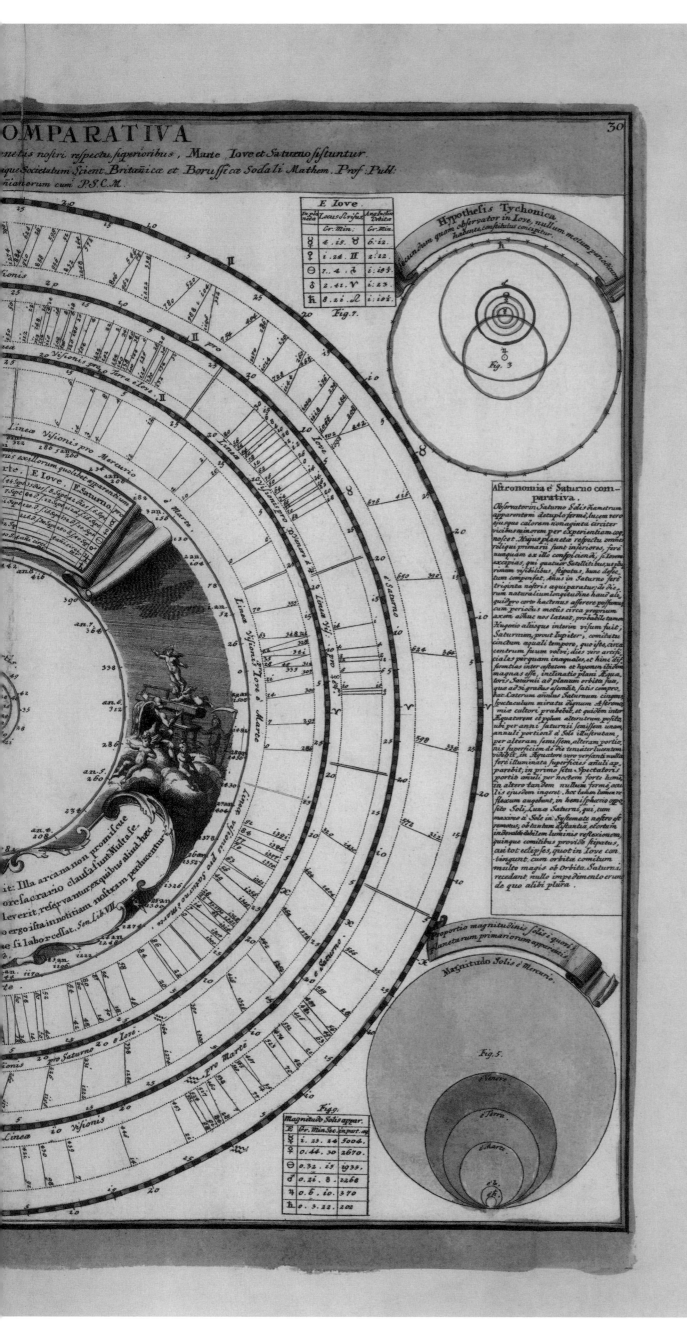

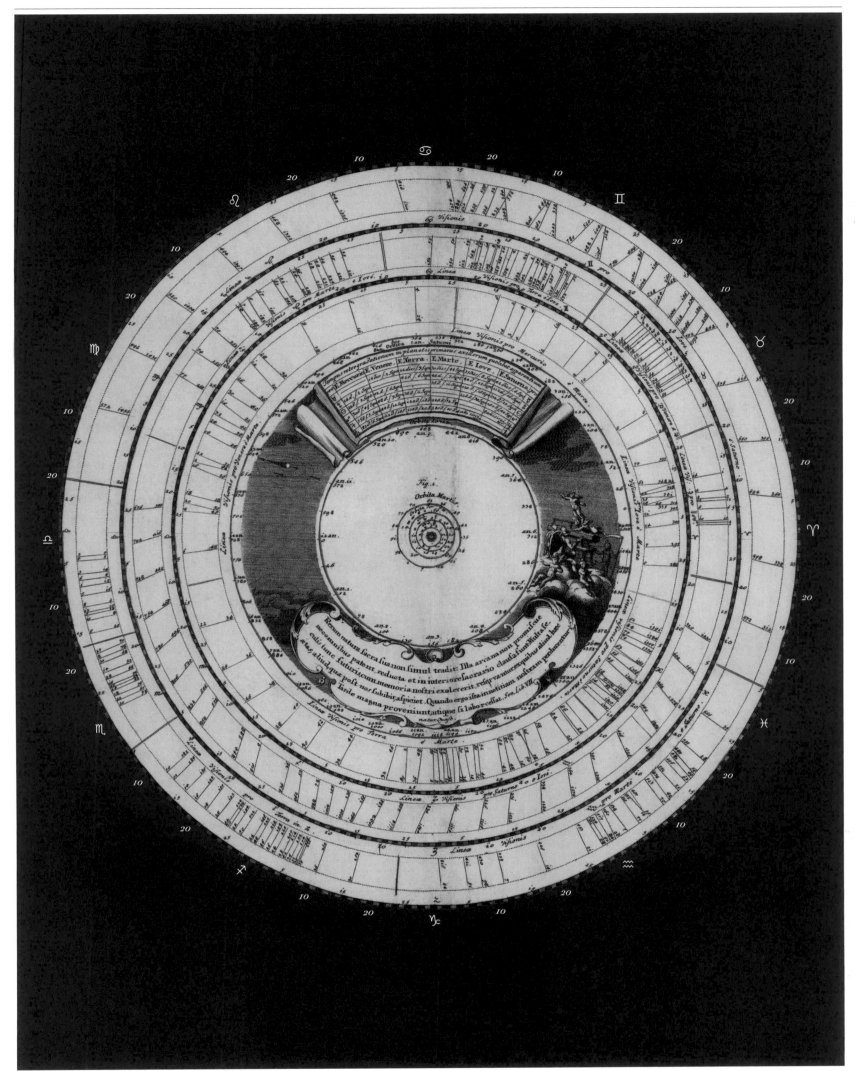

MOTIONS OF THE OTHER PLANETS SEEN FROM
MARS, JUPITER AND SATURN. (FIG. 1.)

The central diagram of Plate 30 extends the concept of the previous
plate, marking regular divisions around the planetary orbits in the
centre and using these to construct lines of sight around the outside,
revealing the behaviour of the planets as seen from Mars, Jupiter and
Saturn. At the top, a table provides the average periods of retrograde
motion that each planet will display when viewed from the vantage
point listed at the top of the column.

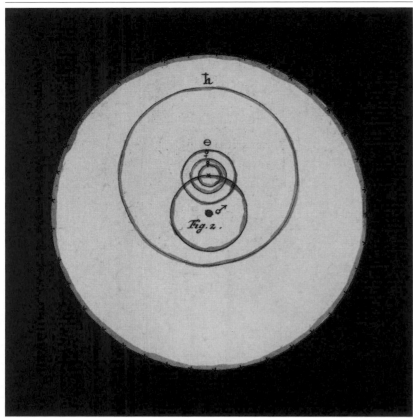

TYCHONIC SYSTEM AROUND A FIXED MARS. (FIG. 2.)

Here, Doppelmayr visualizes a system in which Mars is the fixed centre of the universe, orbited by the Sun, which is, in turn, circled by the other planets. The table details how the other planetary orbits appear if measured from Mars.

from MARS.

Planet	Direction of ascending node	Orbital inclination
	Grad. Min.	Grad. Min.
☿	12. 41. ♉	5. 4.
♀	10. 4. ♋	1. 54.
⊕	18. 22. ♏	1. 50½.
♃	2. 41. ♎	1. 23.
♄	6. 21. ♍	2. 24.

TYCHONIC SYSTEM AROUND A FIXED JUPITER. (FIG. 3.)

This diagram shows a Tychonic system in which Jupiter is fixed at the centre. The Sun orbits around it on a circular path, and is itself orbited by the other planets. The table shows the relationship of the other planets' orbits with that of Jupiter.

from JUPITER.

Planet	Direction of ascending node	Orbital inclination
	Grad. Min.	Grad. Min.
☿	4. 15. ♉	6. 12.
♀	1. 24. ♊	2. 12.
⊕	7. 4. ♑	1. 19⅔.
♃	2. 41. ♈	1. 23.
♄	8. 21. ♌	1. 19½.

TYCHONIC SYSTEM AROUND A FIXED SATURN. (FIG. 4.)

One last Tychonic diagram shows a system in which Saturn becomes the fixed centre point, orbited by the Sun in a wide circle, while the Sun is orbited by the other planets. The standard table shows the characteristics of the other planetary orbits from a Saturnian point of view.

from SATURN.

Planet	Direction of ascending node	Orbital inclination
	Grad. Min.	Grad. Min.
☿	22. 32. ♈	6. 25½.
♀	25. 59. ♈	2. 4½.
⊕	22. 38. ♑	2. 32.
♃	6. 21. ♓	2. 24.
♄	8. 21. ♒	1. 19½.

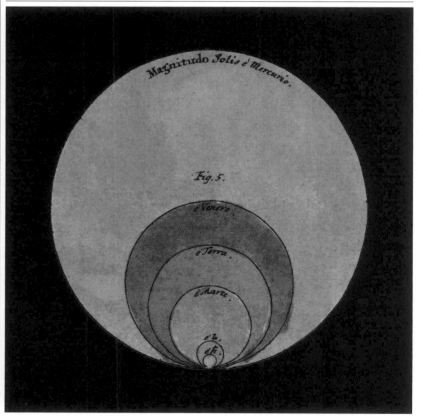

SIZE OF THE SUN FROM THE PRIMARY PLANETS. (FIG. 5.)

Doppelmayr visualizes the relative dimensions of the Sun, as seen from each of the major planets. The tiny discs seen from Jupiter and Saturn are represented at larger scale on Plate 29.

APPARENT SIZE OF THE SUN.

E	Gr. Min. Sec.	In equal parts
☿	1. 23. 24.	5004.
♀	0. 44. 30.	2670.
⊕	0. 32. 15.	1935.
♂	0. 21. 8.	1268.
♃	0. 6. 10.	370.
♄	0. 3. 22.	202.

THE GREAT COMET, 1881.

PART OF THE MILKY WAY.

THE NOVEMBER METEORS.

THE PLANET SATURN.

MARE HUMORUM.

AURORA BOREALIS.

TOTAL ECLIPSE OF THE SUN.

THE ZODIACAL LIGHT.

PARTIAL LUNAR ECLIPSE.

STAR CLUSTER IN HERCULES.

GROUP OF SUNSPOTS.

THE GREAT NEBULA IN ORION.

LEGACY

The compilation of the *Atlas Coelestis* in the decades prior to its 1742 publication coincided with a unique period in the story of astronomy. While most astronomers were by now convinced that the Copernican model of a Sun-centred solar system was true, the clinching evidence that Earth was in motion was only just being discovered, while the breakthroughs that would reveal the true scale of the cosmos lay decades ahead. In closing the story, it is fitting to look at the various ways in which Doppelmayr's magnum opus slots into the wider narratives of both our understanding of the cosmos and the artistic depiction of the heavens, as well as the complex ways in which these interrelate.

*   *   *

From a historical perspective, the *Atlas Coelestis* presents something of a puzzle in at least one respect; by the time it was published in 1742, the final clue confirming Earth's motion through space had been recognized elsewhere for more than a decade. Today, this phenomenon is known as the 'aberration of starlight'.

As previously noted, one of the most reasoned objections to the Copernican theory of a moving Earth was the absence of parallax – the expected annual shifts in the apparent directions of stars as they were viewed from different points on our planet's supposedly huge orbit. The principal Copernican response to the lack of parallax was to suggest that the stars were so far away that their parallax was undetectable – even using the most powerful telescopes and accurate measuring devices. This in turn implied that the stars must be suns in their own right, viewed over great distances. Even Aristarchus of Samos (*c.* 310–230 BCE), the earliest known proponent of a heliocentric system, recognized the parallax issue and its implications.

By the 16th century, however, measurement techniques had improved so much that the so-called 'star size problem' (see pages 88–9) added a significant barrier to the acceptance of Copernicanism. Now, the small but apparently measurable angular diameters of stars, coupled with their supposedly vast distances, implied huge size – even the faintest and smallest stars must, in fact, be monsters far larger than the Sun. The issue was not resolved until the 19th century, when astronomers recognized the various optical effects that cause the point source of light from a distant star to appear as a small but measurable disc. In the meantime, however, many astronomers felt that upholding Copernicanism meant accepting the contrary steps of promoting the Sun to central status in the universe while rendering its size insignificant compared to every other star in the sky. The only way of resolving this problem, it seemed, was to confirm parallax through sufficiently accurate measurements.

Robert Hooke (1635-1703), curator of experiments at London's Royal Society, made an early attempt at measuring parallax in 1669. Aware that calculating the true direction of any star is complicated by the way that Earth's atmosphere bends or refracts its light, he focused his attention on Eltanin (Gamma Draconis), which passed directly overhead from the latitude of London and could be assumed to be free of refraction as it did so. Careful measurements suggested that Eltanin lay twenty-three seconds of arc (about 1/160th of a degree) closer to the sky's north celestial pole in July than it did in October. In 1680, French astronomer Jean Picard (1620–82) reported a similar effect influencing the pole star Polaris. The wider scientific community voiced doubts about these observations, but they became harder to dismiss when British Astronomer Royal John Flamsteed (1646-1719) reported the detection of a similar wobble affecting Polaris in 1698. Based on a sequence of observations going back to 1689, Flamsteed was convinced that he had detected parallax and described the way in which it manifested.

FIG. 1.
This 18th-century engraving by Sutton Nicholls depicts The Monument, built to commemorate the Great Fire of London (1666). In an effort to refine his measurements of the star Eltanin, Hooke collaborated with Sir Christopher Wren on the design of the 62-m (202-ft) tower. It was built with a hollow shaft at the centre to act as a vast zenith telescope. Sadly, vibrations from nearby traffic rendered it unusable for this purpose.

FIG. 1.

PROSPECTUS INTRA CAMERAM STELLATAM.

FIG. 2.

FIG. 2.
This view of the Great Star Chamber at the Royal Observatory in Greenwich by Francis Place (1712) shows the huge windows designed by Wren to allow the mounting of telescopes and other instruments. However, because it was not oriented to the points of the compass, it proved impossible to incorporate a transit telescope aligned to the north–south meridian (necessary to measure objects at their highest point in the sky). Instead, Flamsteed did much of his observing work from other buildings.

While no one dared question the accuracy of Flamsteed's measurements, several astronomers soon expressed misgivings about their theoretical underpinnings. At the time, the observational consequences of parallax were poorly understood, but a number of people pointed out that Flamsteed's cyclical annual motions were awkwardly out of sync with those expected from any plausible theory. Whereas Flamsteed found that the drift of Polaris slowed to a halt and switched directions in June and December, most experts were confident that true parallax would change directions in March and September. Giovanni Domenico Cassini (1625–1712) of the Paris Observatory provided the most detailed critique in a letter of 1702, complete with a somewhat abstruse alternative explanation for the reported motions.

Doppelmayr was certainly not ignorant of this debate. In supplementary diagrams on Plate 28, he explores both the basics of parallax and Flamsteed and Cassini's theories. However, despite clear evidence that he compiled these diagrams in c. 1740, he seems to have mysteriously neglected the crucial resolution to the story.

In the mid-1720s, James Bradley (1692–1762), Savilian Professor of Astronomy at Oxford, set out to repeat Hooke's measurements in collaboration with amateur astronomer Samuel Molyneux (1689–1728). They used a purpose-built, vertically mounted telescope hung down the face of a chimney at Molyneux's house on the outskirts of London. The instrument's tube was suspended by a pivot at the top, allowing the eyepiece end to be swung back and forth through a little over 1 degree using an adjustable screw. A plumb line hanging from the pivot provided a precise vertical line, and the telescope's deviations from this angle could be read from an arc measured in the smallest fractions of a degree. Over the course of almost two years, beginning in December 1725, Molyneux and Bradley went through the laborious process of recording the telescope's view of Eltanin as it reached its highest point in the sky on some eighty occasions. Their results not only put it beyond doubt that the drift in position was real, but also confirmed that it was three months out of step with the predictions of parallax.

At first, the astronomers pondered whether the movement could be down to an annual wobble or 'nutation' in the direction of Earth's axis. Bradley, however – forced to continue the project alone, first by other demands on Molyneux's time and then by his colleague's untimely death – proved that this could not be the case. Using a smaller instrument that could swing across a wider arc of the sky, he made measurements of stars further from the zenith point, taking into account the idealized effects of atmospheric refraction. While a nutation effect could be expected to affect all stars to the same degree, Bradley found that the shifts varied considerably depending on their distance from the celestial pole.

The solution came to him in 1728. While the search for precession was focused on Earth's changing position relative to distant stars, everyone had hitherto overlooked its changing velocity (speed relative to the directions of particular stars). Since light was now known to travel at a finite speed, Earth's motion towards or away from a star's incoming rays of light would affect the apparent direction from which they arrived at Earth, and, therefore, the star's apparent position in the sky. (An everyday analogy is to think about the slight angle at which an umbrella needs to be tilted in order to protect the carrier from rainfall depending on their own direction and speed.)

From his measurements, Bradley estimated that Earth's speed through space was roughly 1/10,200th of the speed of light, and that light propagated from the Sun to Earth in 8 minutes and 12 seconds (very close to the modern figure of 8 minutes and 20 seconds). Thus, he not only proved incontrovertibly that Earth really was moving around the Sun, but he also improved on Ole Rømer's (1644–1710) earlier estimate of the speed of light, helping to refine the scale of the solar system as a whole.

It is a great testament to Bradley's method and insight that his work was widely and rapidly recognized across Europe. He succeeded in taking the stellar misbehaviour that might have seemed a fatal blow to the Copernican theory and transforming it into a vindication. Indeed, Italian mathematician and astronomer Eustachio Manfredi (1674–1739) had been making similar observations in Bologna at around the same time, in the hope of disproving Copernicanism once and for all. Somewhat ironically, Bradley soon adopted Manfredi's term for the effect – 'annual aberration' – as a means of distinguishing it from true parallax.

It is all the more curious, then, that Doppelmayr fails to properly describe Bradley's work in the *Atlas*. His connections across Europe and fellowship of England's Royal Society itself could hardly have left him in ignorance of such an important development, and indeed he even reproduces a diagram that could be used to explore the phenomenon on Plate 28, but he neglects to include any explanatory text. Whether simple pressure of space or some other motivation was at work, it seems the *Atlas* leaves us with one last mystery.

* * *

Once it became widely known and accepted, aberration seemed to settle the question of Earth's motion beyond doubt. From then on, the elusive parallax would retain its importance as a potential tool for measuring the scale of the universe, but it would no longer be the fulcrum of an existential debate between rival systems of astronomy. Yet by the time Bradley concluded more than two decades of meticulous attempts at measurement in 1747, it seemed clear that the real parallax effect remained far too small for any telescope of the period to distinguish. Although there were further intermittent attempts to measure parallax in subsequent decades, it was not until the 1830s that improvements in telescopes and other observing instrumentation allowed them to bear fruit. In 1838, the credit for the first successful parallax measurement went to German astronomer Friedrich Bessel (1784–1846).

Bessel's measurements targeted 61 Cygni: a faint naked-eye star (or, rather, a binary pair of stars) in the constellation of Cygnus, the Swan. 61 Cygni was renowned for the speed of its 'proper motion' across the sky – a drift in comparison to other stars that amounted to the width of the Full Moon every 150 years. When the phenomenon was discovered by Giuseppe Piazzi (1746–1826) in 1804, it earned 61 Cygni the nickname the 'Flying Star'.

FIG. 3.
This diagram from Bessel's collected essays (1876) shows the 'Königsberg heliometer'. Bessel's precise measurements of 61 Cygni made use of this unusual telescope, whose design employs a main objective lens divided into two halves to create a pair of independent images. A fine screw allows the alignment of the lens segments to be minutely adjusted, overlapping the images of separate stars and revealing the precise angle between them.

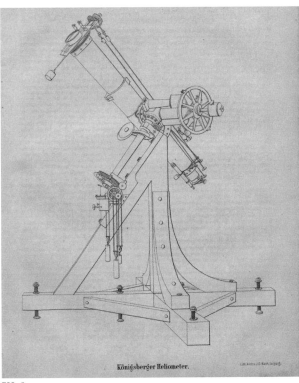

Königsberger Heliometer.

FIG. 3.

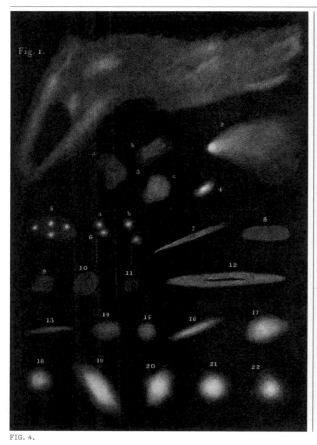

FIG. 4.

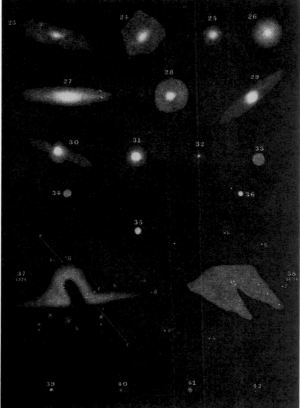

FIG. 5.

Two pages from
the 1811 catalogue
of William Herschel
showcase nebulae
and diffuse objects
in the sky, with the
intention of displaying
their evolution from
one kind to another
over time. Following
the discovery of
large numbers of
nebulae in the 18th
and 19th centuries,
many astronomers
attempted to
understand them as
a single class of object,
rather than as the mix
of unresolved star
clusters, interstellar
gas clouds and remote
galaxies they really are.

As early as 1718, Edmond Halley (1656–1742) had demonstrated the existence of proper motion, but most stars showed far slower drifts. It was logical to reason, however, that if all stars moved at broadly similar speeds in random directions through space, then the nearest ones would show the largest proper motion in Earth's skies.

Bessel, therefore, chose 61 Cygni as his study subject because it was the star with the largest known proper motion of all. At the same time, in St Petersburg, Friedrich Struve (1793–1864) – following an equally plausible chain of reasoning – picked out one of the brightest stars in the sky: Vega, in the constellation of Lyra. With little reason to believe stars were not all broadly similar in their light output, many assumed that the brightest stars would also be the ones closest to Earth and showing the greatest parallax.

As it happened, Bessel had made the better choice. Although Vega is, indeed, nearby in cosmic terms, it lies about twice as far away as 61 Cygni and, therefore, has half the parallax. Vega's dazzling brightness also made it more difficult to measure its central point precisely, in comparison to the much fainter 61 Cygni. As a result, while Struve published a preliminary (and accurate) measurement of Vega's parallax in 1837, his final result of 1840 was actually far less accurate. Bessel's 1838 measurement for 61 Cygni, meanwhile, was within 10 per cent of the modern value. His calculation of the star's parallax as

0.314 seconds of arc, or roughly 1/11,500th of a degree, indicated a distance equivalent to 10.3 light years in more recent terminology. When Bessel announced his findings in a letter to London's Royal Astronomical Society, its president, John Herschel (1792–1871), proclaimed that the 'sounding line in the universe of stars has at last touched bottom'.

The ability to measure parallax, however, did not exactly open the floodgates to mapping the universe. Bessel and Struve's work pushed the limit of what could be achieved with the telescopes of the time, and after their initial two stars, only a couple of dozen others, all very close to Earth, were measured in the 19th century. (Measurements of much smaller angles – indicating stars at far greater distance – had to wait until the era of orbital observatories capable of observing from above Earth's atmosphere.) Nevertheless, knowledge of the true brightness of this handful of stars allowed calculation of their intrinsic light output or luminosity for the first time. From this, other physical properties could be found by following various lines of evidence. The result was a model of stellar physics that could be applied to many more stars.

If Bessel's measurements helped map out distances in the solar neighbourhood, then more distant stars remained frustratingly mysterious. William Herschel (1738–1822, father of the aforementioned John) had tried to map the structure of the Milky Way in the 1780s by counting the number and brightness of stars in different parts of the sky. He concluded that the Milky Way was a somewhat flattened cloud of stars, with the Sun and solar system embedded inside its central plane a little way from the centre.

While astronomers debated the accuracy of Herschel's model, most still assumed, for the majority of the 19th century, that the Milky Way was the one and only, all-encompassing star system. The alternative – that it is but one among many vast clouds of stars in space – had been proposed by Thomas Wright (1711–86) as early as 1750. Wright noted that, while most stars concentrated along the band of the Milky Way, many of the nebulae lay out of the plane and, therefore, seemed separate from it. A few years later, German philosopher Immanuel Kant (1724–1804) was prompted by Wright's ideas to propose that the Milky Way and these distant nebulae were distinct 'island universes'.

Intriguing though the idea was, a second great Copernican Revolution, demoting our galaxy to one among countless billions, only took hold in the late 19th and early 20th centuries. In 1864, British astronomer William Huggins (1824–1910) analysed the light from the various nebulae and found that they fell into two distinct classes: some were formed of glowing gas, while others –

including those that showed spiral structures – were apparently vast unresolvable clouds of stars. Around 1915, further studies by Vesto Slipher (1875–1969) at the Lowell Observatory in Flagstaff, Arizona, revealed that many of the spiral nebulae were moving relative to Earth at speeds of hundreds of kilometres per second – far faster than the typical speeds of stars, and suggesting that they were not bound by the Milky Way's gravity.

The question was finally settled in 1925, when Edwin Hubble (1889–1953) used images from what was then the world's largest telescope, at Mount Wilson Observatory in California, to capture multiple photographs of individual spiral nebulae and reveal their individual stars. Comparing the photographs, Hubble identified a number of variable stars that changed their brightness over time. While there were many types of variable, some proved to belong in a class known as Cepheids, which Harvard astronomer Henrietta Swan Leavitt (1868–1921) had shown pulsated with a period in proportion to their average intrinsic brightness. Using Leavitt's 'period-luminosity relation', Hubble worked out the true luminosity of the Cepheids in his images, and then calculated their distance based on their brightness as seen on Earth. This proved, beyond doubt, that the spiral

FIG. 6.
This engraving – *The Newtonian System of the Universe* – by Isaac Frost (*c.* 1846) depicts the solar system surrounded by orbiting star systems. Just as the Enlightenment astronomer could dare to imagine each star as the centre of its own unique solar system, so today's cosmologists can visualize each galaxy as the centre of its own vastly bigger 'observable universe', extending in every direction for as far as the most distant light can have travelled since the Big Bang.

FIG. 6.

nebulae really were Kant's 'island universes', and that our galaxy was just one among many.

Hubble's finding was the last great breakthrough in uncentring Earth from its special place at the centre of the universe. When, a few years later, he discovered his famous law – that galaxies at greater distances in every direction are moving away from us at higher speeds – astronomers did not rush to assume some special point of view for Earth. Following the principles of relativity laid down by Albert Einstein (1879–1955) at the beginning of the 20th century, they assumed that there are no such privileged viewpoints. The universe would appear the same from any galaxy picked at random, and so the proper interpretation of galactic motions is that the entire universe is expanding, and pulling every galaxy away from its fellows like raisins in a rising cake.

In one sense, however, Earth does remain at the centre of everything even now. The discovery of cosmic expansion (and its corollary that the expansion could be traced back to a single point of origin, now known as the Big Bang) places limits on how far we can see, and puts Earth at the centre of its own unique 'observable universe'. The limited speed of light means that the further away we look in space, the longer light has travelled to reach us, and the further back we are looking in time. Thus there is a boundary where light has only just had time to reach us in the 13.8 billion years since the birth of the universe.

Every point in space sits at the centre of its own observable universe, with the true all-encompassing universe stretching far beyond. As this universe ages, each observable universe grows along with it. The ancient Greek astronomers might be comforted to know there is, indeed, one last ineluctable heavenly sphere that can never be penetrated.

* * *

Astronomy is both the oldest and the most visual of the sciences, so it is little wonder that attempts to understand the universe – and to transmit that knowledge to others – have overlapped with artistic endeavours.

The great European star atlases of the 17th and 18th centuries mark the culmination of a tradition dating back to prehistory, in which the patterns of stars in the night sky, and other celestial phenomena, were associated with powerful deities, symbols and mythical figures. The influence of such star atlases – principally *Uranometria* (1603) by Johann Bayer (1572–1625), and *Uranographia* (1687) by Johannes Hevelius (1611–87) – is traced elsewhere in this book.

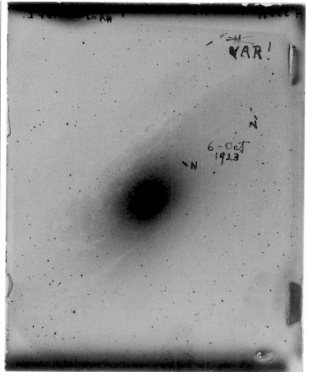

FIG. 7.

Doppelmayr's stellar plates in the *Atlas Coelestis* were particularly influenced by *Uranographia*. But by the time the *Atlas* was completed, a major new work had emerged in the form of Flamsteed's *Historia Coelestis Britannica* (1725) and its accompanying atlas – the long-awaited result of some forty years of meticulous observations. Flamsteed's accurate measurement of the positions of some 2,935 stars in fifty-four constellations set a new standard that prevailed through the 18th century. Although Doppelmayr's atlas (and other works derivative of earlier catalogues) remained popular, the most serious observers relied on Flamsteed. Throughout this era, charts continued to give broadly equal prominence to both the stars and the constellation figures, but as telescopes improved and astronomers catalogued ever more objects with increasing precision, this meant that star atlas plates became increasingly crowded.

The last great atlas of the golden age – *Uranographia, sive Astrorum Descriptio* (1801) – was compiled by Johann Elert Bode (1747–1826), director of the Berlin Observatory. Featuring more than 17,000 stars plotted to well below naked-eye visibility fighting for attention with elaborate depictions of the constellation figures, it pushed the limits of usability.

In the 19th century and beyond, star charts became increasingly utilitarian, as the number of celestial objects burgeoned and the constellation figures retreated into the background and ultimately disappeared.

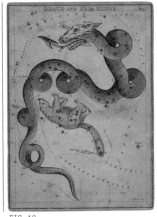

FIG. 8.   FIG. 9.   FIG. 10.   FIG. 11.

FIGS. 8–11. By the 19th century, constellation figures had begun to fade from view on the increasingly crowded charts of serious astronomers. They lingered on as a decorative nicety and aide-memoire for amateur astronomers: for example, on *Urania's Mirror*, a set of thirty-two astronomical cards published in 1824. Each card was punched with holes that allowed light to shine through, mimicking the appearance of the stars in the sky.

Bode pointed the way forward with his tentative introduction of boundaries between the constellations, enabling any new objects that were found to be allocated to a distinct and unique region of the sky without argument. Ultimately, this idea was systematized in the 20th century, with a final array of eighty-eight constellations – covering every area of the sky – being adopted by the International Astronomical Union in 1928.

Despite this, the romance of the constellation figures never quite disappeared. Their ghosts have remained as valuable teaching aids for amateurs navigating the sky, in incarnations ranging from the beautiful cards of *Urania's Mirror* (1824) to the switchable layers of modern stargazing apps. The *Atlas Coelestis*, therefore, captures just one beguiling moment in the ongoing story of the way we chart the stars.

*   *   *

Well beyond Doppelmayr's time, draftsmanship remained a key astronomical skill and the only way of preserving the ephemeral view through a telescope eyepiece to share with others. While the maps of the *Atlas* owe their stylistic heritage largely to Renaissance cartographers and pioneering stellar maps, such as those of Bayer and Hevelius, realistic renderings of individual objects, such as the nebulae featured at the foot of Plate 26, hint at the relatively new challenge of representing what was actually seen in the sky.

Early telescopic astronomers were largely content to record the positions and shapes of objects in ink on paper. Stars were either present or they were not, and the conceit that brighter ones were visually 'larger' was already well established. More subtle methods were required to record objects that showed variations in their intensity or light and shade: for example, spots on the face of the Sun, or features on the Moon. The representation of comets is another area that has its own rich history – beginning

with the line sketches on the Chinese *Silk Atlas of Comets* (*c*. 185 BCE) and extending through woodcuts and imaginative medieval illuminations into the telescopic era.

The first attempts to represent the sky itself somewhat realistically, however, are found in Giovanni Battista Hodierna's (1597–1660) book *On the System of Cometary Orbits, and on the Wonderful Things in the Heavens* (1654). Here, Hodierna wrote one of the first detailed considerations of what today's astronomers call 'deep sky' objects: nebulae and other phenomena with an apparently non-stellar appearance. Hodierna used simple woodcuts to create blocks of black space out of which stars emerged, and the approximate limits of nebulae could be hatched.

Dutch astronomer Christiaan Huygens (1629–95) improved on both the observation of the sky and its rendition into print in his book *The System of Saturn* (1659), which used engraving to tint an entire area of the Orion Nebula, and to indicate the brightness of features on Mars and Jupiter. Although the technique introduces only one shade of grey intermediate between the stars and the blackness of space, it hints at the beginnings of the astronomical sketching tradition that uses crayon on black paper to render the sky more accurately, instead of the cruder form of black stars against white.

Huygens's sketch is one of the two reproduced by Doppelmayr on Plate 26; the other is a more detailed view captured by Picard in 1673, which added a handful of new stars to the region and rendered the shape of the nebulosity rather differently. Some modern researchers have pondered whether the absence of nebulosity in earlier and even contemporary records indicates that this nebula has changed significantly in recent history. It seems more probable that the variations are due to the foibles of early telescopes and subjective interpretation.

The need for accurate representation became more pressing in the late 18th century as improving telescope technology brought a range of new and often nebulous objects within reach. Comet hunter Charles Messier (1730–1817) published the first version of his catalogue of such objects (intended to help his fellow astronomers pinpoint new discoveries more easily) in 1774, and expanded it from some 45 to 103 entries the following decade. While many of Messier's objects could be resolved into individual stars, others could not. His depiction of the Orion Nebula marks a huge leap forward in the visualization of the heavens, using subtle graduations of light and shadow to capture an apparently three-dimensional object within the circle of a telescopic field of view. Overlaid grid lines, meanwhile, not only reveal the working method by which Messier achieved such accuracy, but also serve as signifiers to readers that this was worthwhile and scientific work.

Messier's contribution to cataloguing the nebula was soon improved upon, however, by William Herschel from the 1780s onwards. Herschel was a self-taught telescope maker, who produced the finest instruments of his day. Working alongside his sister Caroline (1750–1848), he used them to make groundbreaking observations of the heavens. After his discovery of a new planet (now known as Uranus) in 1781, he was renowned as the greatest observer in Europe. Among his many discoveries, he found countless new nebulae and became satisfied that some of them were truly gaseous in nature.

Faced with the problem of discussing objects that only he could properly observe, in 1811 he illustrated the various forms of nebulae in an influential engraving for his paper 'On the construction of the heavens' (see page 239).

Herschel's son, John, inherited his father's mantle as the pre-eminent observer of the early 19th century. A skilled draftsman from his youth, he and many others benefited from the invention in 1795 of pencils that combined graphite and clay in varying amounts to produce drawing tools of different hardnesses. Using his father's telescopes and some of his own construction, John Herschel produced meticulous charts of many diffuse objects in the sky. He developed an ingenious way of accurately recording faint stars and nebulosity by first plotting the precise positions of a few key stars, then constructing a polygonal skeleton of lines between them, and finally adding fainter features with reference to these imaginary lines in the sky. He was concerned about accuracy because he wanted to look for movement and evolution in the shape of the nebulae, particularly in the hope of settling the widespread debate about whether they were nascent solar systems. Around 1839, meanwhile, a young astronomer at Yale College, Ebenezer Porter Mason (1819–40), pioneered a different approach to mapping nebulae, using isophotes – lines of equal brightness analogous to the elevation lines on a map.

The most influential astronomical illustration of the 19th century, however, emerged from a different quarter. In 1845, William Parsons, 3rd

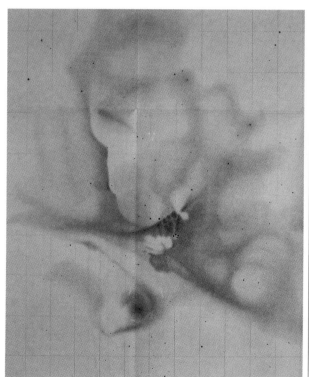

FIG. 12.

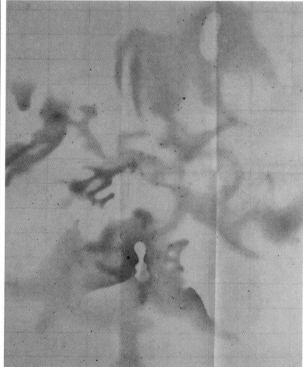

FIG. 13.

FIGS. 12–13.
Two of John Herschel's astonishing astronomical sketches from the 1830s show the Orion Nebula (far left) and the Keyhole Nebula in the southern constellation of Carina (near left). Herschel's combination of meticulous plotting with an artistic eye and the latest in drawing tools allowed him to capture extraordinary levels of detail.

Earl of Rosse (1800–67), completed construction of the largest telescope to date, and began his own research into the nebulae. Fittingly known as the Leviathan, Rosse's telescope had a 1.8-m (6-ft) mirror and a tube so long that it had to be supported between two parallel walls. This limited horizontal movement and kept individual objects in view for only an hour at best. Almost immediately after the telescope was installed, Parsons used it to sketch a view of Messier 51, a nebula in the constellation Canes Venatici. Where Herschel had previously seen only a ring-shaped structure, Parsons found a wealth of detail suggesting a spiral in space.

Parsons's illustration was reproduced in 1846 as a mezzotint – a form of engraving that relies on the roughened surface of the plate to control the distribution of ink. The method avoids the need for larger and more deliberate marks to produce the impression of half-tones, such as hatching or dots. As the 19th century progressed, there was an increasing recognition of the implications of both choice of sketching technique and methods of reproduction. For example, in Plate 26, the traditional techniques such as stippling that are used to create graduated shading may also, in an astronomical context, give the false impression of a field of tiny stars. The structure of the nebulae was particularly misleading, and astronomers went to great efforts to distinguish whether they were truly nebulous gases or whether they consisted of countless unresolvable stars. The fact that the features broadly termed 'nebulae' encompassed several different classes of object (principally glowing gas clouds in areas of star formation, gaseous shells expelled from dying stars and distant, star-packed galaxies) did little to help.

Astronomers, nevertheless, did their best to be objective. Some experimented with artistic media that would avoid ambiguity, such as watercolours and oils. Most notably, William Lassell (1799–1880) produced detailed oil paintings of complex nebulae in an effort to convey his impressions. Parsons, meanwhile, employed several skilled observing assistants, each of whom kept their own book in which they recorded what they saw at the eyepiece using their own distinctive techniques. Despite this, interpretation could not help but seep in – a problem at its most dangerous in the case of Messier 51. In reality, this 'nebula' is a distant galaxy whose spiral structure is an illusory pattern caused by the 'clumping' of its brightest stars at certain locations as they follow broadly circular, concentric orbits. However, the spiral appearance tempted many observers to render it as a celestial whirlpool of material being drawn towards the bright centre, in keeping with theories of how the solar system had been formed. Popularized by French astronomer Camille Flammarion (1846–1936), this idea is

FIGS. 14–15.
This series of sketches of nebulae was published by Rosse in a paper of 1850. The pronounced structure of Messier 51 (top left, drawn in 1848) is echoed to a greater or lesser extent in the other objects, all of which are now known to be distant spiral galaxies. Rosse's iconic drawing of the 'whirlpool' became the standard image of a spiral nebula for more than half a century.

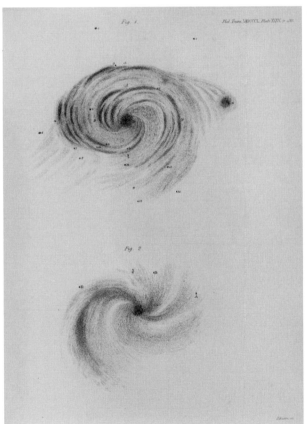

FIG. 14.

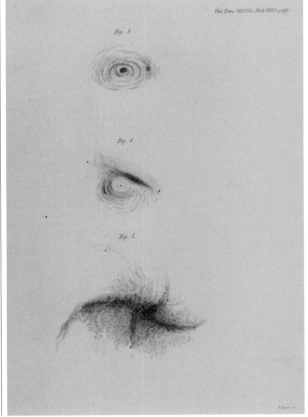

FIG. 15.

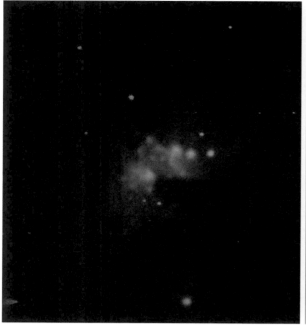

FIG. 16.

FIG. 17.

FIGS. 16–17.
More than a century
of progress in
astronomy is bridged
by two iconic images
of the Orion Nebula.
At far left, Draper's
landmark photograph
(1880) was made using
a new process that
massively increased
sensitivity and brought
a far wider range
of objects within the
reach of photography
for the first time. At
near left is a 2006 view
from the Hubble Space
Telescope, which uses
a panoply of modern
imaging technology
in five separate colours
to create a stunning
rendition that
combines scientific
accuracy with a touch
of artistic flair.

even said to have influenced the vortex-like depiction of the heavens in Vincent van Gogh's (1853–90) painting *The Starry Night* (1889).

The first hints of a coming revolution in astronomical imaging began as early as 1839, when French artist and photographer Louis Daguerre (1787–1851) captured a mottled image of the Moon using his new light-sensitive daguerrotype process. However, it was some four decades before photographic technology had matured sufficiently to rival illustration. Beginning in 1880, amateur astronomer Henry Draper (1937–1882) used the innovations of fast-reacting 'dry plate' photography to capture images of the Orion Nebula, with results so impressive that others soon followed in his wake.

Photography was not without its own drawbacks, of course, and the techniques of developing a print could themselves require some subjective judgment calls. However, the ability of long exposures to outperform even the most skilled observer's eye had obvious advantages. Coupled with new techniques such as spectroscopy (the splitting-up of starlight into rainbow-like spectra to study the intensity of different wavelengths), the first large-scale photographic surveys ushered in an era of astronomical 'big data'. Analysing and comparing the properties of stars en masse revealed patterns and relationships among them that were ultimately crucial to discovering their physical properties and uncovering their life stories through a new branch of science – astrophysics.

Thanks to the advances of photography and, more recently, of digital imaging, the professional astronomers of the 20th and 21st centuries have stepped back from the eyepieces of their telescopes. Nowadays, computers do the work of controlling instruments and gathering data from mountain-top observatories thousands of kilometres away. The element of human talent and inspiration survives, however, in formulating the most perceptive questions to ask of the universe, discovering the best ways of obtaining the clearest answers, and interpreting what they all mean.

The ancient partnership between art and astronomy, of which the *Atlas Coelestis* forms a vivid historical highlight, lives on today in many different forms. For example, astronomical sketching still has its devoted practitioners among amateur astronomers. And what are the carefully processed images from instruments such as the Hubble Space Telescope, but works of art designed to coax the maximum detail from electronic sensors and combine them to the greatest effect?

The very nature of our place in the cosmos limits our ability to see it clearly. Even when viewed through the most powerful telescope, the stars remain mere points of light, and our location within the vast spiral galaxy of the Milky Way clouds our view. However, astronomers are nothing if not ingenious in finding new ways to learn about the phenomena of the universe, and very often, when they want to communicate their ideas, they turn to the talent of artists to bridge the gap and transform equations, digits and data into visions of other stars and other worlds. Just as Doppelmayr discovered some 300 years ago, the enticement of visual splendour can be a powerful tool for conveying scientific arguments.

MILESTONES OF ASTRONOMY.

Eudoxus of Cnidus, a former student of Plato, develops a mathematical system that describes the movement of celestial objects through the rotation of concentric spheres on varying axes around a central Earth.

Aristotle develops a cosmology, based on the Eudoxan spheres, proposing physical connections to transmit motive force from one sphere to the next. He argues that planets, stars and celestial spheres are made from aether, a substance with different properties to the terrestrial elements of earth, air, fire and water.

Aristarchus of Samos considers a Sun-centred theory of the universe as a possible alternative to the geocentric ideas of Aristotle, but is unable to prove it.

Ptolemy of Alexandria's astronomical treatise the *Almagest* refines the model of celestial spheres using subsidiary circles called deferents along with other mathematical tools that enable it to survive unchallenged for 1,400 years.

Indian mathematician Aryabhata makes the earliest known argument that the daily motion of the sky is due to Earth's rotation on its axis. However, his ideas do not reach European astronomers.

Ibn al-Haytham writes *Doubts on Ptolemy*, one of several works by Islamic astronomers expressing concerns about the accuracy of the geocentric system.

Alfonso X of Castile sponsors the creation of the *Alfonsine Tables*, a series of tables using the Ptolemaic model (with refinements by Islamic astronomers) to predict the positions of celestial bodies.

French philosopher Nicolas d'Oresme's *Book of the Heavens and the Earth* demonstrates that many of the arguments used against a moving rotating Earth can be disregarded, but nevertheless it retains support for an Earth-centred cosmology.

Regiomontanus publishes *New Theory of the Planets* by his former teacher Georg von Peuerbach, a simplified description of the Ptolemaic model. It is the first printed astronomical textbook. Regiomontanus later completes a translation of Ptolemy's work, *The Epitome of the Almagest*, printed in 1496.

The *Alfonsine Tables* are printed for the first time (with a second edition in 1492). Improved access to both data and the underlying theories helps to highlight shortcomings in the Ptolemaic model.

Nicolaus Copernicus privately circulates manuscript copies of his *Little Commentary* – an early summary of his heliocentric theory. Word rapidly spreads through Europe's natural philosophical community.

Albrecht Dürer, Johannes Stabius and Conrad Heinfogel produce a pair of celestial hemispheres – the first star charts to be printed in Europe.

Copernicus's full explanation of his theory, *On the Revolutions of the Heavenly Spheres*, is published, thanks in part to pressure from his friend and pupil, German mathematician Georg Joachim Rheticus.

German astronomer Erasmus Reinhold publishes the *Prutenic Tables*, a series of predictive tables based on the Copernican model that are soon found to be more useful than the *Alfonsine Tables*.

A supernova in Cassiopeia is the first inarguable sign of change in the heavens since antiquity. Tycho Brahe makes measurements that show it must be more distant than the Moon, undermining the Aristotelean view of an unchanging celestial realm.

Tycho measures the path of a brilliant comet, showing that this, too, lies beyond the orbit of the Moon.

Inspired by scientific arguments against the Copernican theory, Tycho publishes his own model of the cosmos, in which the planets orbit the Sun, but the Sun orbits Earth.

Dutch cartographer Petrus Plancius creates a globe that introduces new constellations to the far southern skies, mapping stars reported by sailors during a recent Dutch trading expedition.

Johann Bayer publishes *Uranometria*, a star atlas that covers the entire celestial sphere. It introduces the convention of designating stars according to their brightness within an individual constellation using Greek letters.

The telescope is invented in the Netherlands, probably by Dutch spectacle maker Hans Lippershey. Word of the new invention spreads rapidly across Europe.

Johannes Kepler publishes *The New Astronomy*, arguing for a Copernican system in which planets follow elliptical, rather than perfectly circular, orbits, and vary in speed according to their distance from the Sun.

Galileo publishes *The Starry Messenger*, a summary of his early telescopic discoveries, including the moons of Jupiter. He becomes a convinced Copernican, but is soon enjoined from teaching the theory as a physical reality by church authorities.

1619

Kepler elaborates on the theory of elliptical orbits in *Harmony of the World*, adding a third law that links a planet's orbital period to its mean distance from the Sun.

1632

Galileo publishes *Dialogue Concerning the Two Chief World Systems*, a comparison between the Ptolemaic and Copernican models. His satirical attacks on the geocentric view and its followers lead to a trial before the Inquisition.

1651

Giovanni Domenico Riccioli publishes the *New Almagest*, an encyclopaedic treatise on astronomy that includes detailed maps of the Moon and an analysis of arguments around the motion of Earth. They ultimately lead him to favour a modified Tychonic system.

1655

Christiaan Huygens uses a powerful new telescope of his own design to observe Saturn, identifying the structure of its rings and discovering its largest satellite, Titan.

1660

The first national scientific academy is founded in London when the Royal Society receives its charter from Charles II. In 1665, it begins publishing the *Philosophical Transactions*, the world's first scientific journal.

1667

The Paris Observatory is founded under a charter from Louis XIV, with the aim of improving understanding of astronomy and navigation.

1671–2

Giovanni Domenico Cassini, working at the Paris Observatory, uses surface markings to determine the rotation of Mars and Jupiter and discovers four new moons of Saturn.

1675

A Royal Observatory is established by Charles II at Greenwich, London, with similar aims to the Paris Observatory. John Flamsteed is appointed as England's first Astronomer Royal.

1676–7

Edmond Halley maps southern hemisphere skies from the South Atlantic island of St Helena and records a transit of Mercury across the face of the Sun.

1681

The reappearance of a bright comet (discovered by Gottfried Kirch and previously seen heading towards the Sun) following an apparent change in direction provides convincing evidence that comets move along parabolic curves.

1686

French author Bernard de Fontenelle publishes *Conversations on the Plurality of Worlds*, an influential book exploring the evidence that the stars are other Suns with solar systems of their own.

1687

Isaac Newton publishes the *Principia* (*Mathematical Principles of Natural Philosophy*), outlining laws of motion and universal gravitation that explain the force behind planetary motions and elliptical orbits.

1690

Johannes Hevelius's *Prodromus Astronomiæ*, an influential star catalogue and atlas, is published posthumously, introducing ten new constellations to northern skies.

1698

Huygens's *Cosmotheoros*, an influential book speculating on extraterrestrial life and including the first attempt to estimate stellar distances, is published posthumously.

1700

Frederick III of Prussia grants a patent for the building of the Berlin Observatory, with Gottfried Kirch as its first director.

1705

Halley publishes the *Synopsis of the Astronomy of Comets*, arguing that these celestial objects follow elliptical orbits around the Sun and predicting the 1758 return of a comet seen in 1531, 1607 and 1682.

1728

James Bradley announces his discovery of a small annual shift in the positions of stars (now known as 'aberration'), caused by the changing angle at which light approaches Earth as Earth orbits the Sun.

1729

John Flamsteed's *Atlas Coelestis*, the most detailed and accurate star atlas so far compiled, is published posthumously.

1742

The heirs of Johann Baptist Homann publish Doppelmayr's own *Atlas Coelestis*, combining celestial charts with a detailed summary of contemporary ideas about the universe as well as historical and alternative theories.

1750

Nicolas Louis de Lacaille begins four years of astronomical observations from the Cape of Good Hope, compiling detailed catalogues of southern skies and introducing fourteen new constellations.

1758

Amateur astronomer Johann Georg Palitzsch uses charts from Doppelmayr's *Atlas Coelestis* to confirm the return of the comet predicted by Halley.

1774

Charles Messier publishes the first version of his catalogue of diffuse objects in the sky – the beginnings of detailed study of such objects, including star clusters, nebulae and galaxies.

1781

William Herschel uses a self-built telescope to discover Uranus, the first new planet added to the solar system since prehistoric times.

1785

Herschel and his sister Caroline make the first attempt to map the structure of the Milky Way, based on the number of stars seen in various directions.

1801

Johann Elert Bode publishes *Uranographia*, an exhaustive atlas cataloguing more than 17,000 stars to well below naked-eye visibility.

1838

Friedrich Wilhelm Bessell successfully measures the annual parallax of the star 61 Cygni, confirming the vast scale of interstellar distances for the first time.

1846

French mathematician Urbain Le Verrier uses unexpected variations in the motion of Uranus to predict the existence of an eighth planet, Neptune, which is soon discovered.

1846

William Parsons, 3rd Earl of Rosse, uses a giant telescope to discover the whirlpool-like structure of the object Messier 51. Many other nebulae soon prove to have similar spiral forms.

1864

William Huggins uses the new technique of spectroscopy to distinguish between gaseous nebulae and those whose light is composed of countless individual stars.

1886

Edward Pickering of Harvard College Observatory begins a comprehensive photographic and spectroscopic survey of the sky that will ultimately unlock the secrets of stellar chemistry and physics.

1912

Henrietta Swan Leavitt identifies a relationship between the brightness and period of Cepheid variable stars in the Small Magellanic Cloud – a companion galaxy of the Milky Way. This relationship unlocks the key to measuring vast cosmic distances.

1925

Edwin Hubble identifies Cepheid variables in a number of spiral nebulae, allowing him to estimate their distance and proving that they are independent galaxies far beyond the Milky Way.

1929

Hubble discovers a broad rule that the further away a galaxy is from the Milky Way, the faster it is receding – a relationship that reveals the universe as a whole is expanding from an origin at a single point in the distant past.

DRAMATIS PERSONÆ.

ARISTOTLE
(384–322 BCE)

Ancient Greek philosopher who developed a cosmology in which stars and planets are carried on interlocking concentric spheres around a central Earth.

PTOLEMY
(c. 100–170 CE)

Greek-Egyptian astronomer and geographer who refined the geocentric model of the universe to create a theory that persisted for 1,500 years.

HASAN IBN AL-HAYTHAM
(965–1040)

Arab scholar who investigated the principles of optics and vision, published an influential critique of Ptolemy and developed a model of planetary motions.

NICOLAUS COPERNICUS
(1473–1543)

Polish priest and astronomer who put forward a Sun-centred model of the universe to explain problems in the Ptolemaic model.

TYCHO BRAHE
(1546–1601)

Danish astronomer who proposed a hybrid theory of the universe to resolve issues with the models of both Ptolemy and Copernicus.

GALILEO GALILEI
(1564–1642)

Italian physicist whose early telescopic observations supported the Copernican model, bringing him into conflict with the Catholic Church.

JOHANNES KEPLER
(1571–1630)

German astronomer who refined the Copernican system with elliptical, rather than circular, orbits to make its predictions accurate.

JOHANN BAYER
(1572–1625)

German cartographer who produced the first star atlas covering the entire celestial sphere, the frontispiece of which is pictured here.

JOHANNES HEVELIUS
(1611–87)

German-Polish astronomer who produced maps of the Moon, devised new constellations and published a star atlas that heavily influenced Doppelmayr.

GIOVANNI DOMENICO CASSINI (1625–1712)

Italian astronomer who made important observations of the planets and switched from a geocentric to a Copernican model of the universe.

CHRISTIAAN HUYGENS (1629–95)

Dutch astronomer and inventor who first accurately described the rings of Saturn, discovered its largest satellite Titan and invented the pendulum clock.

ISAAC NEWTON (1643–1727)

English physicist who identified laws of motion and gravitation that describe the behaviour of the planets and other celestial objects.

JOHN FLAMSTEED (1646–1719)

English Astronomer Royal, who charted some 3,000 stars with unprecedented accuracy and produced an influential star catalogue and atlas.

EDMOND HALLEY (1656–1742)

English astronomer and geophysicist who produced the first detailed map of southern stars and predicted the return of the comet named after him.

JAMES BRADLEY (1692–1762)

English astronomer whose precise measurements of stellar positions gave conclusive evidence that Earth orbits the Sun.

FRIEDRICH BESSEL (1784–1846)

German astronomer who made the first measurement revealing the distance to another star, beginning to establish the true scale of the universe.

HENRIETTA SWAN LEAVITT (1868–1921)

US astronomer who discovered relationships between the brightness and pulsation period of certain measurable stars.

EDWIN HUBBLE (1889–1953)

US astronomer who established the existence of distant galaxies beyond the Milky Way and subsequently discovered the expansion of the universe.

ABERRATION
A shift in the apparent direction of stars caused by the changing direction of Earth in its orbit over the course of each year (contrast with *parallax*).

AETHER
The 'fifth element' in Aristotelean cosmology - an incorruptible material prone to perfect circular motion, from which the spheres of the planets and stars were composed.

ALTITUDE
An astronomical object's angular separation from the horizon, measured along a great circle between the horizon and the zenith.

ANGULAR DIAMETER
The apparent size of an object in the sky, measured in terms of the angle between its opposite edges as seen by a distant observer.

APHELION
The point on a planet's orbit where it lies furthest from the Sun. Similarly, apogee is the point on the Moon's orbit where it lies furthest from Earth.

ARMILLARY
A model of the celestial sphere, centred on the Earth or Sun, used for understanding and predicting the motions of stars and planets.

ASTROLABE
A traditional astronomical instrument combining tools for measuring angles with a planisphere for predicting the positions of celestial objects.

ASTROLOGY
A range of ancient traditions linking the movement of objects in the sky such as the Sun, Moon and planets, to events on Earth.

AZIMUTH
An astronomical object's direction relative to an observer, measured as the angle between due north or south on the horizon and the point where a line from the zenith, passing through the object, touches the horizon.

CARTOGRAPHY
The art and science of map making, principally concerned with the accurate translation of positions on the surface of the Earth (or on the 'celestial sphere' of the sky) onto flat charts or spherical globes.

CELESTIAL COORDINATES
Any system of coordinates for pinning down the position of an object in the sky - examples include horizontal coordinates (altitude and azimuth), equatorial coordinates (right ascension and declination) and ecliptic coordinates.

CELESTIAL EQUATOR
An imaginary line running around the sky midway between the north and south celestial poles, corresponding to the plane of Earth's equator extended into the sky.

CELESTIAL POLE
One of two points in the sky, directly in line with Earth's north and south poles, about which the sky appears to spin due to Earth's daily rotation.

CELESTIAL SPHERE
In modern astronomy, an imaginary sphere around the Earth on which the positions of celestial objects can be determined by a coordinate system.

COMET
A small, icy object in an often highly elliptical orbit, typically returning to the inner solar system within a period of decades, centuries or longer, when heat from the Sun causes it to develop a fuzzy coma and extended tail.

CONJUNCTION
Any alignment of two celestial objects in the sky.

CONSTELLATION
Traditionally, a pattern created by the chance alignment of stars in the sky, associated by observers with mythological figures, animals or cultural artefacts. Modern astronomers define eighty-eight constellations as distinct areas that fit together to encompass the entire sky.

COPERNICANISM
See *Heliocentrism*.

DECLINATION
A measure of the angle between the celestial equator and a star or other object, along a line passing through both celestial poles.

DEFERENT
In complex geocentric models of the solar system, the larger circle upon which an epicycle turns.

DEGREE
A common measure of angle; a full circle is divided into 360 degrees, each of which are subdivided into 60 minutes of arc, themselves further divided into 60 seconds of arc.

DIGIT
A traditional unit of astronomical angle or extent, corresponding to one-twelfth the diameter of the Sun or Moon.

ECCENTRICITY
A measure of how much an ellipse such as an orbit diverges from a perfect circle. A circle has an eccentricity of 0, and highly elongated ellipses can approach but not reach an eccentricity of 1.

ECLIPSE
An astronomical event in which one object passes in front of another.

ECLIPTIC
An imaginary line around the sky, tilted at 23.5° to the celestial equator and marking the apparent track of the Sun through the constellations of the zodiac over the course of the year. Since the Sun's apparent motion is in fact created as Earth moves along its orbit, the ecliptic is also an extension of the plane of that orbit – and often treated as the de facto plane of the entire solar system.

ECLIPTIC LATITUDE
A measure of the angle between the ecliptic and a star or other object, along a line passing through both ecliptic poles.

ECLIPTIC LONGITUDE
A measure of an object's position along lines parallel to the ecliptic, measured in degrees eastwards from a line perpendicular to the ecliptic at the First Point of Aries.

ECLIPTIC POLE
The points at 90 degrees north and south of the ecliptic, about which the celestial poles appear to very slowly circle due to precession.

ELEMENTS (ARISTOTELEAN)
The traditional materials comprising the universe in the physics of Aristotle: earth, water, air and fire in the sublunary world, and aether in the heavens. The tendency of substances to rise or sink to their appropriate level within this scheme was used to explain natural motions.

ELLIPSE
A closed uniform curve resembling a circle stretched along one axis, with two focal points equidistant on either side of its centre. Objects follow elliptical orbits under the influence of gravity, with the parent body around which they orbit lying at one of the two foci.

ELONGATION
The angular separation of an inferior planet (Mercury or Venus) to the west or east of the Sun.

EPICYCLE
A smaller circle whose centre lies on the circumference of a larger one; in later classical and medieval geocentric models, the positioning of the planets on rotating epicycles helped to explain their complex motions.

EQUANT
A concept introduced in the geocentric system of Ptolemy, about which a planet moves with steady angular speed. The equant allowed Ptolemy's system to preserve the classical ideals of uniform and circular motion while more accurately predicting the paths of planets in the sky.

FIRST POINT OF ARIES
The point where the Sun's path on the ecliptic crosses the celestial equator on its way from the southern to the northern hemisphere at the start of northern spring. The First Point (denoted by the symbol ♈) is used as reference point for both the ecliptic and equatorial coordinate systems.

FIXED STARS
The apparently unchanging points of light that form the constellations and appear to spin around the sky once each day, with no additional complicating motions. Pre-Copernican astronomy imagined the stars as points on an outer Earth-centred sphere beyond those of the planets.

GEOCENTRISM
The traditional and once widely held view that Earth is static at the centre of the universe, with the Sun, Moon, planets and stars moving around it at varying rates.

GRADIAN
A measure of angle used today mostly in surveying; a gradian is 1/100th of a right angle, so 100 gradians correspond to 90 degrees.

GRAVITATION
An attractive force between objects with mass, linked to both their masses and the distance between then. The accelerating attractive force one object exerts on those nearby as a result is known as gravity.

GREAT CIRCLE
Any circle drawn around a sphere whose plane passes through its centre, and which therefore has the largest possible circumference. Lines such as the ecliptic, celestial equator and meridians passing through the celestial poles are all great circles.

HELIOCENTRISM
The view that the Sun is at the centre of the solar system, with Earth as the third in a family of planets orbiting around it.

HOUSE (ZODIAC)
A subdivision of the sky traditionally associated with the Sun's passage through a certain zodiac constellation. In Doppelmayr's context, a means of dividing the ecliptic into twelve equal segments, each encompassing some 30 degrees of angular measurement.

INCLINATION
The angle at which one orbit is tilted or inclined relative to another.

INERTIA
The tendency for an object to remain unchanged – either at rest or in constant motion – unless acted upon by an outside force.

INFERIOR CONJUNCTION
A conjunction in which a planet passes between Earth and the Sun (though it may lie above or below it in the sky due to the inclination of its orbit). Such events are only possible for Venus and Mercury.

INFERIOR PLANET
In the astronomy of Ptolemy, a planet whose epicycle remains aligned with the direction of the Sun as seen from Earth (contrast with *superior planet*). In the

Copernican system, one of the planets orbiting closer to the Sun than Earth.

LATITUDE
A positional coordinate on the surface of Earth or another spherical body, measured in degrees north or south of the equator.

LONGITUDE
A positional coordinate on a spherical body, measured along lines parallel to its equator, in degrees east or west of a selected meridian line.

LUNAR ECLIPSE
An event in which Earth passes between the Moon and Sun and casts its shadow on all parts of the lunar surface at Full Moon.

MERIDIAN
A line around the sky passing through both celestial poles, as well as the observer's nadir and zenith. The meridian touches the horizon at points due north and south, and objects reach their highest altitude in the observer's sky as they cross it.

MERIDIONAL
A traditional term for the southern direction, once widely used by European geographers and astronomers.

MOON
Written with a capital 'M', Earth's natural satellite; the natural satellites of other planets are known as moons with a lower-case 'm'.

NADIR
The point directly beneath an observer's feet, the opposite of the zenith.

NODE
One of the two points where the planes of orbits with different inclinations intersect, allowing precise alignments such as those which produce eclipses and transits.

OCCULTATION
An event in which one celestial body passes in front of another and blocks it from view – typically used to describe the disappearance of stars behind planets, or planets and stars behind the Moon.

OPPOSITION
The point at which a superior planet (such as Mars, Jupiter and Saturn) lies directly opposite the Sun in the sky, and is at its closest to Earth.

ORBIT
The typically elliptical path taken by one object around another under the influence of its gravity.

PARALLAX
A shift in the apparent direction of stars caused by the shifting position of Earth at different points on its orbit over the course of each year (contrast with *aberration*). The size of the parallax shift is inversely related to its distance from Earth.

PENUMBRA
The area of partial shadow cast by the Moon during an incomplete solar eclipse; in Doppelmayr's usage, also the shadow on the eclipsed Moon during a lunar eclipse.

PERIHELION
The point on a planet's orbit where it lies closest to the Sun. Similarly apogee is the point on the Moon's orbit where it comes closest to Earth.

PHASE
The apparent change in the shape of a spherical body created as different amounts of its sunlit hemisphere are displayed towards Earth.

PLANET
One of eight large worlds orbiting the Sun, although the outermost two (Uranus and Neptune) were unknown in Doppelmayr's time.

PLANISPHERE
A traditional device for calculating the visibility of stars from a certain location at a certain date and time.

PRECESSION
A slow change in the orientation of the constellations created by a 25,800-year 'wobble' of Earth's poles, which causes the celestial poles to change position and the First Point of Aries to slowly shift along the ecliptic.

PRIMUM MOBILE
The 'prime mover' or instigator of action – the outermost sphere in geocentric systems of the universe, lying beyond the fixed stars and imparting rotational motion to all the inner spheres.

PROJECTION
Any mathematical system used by cartographers for transposing positional measurements measured on a real or apparent sphere (such as stellar coordinates) onto a flat map or chart.

QUADRANT
A traditional device consisting of a swinging sighting bar and a graduated quarter circle, sometimes mounted on a precisely aligned wall and used for measuring the altitude of celestial objects.

REFRACTION
The bending of light as it passes between materials where it moves with different speeds. Refraction affects measured celestial positions due to the slower speed of light in Earth's atmosphere compared to empty space.

RETROGRADE
A type of planetary motion in which a body appears to slow, stop and move backwards against the background stars before resuming its normal 'prograde' (west-to-east) course.

RIGHT ASCENSION
A measure of an object's position along lines parallel to the celestial equator, today measured in hours, minutes and seconds eastwards from a line perpendicular to the equator at the First Point of Aries (and therefore indicating how long after the First Point they cross the meridian line), but previously measured using various angular schemes.

SATELLITE
Any object that orbits another one under the influence of gravitational attraction.

SEPTENTRIONAL
A traditional term for the northern direction, once widely used by European geographers and astronomers.

SOLAR ECLIPSE
An event in which the Moon passes in front of the Sun as seen from Earth's surface, blocking the solar disc either partially or entirely. A total solar eclipse requires a very precise alignment and can be seen from only a limited area.

SPHERES
In geocentric cosmology, a series of rotating concentric spheres composed of invisible crystalline material, upon which the planets and other celestial objects are mounted.

SUBLUNARY
In Aristotelean cosmology, the changeable realm beneath the sphere of the Moon, filled with the four classical elements of earth, water, air and fire, within which objects tend to move in straight lines.

SUN
The incandescent body at the centre of our solar system, which illuminates all the bodies around it. The Copernican Revolution led rapidly to acceptance that the Sun is just one of countless stars.

SUNDIAL
A device that uses the Sun's direction to find the local time at a location, based on the shadow cast by a bar called a gnomon.

SUNSPOT
A dark marking on the face of the Sun, carried across its face with its rotation.

SUPERIOR CONJUNCTION
A conjunction in which a planet lies exactly on the opposite side of the Sun as seen from Earth (though it may lie above or below it in the sky due to the inclination of its orbit).

SUPERIOR PLANET
In the astronomy of Ptolemy, a planet whose epicycle can change its alignment with Earth and Sun (contrast with *inferior planet*). In the Copernican system, a planet orbiting further from the Sun than Earth.

TELESCOPE
A devices that uses lenses or mirrors to capture more light than the human eye alone, in order to produce a magnified image that shows objects in more detail.

TRANSIT
An event in which a smaller object passes across the face of a larger one – for instance the passage of Mercury or Venus in front of the Sun, or a satellite in front of Jupiter.

TYCHONIC SYSTEM
A system of the universe in which the planets orbit the Sun, but the Sun orbits Earth, devised by Tycho Brahe to address contemporary problems with a purely heliocentric system.

UMBRA
The area of deep shadow cast by the Moon during a total solar eclipse.

ZENITH
The point directly above an observer's head, opposite the nadir.

ZODIAC
A band of twelve ancient constellations around the path of the ecliptic through the sky, through which the Sun appears to move over the course of each year, and where the planets are also most often seen. At present, the effects of precession mean that the ecliptic also crosses a thirteenth constellation, Ophiuchus.

WORKS OF JOHANN GABRIEL DOPPELMAYR.

— Doppelmayr, Johann Gabriel and Johann Christoph Sturm, *Optico-physical dissertation on the camera obscura* (Altdorf: Meyer, 1699)

— Doppelmayr, Johann Gabriel, *Detailed explanation of two new Homann charts, on the Copernico-Huygenian solar system and the European eclipse* (Nuremberg: Johann Baptist Homann, 1707)

— Doppelmayr, Johann Gabriel, *Gnomonica, or thorough instruction of how to set up all regular sundials* (Nuremberg: Johann Christoph Weigel, 1708)

— Doppelmayr, Johann Gabriel, *Short introduction to the noble astronomy* (Nuremberg: Johann Baptist Homann, 1708)

— Doppelmayr, Johann Gabriel, *A short summary of the physical experiments and artistic tests presented and demonstrated in the so-called Collegium Curiosum* (Nuremberg: 1716)

— Doppelmayr, Johann Gabriel, *New and thorough instructions on the placement of sundials* (Nuremberg: Johann Christoph Weigel, 1719)

— Doppelmayr, Johann Gabriel, *Historical account of the Nuremberg mathematicians and artists* (Nuremberg: Peter Conrad Monath, 1730)

— Doppelmayr, Johann Gabriel, *Illustrated physical experiments, or natural science promoted through many experiments and artistic tests demonstrated in the so-called Collegium Curiosum* (Nuremberg: Rüdiger, 1731)

— Doppelmayr, Johann Gabriel, *The Atlas of the Heavens, in which the world is to be seen, and in the same place the phenomena of the wandering and fixed stars* (Nuremberg: Heirs of Homann, 1742)

— Doppelmayr, Johann Gabriel, *Newly discovered phenomena due to admirable effects of nature arising from electrical power that accrues to nearly all bodies* (Nuremberg: Endter & Engelbrecht, 1744)

OTHER WORKS.

— Allen, Richard Hinckley, *Star Names: Their Lore and Meaning* (New York: Dover, 1963)

— Astronomical Society of Malta and National Library of Malta, *Respicite Astra: A Historic Journey in Astronomy through Books* (Valetta: National Library of Malta, 2010)

— Bernardini, Gabriella, *Giovanni Domenico Cassini: A Modern Astronomer in the 17th Century* (New York: Springer, 2017)

— Bernhard Cantzler et al., *A summary of practical geometry*, trans. Johann Gabriel Doppelmayr (Nuremberg: Wolfgang Moritz Endter, 1718)

— Bion, Nicolas, *A further opening of the new mathematical work-school*, trans. Johann Gabriel Doppelmayr (Nuremberg: Peter Conrad Monath, 1717)

— Bion, Nicolas, *A third opening of the mathematical work-school of Nicolas Bion, describing the preparation and use of various astronomical instruments*, trans. Johann Gabriel Doppelmayr (Nuremberg: Peter Conrad Monath, 1721)

— Bion, Nicolas, *The newly opened mathematical work-school, or thorough instructions of how to use mathematical instruments not only properly and correctly, but also to make them in the best and most accurate manner*, trans. Johann Gabriel Doppelmayr (Frankfurt, Leipzig & Nuremberg: Hofmann, 1712)

— Borbrick, Benson, *The Fated Sky: Astrology in History* (New York: Simon & Schuster, 2005)

— Buchwald, Zed J. and Robert Fox, *The Oxford Handbook of the History of Physics* (Oxford: Oxford University Press, 2013)

— Chapman, Allan, *Stargazers: Copernicus, Galileo, the Telescope and the Church* (Oxford: Lion Books, 2014)

— Christianson, John Robert, *Tycho Brahe and the Measure of the Heavens* (London: Reaktion Books, 2020)

— Drake, Stillman, 'Galileo's First Telescopic Observations', *Journal for the History of Astronomy*, 7, 3 (1976), 153–68

— Drake, Stillman, *Galileo at Work: His Scientific Biography* (Chicago: University of Chicago Press, 1978)

— Duerbeck, Hilmar W., *A Reference Catalogue and Atlas of Galactic Novae* (Chicago: University of Chicago Press, 1987)

— Ferguson, Kitty, *Tycho and Kepler: The Unlikely Partnership that Forever Changed our Understanding of the Heavens* (New York: Walker & Company, 2002)

— Gaab, Hans, 'Johann Gabriel Doppelmayr (1677–1750)', *Beitrage zur Astronomiegeschitchte*, band 4, in *Acta Historia Astronomiæ*, 13 (Frankfurt: Verlag Harri Deutsch, 2001), 46–99

— Gingerich, Owen, *The Book Nobody Read: Chasing the Revolutions of Nicolaus Copernicus* (London: William Heinemann, 2004)

— Graney, Christoper M., *Setting Aside All Authority: Giovanni Battista Riccioli and the Science against Copernicus in the Age of Galileo* (Notre Dame, IN: University of Notre Dame Press, 2015)

— Heilbron, John L., *Galileo* (Oxford: Oxford University Press, 2010)

— Hirshfeld, Alan, *Parallax: The Race to Measure the Cosmos* (New York: Henry Holt and Company, 2001)

— Hoskin, Michael (ed.), *Cambridge Illustrated History of Astronomy* (Cambridge: Cambridge University Press, 1997)

— Inwood, Stephen, *The Man Who Knew Too Much: The Strange and Inventive Life of Robert Hooke* (London: Macmillan, 2003)

— Irby-Massie, Georgia L. and Paul T. Keyser, *Greek Science of the Hellenistic Era: A Sourcebook* (London: Routledge, 2002)

— Kanas, Nick, *Star Maps: History, Artistry, and Cartography* (3rd edn) (New York: Springer, 2019)

— Koestler, Arthur, *The Sleepwalkers: A History of Man's Changing Vision of the Universe* (London: Pelican, 1970)

— Lankford, John (ed.), *History of Astronomy: An Encyclopedia* (New York: Garland Publishing, 1997)

— Nasim, Omar, *Observing by Hand: Sketching the Nebulae in the Nineteenth Century* (Chicago: University of Chicago Press, 2014)

— Pilz, Kurt, *600 Jahre Astronomie in Nürnberg* (Nuremberg: Verlag Hans Carl, 1977)

— Rogers, John H., 'Origins of the Ancient Constellations I and II', *Journal of the British Astronomical Association*, 108, 1, 2 (1998), 9–28; 79–89

— Rousseau, A. and S. Dimitrakoudis, 'A Study of Catasterisms in the "Phaenomena" of Aratus', *Mediterranean Archaeology and Archaeometry*, 6, 3 (2006), 111–19

— Sandler, C., *Johann Baptista Homann, die Homännischen Erben, Matthäus Seutter und ihre Landkarten* (Amsterdam: Meridian Publishing Co., 1979)

— Savage-Smith, Emilie, 'Celestial Mapping', in *The History of Cartography, Volume 2, Book 1: Cartography in the Traditional Islamic and South Asian Societies* (Chicago: University of Chicago Press, 1992), 12–70

— Seargent, David, *The Greatest Comets in History: Broom Stars and Celestial Scimitars* (New York: Springer, 2009)

— Sobel, Dava, *Longitude* (London: 4th Estate, 1995)

— Streete, Thomas, *Astronomia Carolina, a new theory of celestial motion*, trans. Johann Gabriel Doppelmayr (Nuremberg: Andreas Otto, 1705)

— Toomer, G. J., *Ptolemy's Almagest* (Princeton, NJ: Princeton University Press, 1998)

— Walker, Christopher (ed.), *Astronomy Before the Telescope* (London: British Museum Press, 1996)

— Wilkins, John, *John Wilkins's defence of Copernicus, in two parts*, trans. Johann Gabriel Doppelmayr (Leipzig: Peter Conrad Monath, 1713)

— Williams, M. E. W., 'Flamsteed's Alleged Measurement of Annual Parallax for the Pole Star', *Journal for the History of Astronomy*, 10, 2 (1979), 102–16

— Wilson, Curtis, 'On the Origin of Horrocks's Lunar Theory', *Journal for the History of Astronomy*, 18, 2 (1987), 77–94

— Woolard, Edgar W., 'The Historical Development of Celestial Coordinate Systems', *Publications of the Astronomical Society of the Pacific*, 54, 318 (1942), 77–90

— Wulf, Andrea, *Chasing Venus: The Race to Measure the Heavens* (London: William Heinemann, 2012)

All images from the *Atlas Coelestis* from David Rumsey Map Collection www.davidrumsey.com, unless otherwise stated below.

Every effort has been made to locate and credit copyright holders of the material reproduced in this book. The author and publisher apologize for any omissions or errors, which can be corrected in future editions.

a = above
c = centre
b = below
l = left
r = right

4–5 Universitätsbibliothek Freiburg i. Br., J 8564, b, S. 18
8 Fine Art Images/Heritage Images/Getty Images
9 ESA/Hubble & NASA, J. Lee and the PHANGS-HST Team
12–13 *Harmonia macrocosmica seu atlas universalis et novus*, Andreas Cellarius, 1708
14l © The Trustees of the British Museum
14c © History of Science Museum, University of Oxford, inventory no. 40,443
14r © History of Science Museum, University of Oxford, inventory no. 49296
15l *Templum S: Egidy cum Gymnasis*, Johann Andreas Graff, Jeremias Wolf, 1695–1725
15r *Maxplatz*, Johann Andreas Graff, 1693
16 Wellcome Library, London
17 Eimmart Observatory at Vestnertor Bastion of Nürnberg Castle, Johann Adam Delsenbach, 1716
18 *Johann Wilhelm Winter*, Johann Baptist Homann, 1728–1750
19 *1706 map of Europe during the Solar eclipse of May 12*, Johann Gabriel Doppelmayr, Johann Homann, 1706
20–21 © Mathematisch-Physikalischer Salon, Staatliche Kunstsammlungen Dresden, Photo: Peter Müller
24–25 Courtesy of The Linda Hall Library of Science, Engineering & Technology
26 *Harmonia macrocosmica seu atlas universalis et novus*, Andreas Cellarius, 1708
27 Bibliothèque nationale de France, département Cartes et plans, GE EE–266 (RES)
28l © Bibliotheque Mazarine/ © Archives Charmet/Bridgeman Images
28r Bibliothèque nationale de France. Département des Manuscrits. Grec 1401
29 © Bodleian Libraries, University of Oxford
32 Private Collection
33 *Sphaera armillaris; Instrumentum artificiale orrery ab inventore appellatum*, T. C. Lotter, 1774
39, 45 *Harmonia macrocosmica seu atlas universalis et novus*, Andreas Cellarius, 1708

50–51 *Tychonis Brahe Astronomiæ instauratæ mechanica*, Tycho Brahe, 1598
52 Photo © RMN-Grand Palais (domaine de Chantilly) © Michel Urtado
53 © British Library Board. All Rights Reserved/Bridgeman Images
54 *Tabulae Rudolphinae: quibus astronomicae*, Johannes Kepler, 1571–1630
55 Courtesy of The Linda Hall Library of Science, Engineering & Technology
60–61 Images owned by the Complutense University of Madrid
62–63 Library of Congress, Washington DC
68 The Picture Art Collection/Alamy Stock Photo
69 © British Library/Bridgeman Images
74–75 Photos © RMN-Grand Palais (domaine de Chantilly) © René-Gabriel Ojeda
76 Wellcome Library, London
77 Artokoloro/Alamy Stock Photo
83 The Digital Walters, W. 73, Cosmography
88 Metropolitan Museum of Art, New York, The Elisha Whittelsey Collection, The Elisha Whittelsey Fund, 1959
89 *Tychonis Brahe Astronomiæ instauratæ mechanica*, Tycho Brahe, 1598
94 *Description de l'Univers*, Alain Manesson Mallet, Paris, 1683
95l *Astronomia Mova Aitiologetos Sev, Physica coelestis*, Johannes Kepler, Tycho Brahe, 1609
95r *Astronomy explained upon Sir Isaac Newton's Principles*, James Ferguson, 1794
100–101 Leemage/UIG via Getty Images
102l National Archives of Belgium, Hand drawn maps and plans, Inv. Series II, no. 7911
102r *Mundus subterraneus, in XII libros digestus*, Athanasius Kircher, 1678
103 *Ars Magna Lucis et Umbrae*, Athanasius Kircher, 1671
108 Maria Clara Eimmart (Nuremberg 1676–1707), Lunar phase observed on 29 August 1697, Nuremberg, late XVII cen., (Inv. MdS-124e) mixed media on paper, Alma Mater Studiorum Università di Bologna, Sistema Museale di Ateneo, Museo della Specola
109 Maria Clara Eimmart (Nuremberg 1676–1707), Full Moon, Nuremberg, late XVII cen., (Inv. MdS-124c), mixed media on paper, Alma Mater Studiorum Università di Bologna, Sistema Museale di Ateneo, Museo della Specola
110 Wellcome Library, London
111 Historic Illustrations/Alamy Stock Photo
116 The Metropolitan Museum of Art, The Elisha Whittelsey Collection, The Elisha Whittelsey Fund, 1960
117 *Johannis Hevelii Selenographia, sive, Lunae descriptio*, Johannes Hevelius, 1647
118 *La cosmographie de Pierre Apian*, Peter Apian, 1544
119 *A Description of the Passage of the Shadow of the Moon over England, In the Total Eclipse*

of the Sun, on the 22d Day of April 1715 in the Morning, Edmund Halley, I. Senex and William Taylor, 1715
124–125 Copyright of the University of Manchester
126 Library Of Congress/Science Photo Library
127l *Cristiani Hugenii Zulichemii, Const. f. Systema Saturnium*, Christiaan Huygens, 1659
127r Eth-Bibliothek Zürich/Science Photo Library
132 *Introductio geographica Petri Apiani in doctissimas Verneri annotationes*, Peter Apian, 1533
133 Cambridge University Library, RGO 14/30, fol. 504
138 Metropolitan Museum of Art, New York, Rogers Fund, 1919
144–145 Leiden University Libraries, VLQ 79
147l *Harmonia macrocosmica seu atlas universalis et novus*, Andreas Cellarius, 1708
147r *Atlas Celeste de Flamsteed*, John Flamsteed, M. J. Fortin, 1776
152–153 The Metropolitan Museum of Art, New York, Rogers Fund, 1913
155 Österreichische Nationalbibliothek Wien, Cod. 5415, fol. 168r (E 4561-C)
160 Wellcome Library, London
161l *Blaeu Atlas Maior*, Joan Blaeu, 1665
161r Utrecht University Library, Mag: AC 62 Rariora (Vol. 1)
167 David Rumsey Map Collection www.davidrumsey.com
172 © Interfoto Scans www.agefotostock.com
173 Interfoto/Alamy Stock Photo
178–179 *Uranometria*, Johann Bayer, 1661
184–185 *Firmamentum Sobiescianum sive Uranographia*, Johannes Hevelius, 1687
190–191 *Atlas Celeste de Flamsteed*, John Flamsteed, M. J. Fortin, 1776
196–197 Courtesy of The Linda Hall Library of Science, Engineering & Technology
202–203 David Rumsey Map Collection www.davidrumsey.com
205 Library of Congress, Washington DC
210 *The Augsburg Book of Miracles*, c. 1550
211 *Theatrum cometicum, duabus partibus constans*, Stanislaw Lubieniecki, 1668, Biblioteka Narodowa, Poland
216–217 *Kometenbuch*, 1587. Nordostfrankreich bzw. Flandern.
218 *De nova et nullius ævi memoria prius visa stella jam pridem Anno a nato Christo 1572 mense Novembr*, Tycho Brahe, 1573
219 *De stella nova in pede serpentarii, et qui sub eius exortum de novo iniit, Trigono Igneo*, 1606/Private Collection/Photo © Christie's Images/Bridgeman Images
224 *Somnivm seu opvs posthvmvm de astronomia lvnari*, Johannes Kepler, 1634
225 Antiqua Print Gallery/Alamy Stock Photo
234 Rare Book Division, The New York Public Library
236 The Monument, Sutton Nicholls, 1725–1728, British Library, London

237 © National Maritime Museum, Greenwich, London
238 *Abhandlungen von Friedrich Wilhelm Bessel*, Vol. 2, Friedrich Wilhelm Bessel, Leipzig, 1872
239 *Philosophical Transactions of the Royal Society of London*, 1811, Wellcome Library, London
240 David Rumsey Map Collection www.davidrumsey.com
241 Courtesy of Carnegie Institution for Science
242 Library of Congress, Washington DC
243 *Results of astronomical observations made during the years 1834, 5, 6, 7, 8, at the Cape of Good Hope*, John F. W. Herschel, 1847
244 *Philosophical Transactions of the Royal Society of London*, Vol. 140, Royal Society, 1850
245l *Monograph of the central parts of the nebula of Orion*, Edward Singleton Holden, 1882
245r STScI, Hubble Heritage
248al Wellcome Library, London
248ac *Les vrais pourtraits et vies des hommes illustres grecz, latins et payens*, André Thevet, 1584
248ar *Johannis Hevelii Selenographia, sive, Lunae descriptio*, Johannes Hevelius, 1647
248cl Hulton Archive/Getty Images
248cc The Museum of Fine Arts, Sarah Campbell Blaffer Foundation, Houston
248cr ZU_09
248bl Wellcome Library, London
248bc *Uranometria*, Johann Bayer, 1603
248br Ian Dagnall Computing/Alamy Stock Photo
249al Chronicle/Alamy Stock Photo
249ac Wellcome Library, London
249ar Georgios Kollidas/Alamy Stock Photo
249cl Wellcome Library, London
249cc Georgios Kollidas/Alamy Stock Photo
249cr Pictorial Press Ltd/Alamy Stock Photo
249bl Lebrecht Music & Arts/Alamy Stock Photo
249bc GL Archive/Alamy Stock Photo
249br Archive PL/Alamy Stock Photo

ACKNOWLEDGMENTS.

Putting this book together involved an exhaustive dive into the world of Johann Gabriel Doppelmayr and Renaissance and Enlightenment astronomy in general. It would have been impossible without the work of the academic researchers whose writings I've devoured along the way, many of whom are acknowledged in the Further Reading. With much of the original literature in German, I also owe my wife Katja Seibold a great debt of gratitude for her unfailing patience in helping me untangle some of the knottier passages.

Particular thanks are also due to the reviewers who agreed to cast an eye over the book for their invaluable feedback and suggestions for improvements and additions – and especially to Doppelmayr authority Hans Gaab for his insights.

Finally, thanks to the publishing team: Isabel Jessop, Florence Allard, Phoebe Lindsley and Susanna Ingram at Thames & Hudson, as well as copy-editor Becky Gee, for your efforts throughout such a lengthy and detail-intensive project; Jane Laing and Tristan de Lancey for help in formulating the concept and careful watching over its gestation; and Karen Merikangas Darling for taking the book under her wing at Chicago.

With special thanks to Martin Rees for the foreword.

ABOUT THE AUTHOR.

Giles Sparrow is a writer and editor specializing in astronomy and physics. He studied astronomy at University College London and science communication at Imperial College and is a Fellow of the Royal Astronomical Society. He has written for books, magazines and multi-volume encyclopaedias on a wide range of topics, from cutting-edge space technology to the history of science, and from distant constellations to ancient archaeology.

He regularly contributes to magazines, including *All About Space* and *Sky at Night*, and is the author of the bestselling *Cosmos*, *A History of the Universe in 21 Stars*, *Spaceflight* and *What Shape is Space? A primer for the 21st century*, from Thames & Hudson's The Big Idea series.

ABOUT MARTIN REES.

Martin Rees (Lord Rees of Ludlow, OM FRS) is an astrophysicist and cosmologist, and the United Kingdom's Astronomer Royal. He was Master of Trinity College, Cambridge, from 2004 to 2012 and President of the Royal Society between 2005 and 2010. In 2012 he co-founded the Centre for the Study of Existential Risk, an interdisciplinary research centre dedicated to studying and reducing the threat of global catastrophes.

He has written more than 500 research publications, with topics ranging from black holes and gamma ray bursts to the multiverse and the origin of galaxies. Alongside his academic research, he is the author of many popular books, including *On the Future: Prospects for Humanity*, *Our Cosmic Habitat* and *From Here to Infinity: Scientific Horizons*, which was expanded from a series of BBC Radio 4 Reith Lectures.

COVER & SPINE Images taken from the *Atlas Coelestis*, courtesy David Rumsey Map Collection www.davidrumsey.com

ENDPAPERS Marbled paper: RTRO/Alamy Stock Photo

PAGE 2 This detail of the Copernican solar system is taken from Plate 2 of the *Atlas*.

PAGES 4–5 This page depicts a pair of celestial hemispheres divided along the line of the ecliptic. It is sometimes included in later editions of the *Atlas Coelestis* as an additional plate. Although supplementary details of the plate vary as the engraving was modified over time, the map is thought to originate with George Christoph Eimmart, founder and first director of the Nuremberg Observatory.

PAGE 10 These details of the constellations are taken from Plate 2 of the *Atlas*.

PAGE 22 These diagrams are excerpted from Plates 3, 4 and 6 within the *Atlas* and depict the relative sizes of celestial bodies, hypothetical models of the solar system and the motions of the planets.

PAGE 234 Details from a series of stunning illustrations of celestial phenomena by French artist and astronomer Étienne Léopold Trouvelot (1827–95). After emigrating to the United States, Trouvelot worked on the staff at Harvard College Observatory and the US Naval Observatory, producing thousands of drawings, including these superb pastels.

736 illustrations

First published in the United Kingdom in 2022 by Thames & Hudson Ltd, 181A High Holborn, London WC1V 7QX

Phænomena: Doppelmayr's Celestial Atlas © 2022 Thames & Hudson Ltd, London

Foreword © 2022 Martin Rees

Text © 2022 Giles Sparrow

For image copyright information see p. 253

Designed by Daniel Streat, Visual Fields

British Library Cataloguing-in-Publication Data
A catalogue record for this book is available from the British Library

ISBN 978-0-500-02429-4

Printed in China by Reliance Printing (Shenzhen) Co. Ltd

Be the first to know about our new releases, exclusive content and author events by visiting
thamesandhudson.com
thamesandhudsonusa.com
thamesandhudson.com.au

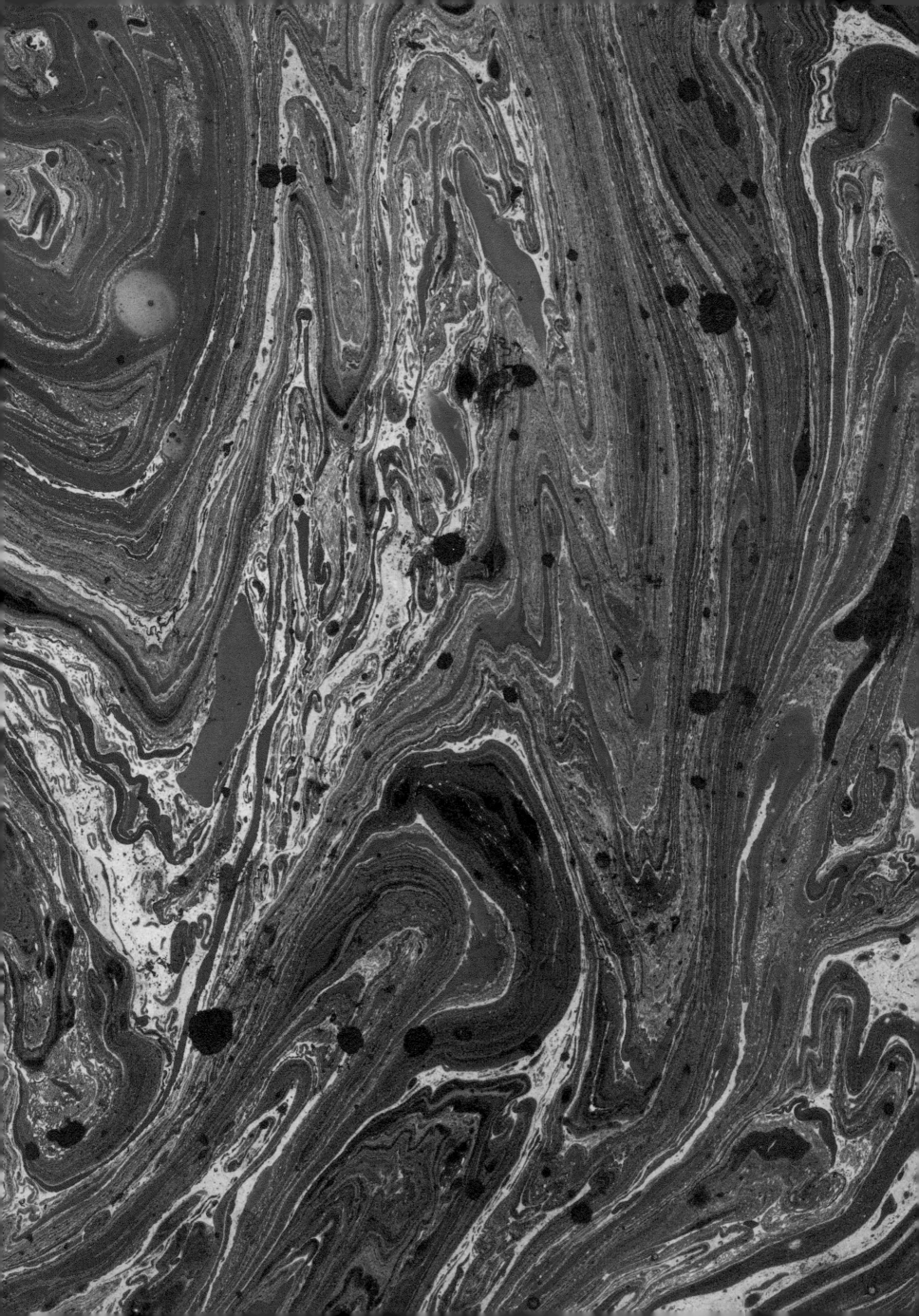